Trends in Environmental Management

The Editor

Dr. A.G. Devi Prasad is presently working as an Associate Professor in the Post Graduate Department of Environmental Science at University of Mysore.

He has 26 years of teaching and research experience. He has published more than 75 research papers in National and International Journals. He has guided successfully eight students for their Ph.D. Degree. He has served as a member of academic bodies of various Universities in India. He is the recipient of International Suvarna Parisara Rathna Award in 1998 conferred by National Pollution Control and Environment Conservation Committee (NPCECC) and Indira Priyadarshini Best Scientist Award in 2011 for Environmental Science conferred by Bose Science Society.

Trends in Environmental Management

–Editor–

Dr. A.G. Devi Prasad

2015

Daya Publishing House®

A Division of

Astral International Pvt. Ltd.

New Delhi – 110 002

Cataloging in Publication Data--DK
Courtesy: D.K. Agencies (P) Ltd. <docinfo@dkagencies.com>

Trends in environmental management / editor, Dr. A.G. Devi Prasad.
pages   cm
Includes bibliographical references and index.

ISBN 978-93-5130-689-4 (International Edition)

1. Environmental management--India.   I. Devi Prasad, A. G., editor.

DDC 363.700954   23

Published by   : **Daya Publishing House®**
                            A Division of
                            Astral International Pvt. Ltd.
                            – ISO 9001:2008 Certified Company –
                            4760-61/23, Ansari Road, Darya Ganj
                            New Delhi-110 002
                            Phone: 011-011-4354 9197, 23278134
                            E-mail: info@astralint.com
                            Website: www.astralint.com

Preface

Environment is the only source of all natural resources for manifestation of life on this biosphere. These resources which are the eco-wealth, acquire significance when they are developed by man. Urbanization, industrialization and the advent of modern materialistic desires have exerted tremendous pressure on our healthy resources. Burgeoning population and their activities have caused increased pollution, waste generation and unscientific management practices of resources have resulted in health risk of organisms, heightened the demand for healthy natural resources, ignoring the maintenance and management of clean and hygienic environment.

The problem of environmental management is a matter of concern all over the world. In this context, the editor had invited the articles related to the current issues of environment and its management from various experts in the field, which are compiled under 15 chapters. The contents of the book are comprehensive and straight forward. Each chapter focuses on specific environmental issues with the recent trends in management aspects. It provides basics and techniques of environmental management. The book will serve as a valuable tool for teachers, scientists, researchers and students providing an insight to scientific environmental management.

Dr. A. G. Devi Prasad
Editor

Acknowledgements

I am indebted to the expertised personalities who have contributed their articles for this book.

I am grateful to Professor Mewa Singh, Ramanna Fellow and Life-long Distinguished Professor, University of Mysore, Mysore-06, for reviewing the book and for writing the foreword.

I sincerely thank Mrs. Sapna Rao, Miss Shwetha and Miss Komala for their meticulous editing work of the manuscripts.

I thank the publishing team of Astral International Pvt Ltd., New Delhi, for giving their best in bringing out this book.

The active support, sustained help and forbearance shown by my wife Dr. Geetha M. J. and children Shravan A. P and Shishir A. P, are, perhaps, the instrumental factors in showing this book the light of the day.

Dr. A. G. Devi Prasad

Foreword

This book is a major contribution to the field of environmental management, especially because it comes at a time when it is most indispensable. The book is an anthology of articles by different authors, who are veritable experts in their own fields. It successfully explores and integrates various perspectives in the field of environmental management, and it is the perspectives and the integration of perspectives that is necessary for tackling the current problems in managing the environment.

So why do we need a right perspective to find solutions for proper management of environment? The answer is rather simple and the need for the same is evidently observable around us. The current environment is drastically degrading, which poses an obstinate threat to mankind and other organisms living in it. The habitat degradation due to several human activities has been a major contributing factor for the threats. This has lead to severe depletion of quantity and quality of several natural resources including energy, water, air and soil. As a consequence of this depletion, the quality of both human and non-human life has been seriously affected apart from driving several species to total or local extinction. Additionally, empirically recorded changes in climate, resulting from this depletion and transformation, are drastically changing the environment, and might make it unsuitable and unsustainable for living organisms. Hence, an understanding about our environment and the right perspective on how to manage it, is very important in the current perilous situation we are in. Chapter nine discusses about the current environmental issues we are facing and the trends on how to manage them.

Development, especially agricultural and industrial, is essential for meeting the needs of present day world population. This development is responsible for overexploitation and hence, depletion in quality and quantity of essential natural resources apart from their misuse and contamination. Sustainable development and responsible usages are the probable solutions to these problems. 'Ecotourism' is one of the quintessential examples of the role of sustainable development in an economic/commercial activity. The inclusive approach of ecotourism entails involvement and support of local workforce who benefits culturally and socio-economically, and this, in turn leads to improved protection, enrichment and conservation of natural resources. A critical evaluation of the benefits and disadvantages of ecotourism are discussed in the first chapter. The fourth chapter discusses how sacred groves, with high degree of species richness are rapidly degrading and how they can be preserved/protected using cultural practices and traditions.

Most of the environmental problems we face in the present day are largely manageable. With the advent of new technologies, long term weather records and availability of comprehensive information on biogeochemical cycles and environmental principles, it is probable for us to save and manage the environment from further degradation. With such an objective in mind, a new field called environmental engineering has gained momentum in the past few decades. It integrates innovative and efficient technology with the knowledge of the environment for the management of the environmental problems. An illustration and evaluation of how technology can mediate environmental management can be found in chapters two, six, seven, eleven and thirteen. Chapter two introduces how 'phytoremediation' technology can be used for degrading, extracting, containing and immobilizing contaminants from soil and water using accumulator plant species. It also discusses how genetically engineered accumulator plants can help in phytoremediation. Chapter thirteen focuses on the phytoremediation capabilities of *Vetiveria zizanoides* through discussion on its unique morphological and physiological characteristics. This species can thus be used for preventing soil erosion and sediment control. Additionally, since it has high heavy metal accumulating capacity, it can be used for treating mine tailings, garbage landfills and industrial waste dumps. Chapter six discusses how nanotechnology can be optimally applied in environmental science studies. This technology can be used for enhancing environmental protection, improve pollution detection and remediation. Nanotechnology can be used for low cost and efficient energy transformation and storage. It can also be used for water treatment and purification and for substituting toxicological hazardous substances. However, the indiscriminate and unscientific use of nanotechnology can compromise standards of safety and health, and this is outlined in the following chapter. Chapter seven discusses how nanotechnology can impact human health and the environment through its toxicity. Chapter eleven discusses how consolidated bioprocessing technology can be used for converting biomass into bioethanol that is renewable, sustainable, efficient and cost effective. Chapter ten discusses how microbial organisms can be used for bio-technical applications, genetic engineering and bio-degradation of lignocelluloses. Chapter fourteen discusses how inoculation of am fungi can help increase seedling vigor of pepper and onion under salt stressed conditions. Chapter twelve critically evaluates the crucial challenge of energy resource management and its sustainable utilization through elucidation of the current trends in energy resource consumption, the immediate need to incorporate sustainability for curtailing harmful impacts on environment, pressing need for utilization of biomass and development of renewable energy resources like wind, solar and tidal powers.

Moving on to the core of the topic being addressed, the authors pivot their discussion on the issue of forest regeneration that apparently, has widespread implications for not only preservation and possible recolonization of native biodiversity but generate NTFP for utilization by native communities. Chapter eight thus, compiles essential information on conditions promoting plant growth and propagation. Followed by this, the final chapter highlights indicators of soil health based on the concept of control chart. The chapter painstakingly evaluates the threshold and critical levels of major soil contaminants for its sustainable management. Finally, very essential to the environment and the secondary producers of the ecosystem are the pollinators, chiefly among them, the honeybees among others. Amidst global crises of colony collapse disorder (CCD), it has become very essential to re-evaluate and reposition the role and function of honey bees in maintenance of floral diversity and hence, in the management of forests and thus, the environment.

In conclusion, the authors of the chapters of this book have done a stupendous job in compiling essential information circumscribing the major tenets of environmental management that can inform and excite both academicians and mangers alike. I have garnered pleasure in reviewing the book and hope the readers do it too.

Prof. Mewa Singh

FASc, FNA, FNASc, FNAPsy
Ramanna Fellow and Life-long Distinguished Professor
University of Mysore, Manasagangotri
Mysore-570006, Karnataka.

Contents

Chapter 1

Ecotourism Management for Sustainable Development

A.G. Devi Prasad

DOS in Environmental Science, University of Mysore, Manasagangotri, Mysore-06, Karnataka

ABSTRACT

The concept of ecotourism is a new-approach in tourism. It creates economic opportunities that make conservation and protection of natural resources advantageous to the local people. Ecotourism has both positive and negative impacts on environment. Ecotourism embraces biological and cultural conservation, preservation, sustainable development, etc.. It offers scope for symbiotic living of man and environment. Sustainable ecotourism promotes the understanding of environment involving local people as well as tourists for sharing socio-economic benefits, enrichment of environmental and cultural knowledge. An ethical evaluation of the above is presented in this chapter.

Key words: Economy, Biodiversity, Cultural activities, Carrying capacity, Solid waste, Conservation

Introduction

India is the second largest country in the world to have maximum natural resources. India, with its vast geographical and climatic diversities, is blessed with rich diversity of flora and fauna. Considering India's wealth of natural resources and rich cultural heritage, tourism can emerge as an important instrument for economic development. Tourism is currently the world's largest industry ($ 3.4 trillion annually) and ecotourism represents the fastest growing sector of this market. Tourism is already the largest source of foreign exchange in countries like Costa Rica and Belize; in Gautemala it is second. Throughout the developing tropics, protected area managers and local communities are struggling to balance the need for economic growth with the preservation of natural resources. Ecotourism may offer one way of striking this critical balance. Well planned ecotourism can benefit both protected areas and residents of surrounding communities by linking long-term biodiversity conservation with local, social and economic development.

The term 'ecotourism' was coined by a Mexican environmentalist, Hector Cebellos-Lascurain in 1983 and brought to light in 1996 by his paper 'Tourism, Ecotourism and Protected Areas'.

The term ecotourism has been used widely as well as interchangeably to refer to sustainable tourism, alternative tourism, ethical tourism, green tourism, special interest tourism, appropriate tourism and responsible tourism.

The term was defined as early as 1965 as that which:

- Respects local culture;
- Optimizes benefits to local people;
- Minimizes environmental impacts; and
- Maximizes visitor satisfaction.

Ecotourism is nature-based tourism, which plays an increasing role in today's environmental management. Ecotourism involves human interaction with nature. It has both positive and negative impacts on the environment. Ecotourism is responsible travel to the natural areas that conserves the environment and sustains the well being of local people. In order to ensure sustainable ecotourism, the United Nations General Assembly (UNGA) declared the year 2002 as the International Year of Ecotourism. Ecotourism aims at conservation of biological resources, enhanced economy of the people, ecological balance of nature, and promotion of socio-cultural development of a region. It also aims at promoting environmental values and ethics and preserving nature in its uninterrupted form. Ecotourism may exert a pressure on environment. It has contributed in its own way to environmental degradation and pollution. Overcrowding, waste and littering, pollution or commercialization resulting from ecotourism have enhanced the destruction of the environment. Thus, there may be several shortcomings and negative impacts due to ecotourism.

Ecotourism, in simple words, means management of tourism and conservation of nature in a way so as to maintain a fine balance between the requirements of tourism and ecology on one hand and the needs of the local communities for jobs-new skills, income generation employment and a better status for women –on the other.

Basic Elements of Ecotourism

The basic elements of ecotourism include conservation, sustainability local involvement of the people-ownership and business opportunities-and biological diversity. Ecotourism follows two important principles of sustainability, namely:

1. To promote conservation of the natural ecosystems; and
2. To support local economies.

The essential key elements of ecotourism are: a well preserved ecosystem to attract tourists, cultural and adventure activities, active involvement of the locals who are able to provide authentic information about nature, culture and their ethnic traditions to the visitors and finally empowering the local populace to manage ecotourism so that they ensure conservation through alternative livelihood opportunities.

Ecotourism involves interaction of human beings with nature. This interaction triggers a pressure on the environment. The growth of ecotourism leads to the modification of the environment. Such a modification may result in either beneficial or harmful effects to the environment itself or to human beings.

Environmental impact of ecotourism

There are both positive and negative environmental impacts due to ecotourism. They are briefly pointed out as follows:

Negative Impacts of Ecotourism

1. The areas that are particularly appealing to the visitors are often places with high biodiversity. The tourist activities may result in loss of habitats and change in composition, destruction of flora and fauna, disruption of breeding and feeding phenomena. Tourists, often unwittingly, can introduce exotic species that are not native to the local environment, which may cause enormous disruption and even destruction of ecosystems. The effects of loss of biodiversity include climatic changes, threatening of sources of wood, food, medicine and energy. Loss of biodiversity destabilizes the ecosystem rendering it weak to deal with natural disasters. It will also result in the stunted growth of economy of a country.

2. The ecological functions are imbalanced-such as species balance, soil formation and greenhouse gas absorption. Productivity of an ecosystem is rendered.

3. Ecotourism may result in land degradation. Important land resources such as minerals, fossil fuels, forests, fertile soil, wildlife, etc. are affected when land is utilized for accommodation, water supplies, restaurants and infrastructure provision for tourists.

4. Forests often suffer because of deforestation caused by fuel wood collection and land clearing by tourists trekking and camp fires.

5. Transport by air, road and rail is continuously increasing in response to the rising number of tourists and their greater mobility. Air pollution from tourist transportation has impacts on the global level, especially CO_2 emissions related to transportation energy use.

6. Noise pollution from vehicles or tourist attractions such as discos, snowmobiles etc. cause annoyance, stress and even hearing loss for humans. It causes distress to wildlife especially in sensitive areas. Visitors can disturb feeding and breeding behaviour of native animals.

7. Tourist activities result in generation of enormous quantities of solid wastes. Waste disposal is a serious problem and improper disposal can be a major despoiler of the natural environment. Trekking tourists and tourists on expedition leave behind untold amounts of garbage that degrade the environment over a period of time.

8. Water pollution through sewage or fuel spillage from pleasure boats, disposal of leftover food or spoiled food, etc. is a matter of concern as it threatens the health of humans and animals.

9. Due to greater demand by tourists for basic amenities like water, fuel, building materials, the resources of the area get depleted.

10. One of the direct impacts of tourism is on the cultural environment of an area. Tourism, especially in rural and undeveloped areas has created a dependence on foreign income among the local population. It has displaced traditional customs and social interactions and made communities vulnerable to foreign economic conditions.

11. Improper land use planning and building regulations in many destinations, introduction of new architectural styles which may clash with the indigenous structural design, sprawling development (such as roads, parking places, housing, waste disposal) have resulted in aesthetic pollution.

12. Trampling tourists impact on vegetation and soil include:

    i.   Breakage and burning of stems;

    ii.   Reduction in regeneration ability;

    iii. Reduction in plant vigor;

    iv.   Species' composition change;

    v.   Loss of ground cover, organic matter and accelerated erosion;

    vi.   Reduction in soil macro-porosity;

    vii. Increase in runoff; and

    viii. Decrease in air and water permeability.

Positive Impact of Ecotourism

1. The essence of ecotourism lies in admiration of nature and outdoor recreation. It encompasses a wide range of activities such as trekking, boating, hiking, mountaineering, bird watching, rafting, biological explorations and visiting wildlife sanctuaries. These recreational benefits are without parallel to any sort of joyous activity and entertainment.

2. Ecotourism stimulates the tourists to respect wild species and envisage conservation schemes.

3. Ecotourism can increase the level of education and activism among travelers making them more enthusiastic and effective agents of conservation.

4. Conservation and revitalization of traditional arts, drama, music, dance, handicrafts, customs, ceremonies and traditional lifestyles directly attract tourists. Thus, ecotourism has found a niche for itself as an effective instrument for preserving culture and tradition.

5. Tourism may provide a new use for formerly unproductive and marginal land.

6. For foreign visitors, ecotourism provides an educational and scientific glimpse of a world not their own, often a world of striking beauty and rich cultural heritage.

7. Ecotourism provides an opportunity for long-term protection of biodiversity, land, water, mineral and other natural resources. It gives scope for designing, developing and promoting conservation strategies of natural resources.

8. Ecotourism plays an important role in rising awareness of the problems facing a particular locale or its people. The more locale people benefit from tourism, the more they will benefit from a commitment to preserve the environmental features, which attract tourists.

9. Ecotourism provides employment opportunities to local people.

10. Ecotourism also encourage the local community to value its natural and cultural assets. It encourages the development of markets in native handicrafts and artwork.

11. Ecotourism generates income including foreign exchange income.

12. Ecotourism offers an opportunity to a visitor to know about the historic sites and monuments offering scope for research activities and protection measures.

13. Ecotourism will help to promote environmental awareness and ethics in the visitors.

14. Above all, visitation to an eco-tourist-spot ensures a tourist to get physically invigorated, mentally rejuvenated, culturally enriched and spiritually elevate.

Characteristics of Sustainable tourism

➤ Sustainable tourism aims at protecting the local culture and tradition

➤ Dissemination of tourists' culture and civilization to the local people

➤ Importance, information and significance of tourist places to the tourists

➤ Promotion of conservation of resources

➤ Provide opportunities for local people to grow socially and economically

➤ Sustainable tourism helps in promoting the integrity of tourist places

Towards Development of Sustainable Ecotourism

Ecotourism has both positive and negative impact on man as well as the environment. While promoting the development of ecotourism one may perceive many difficulties. Hence, some guidelines have to be formulated to achieve sustainable ecotourism, which are as follows:

1. The objectives pertinent to conservation should be clear.

2. A database should be developed on the existing ecological resources of an area. The carrying capacity of the habitat should be evaluated and assessed.

3. The baseline environmental status should be established.

4. Negative impact on the environment, especially degradation and pollution should be assessed.

5. Development of a source of long-term financial stability for the conservation of protected areas. Maintenance of biological diversity should be given prime importance.

6. Identification of tourism activities that are compatible with the area and establishment of standards for quality. The activities should cause least damage to the ecosystem.

7. Integrated approaches should be emphasized which place together emphasis on building local capacity.

8. Ecotourism should surely support a wide range of local economic activities.

9. Development of partnerships with NGOs, especially local communities. This will help to strategize community base ecotourism enterprises.

10. Involvement of local people in planning, management and development programmes. Local people are able to provide authentic information about nature, culture and their ethnic traditions.

11. Visitors' needs and tourists' markets should be analysed. The marketing should focus on environmental, social, cultural and economic sustainability criteria. Basic amenities availability to visitors should be ensured. Over-visitation should be avoided.

12. Establishment of education, training and research programmes for the staff as well as tourist s associated with the ecotourism.

13. Management policies should be established which should help to minimize environmental degradation, such as waste management, zoning for transportation, etc.

14. Government should enforce a legislative framework to regulate tourism trade and industry, create basic infrastructure and health facilities and ensure safety and security of the tourists.

15. Promotion of equity in the distribution of both the economic costs and the benefits of the activity among tourism developers and hosts.

16. Establishment of a programme for monitoring and reviewing the ecotourism activities.

Above all, it is of utmost importance to all of us to cultivate in ourselves a good civic sense with pride in our surroundings. Ecotourism can live up to its promise if it follows the principles of sustainable development, adequately monitors and protect its resources.

Conclusion

Ecotourism, which typically involves nature-based tourism plays a significant role in today's environmental management. Training to develop skills of tour operators, visitors and others to understand what sustainable tourism is and education about best practices are vital activities. Environmental education is, therefore, the best hope in this regard. Environmental education helps in changing practices and behavior, action or participation in environmental issues. To have sustainable ecotourism, it should not be viewed as nature-tourism alone. It must have a wider scope. It must help in eliminating poverty, unemployment, creating new skills, in enhancing the status of women, in preserving cultural heritage, in improving overall environment, in promoting dialogue amongst civilizations and in facilitating the growth of a more just and fair world order. Sustainable ecotourism aids in promoting conservation, sustenance of the well being of humans and helps in the maintenance of a balanced relationship between humanity and the environment.

References

Anonymous, Ecotourism Consultancy Programme. Mexico: IUCN, 1994.

Drumm, A. 'New approaches to community based ecotourism management'. In: Ecotourism-A guide for planners and managers. 2 vols. (Eds) Lindberg, K., Wood, M. E. and Engeldrum, D. Dehradun India: Natraj Publishers, 1999

Hunter, C. and Green, H. Tourism and the Environment: A sustainable relationship. London, U. K.: Routledge, 1995.

Lindberg, K. 'Economic aspect of ecotourism'. In: Ecotourism-A guide for planners and managers. 2 vols. (Eds) Lindberg, K., Wood, M. E. and Engeldrum, D. Dehradun India: Natraj Publishers, 1999.

Panigrahi, S. K. 'Ecotourism in India'. Employment News pp. 21-27, June 2003

Shipra v. Srivastava. 'Tourism as an industry: Tourism linked with pollution and environmental degradation'. In: Advances in environmental biopollution. (Eds). Shukla, A. C. et al. New Delhi, India: Publishing Corporation, 1999.

Hunter: a sustainable relationship'. , C and Green, H. 'Tourism and the environment'. London, U. K.: Routledge, 2005.

Thakur, A. and Bhargava, D. S. 'Environmental equality impacts on tourism'. Ecology 2: 10, 1988.

Chapter 2

Structural and Functional Attributes to Evaluate Lotic Ecosystem Health

K.R. Sridhar

Department of Biosciences, Mangalore University, Mangalagangotri, Mangalore-574 199, Karnataka

ABSTRACT

In this chapter, the ecological integrity of lotic ecosystem have been described which is divided into structural and functional integrity. The structural integrity mainly refers to the qualitative and quantitative composition of biological communities and their resources in pristine conditions. The functional integrity concerns to the rates, patterns and importance of ecosystem-level process under pristine conditions.

A variety of guidelines have been proposed to assess the health of lotic ecosystem. The biological assessment offers cost-effective, rapid results and easy monitoring. Bioassays employing single-species in impact assessment was unsuccessful, while population dynamics, food web and taxonomic structure of communities have been successful. Ecosystem level possesses primary production of benthic algae and macrophytes, community respiration, sediment respiration, secondary production by macro-invertebrates, nitrogen fixation, enzymatic transformations in biofilms and leaf litter decomposition, which are successful depending on the nature of perturbations. Leaf litter decomposition in lotic ecosystem involves a range of organisms such as bacteria, fungi, invertebrates and fishes and will be an ideal criterion for assessment of lotic ecosystem health. The leaf litter breakdown coefficient and diversity and equitability of leaf litter decomposing organisms will be an useful index for comparison of geographical regions, plant species and impact of pollutants. Aquatic hyphomycetes are the major fungal flora involved in plant detritus (mainly leaf litter) decomposition in lotic ecosystem and transfer energy to higher trophic levels. The rate of arrival of conidia, rate of release of conidia, biomass, conidial viability and conidial germination helps to assess the ecosystem potential more precisely, which accounts to structural as well as functional integrity of lotic ecosystem.

Key words: Bioassay, Indicators, Decomposition, Leaf litter

Introduction

Despite freshwater is 0.01% of the global water pool, it supports up to 0.1 million species, which is equivalent to about 6% of described species (Hawksworth and Kalin-Arroyo, 1995; Gleick, 1996; Gilbert and Deharveng, 2002). The rates of decline of biodiversity due to human interference as well as climate change are higher for freshwaters than terrestrial or marine ecosystems (Sala *et al.*, 2000; Jenkins, 2003). Conservation of freshwater biodiversity gained highest priority during the International Decade (2005-2015) for Action 'Water for Life' (Dudgeon *et al.*, 2006). Five major human impacts on freshwater biodiversity include over exploitation, pollution, flow modification, habitat degradation and invasion of exotic species (Dudgion *et al.*, 2006). The contemporary issues in freshwater ecology are to understand the impacts of human interference, methods of assessment and developing appropriate remedial measures for the sustenance of biodiversity.

Human stresses on streams and rivers occur at different magnitudes (Lankerani, 1994). Changes of lotic ecosystem in temporal scales vary from hours to days to decades and centuries depending on the land use pattern and climate changes (Harding *et al.*, 1998; Rosenberg *et al.*, 2000). Spatial scale changes occur in channel, catchment areas, drainage basins at continent or globe through atmospheric deposition and global warming (Gessner and Chauvet, 2002). Ecological integrity of lotic ecosystem refers to the existing situation along the gradient of impairment from impaction to pristine (Karr, 1991). Minshall (1996) subdivided the ecological integrity of lotic ecosystem into two facets as structural and functional integrity. Structural integrity mainly refers to the qualitative and quantitative composition of biological communities and their resources in pristine conditions, while the functional integrity to the rates, patterns and importance of ecosystem-level process under pristine conditions (Hannah *et al.*, 1994; Vitousek *et al.*, 1997; Gessner and Chauvet, 2002). No single indicator or parameter is valuable to assess lotic ecosystem health unequivocally (Boulton, 1999). Although traditional assessment of lotic water quality focused mainly on physical, chemical and biological characteristics, assessment of aquatic biota as a measure of structural or functional integrity of ecosystems has gained importance recently (Norris and Thoms, 1999). Usually fish and macroinvertebrate assemblages are employed for assessment of structural integrity of lotic ecosystem (Norris and Hawkins, 2000; Statzner *et al.*, 2001). However, other alternative communities also serve in assessment of structural integrity which includes algae, protozoa and macrophytes (e.g. Barbour *et al.*, 1999; Norris and Thoms, 1999; Hill *et al.*, 2000). Even though functional integrity complements the structural integrity, it has not been effectively employed or rarely considered for assessment of lotic ecological conditions (Bunn and Davies, 2000). In assessment of lotic ecosystem health, it is difficult understand or to integrate the influence of physical, chemical and biological characteristics (Norris and Thoms, 1999). There are several substantial debates on the importance and sensitivity of indices of lotic ecosystem health. The aim of this chapter is to emphasize some issues on structural and functional attributes of lotic ecosystem health monitoring and assessment based on biodiversity and ecosystem services perspective.

Health of Lotic Ecosystem – the Debate

Ecological criteria of health of lotic ecosystem refer to sustainability, adaptation to stress and ecological integrity (capacity to support and sustain a balanced integrated adaptive full range of elements of biological system) (Karr, 1996). As the lotic water degradation by pollution mainly influences the biota, adoption of biological measures in assessment of lotic ecosystem health has recently gained importance (Karr, 1991; Norris and Norris, 1995; Wright, 1995; Resh *et al.*, 1996; Gessner and Chauvet, 2002). Karr *et al.* (1986) defined healthy biological system as a biological system when its inherent potential is realized, its condition is stable, its capacity for self-repair is restored when perturbed and minimal external support required for management. However, Norris and Thoms (1999) argue that the above definition ignores the use of non-biological parts of the ecosystem on which biota are dependent. Lotic health indicators need to satisfy six criteria: 1) quantify and simplify complex ecological phenomena; 2) provide easily interpretable outputs; 3) respond predictably to damage caused by humans; 4) relate to an appropriate scale; 5) relate to management goals; 6) should be scientifically defensible (Norris and Hawkins, 2000). Lotic ecosystem health depends and is operated by four spatial dimensions: 1) longitudinal (*e.g.* as river continuum, nutrient spiraling, hyporheic corridor and downstream barriers); 2) lateral links with adjacent terrestrial ecosystem (*e.g.*, valley; influence of riparian vegetation and flood pulse); 3) vertical links with river/stream bed (e.g. hyporheic and parafluvial processes); 4) temporal (seasons) (Townsend and Riley, 1999).

To identify structural or functional integrity of river ecosystems, evaluation of aquatic biota have gained acceptance (Norris and Thoms, 1999). A small group of biological ecosystem-level indicators of river ecosystems helps to understand the river condition under stress. However, physical and chemical features of the environment (e.g. flow regime, energy sources and water quality) influence these indicators and ultimately their structure and function. Four methods of habitat assessment have been reviewed by Parsons *et al.* (2004): 1) Index of Stream Condition (ISC) is used as a rating system to assess overall river condition; it is based on the hydrology, physical form, streamside zone, water quality and aquatic life those influence river condition; the aquatic life sub-index is the component of ISC, which reflects biotic condition, and usually macroinvertebrates will be used as it helps in continuous assessment of river environment; it is assumed that degradation of physical, chemical and hydrological features impoverish macroinvertebrate communities; 2) River Habitat Audit Procedure (RHAP), which is composed of 11 data components encompassing hydrology, site description, temporal and spatial reach environments and channel habitats; this method was designed to assess the ecological conditions of rivers rather than faunal surveys (or biological assessment) made at the same sampling reach of lotic ecosystem; 3) The River Styles™ is physical assessment of rivers using geomorphological approaches; a variety of physical field-based variables of catchment area will be collected (catchment/landscape characteristics and reach characteristics) for monitoring; however, River Styles™ does not directly consider the biota, but merges assessment of geomorphological degradation with assessment of ecological degradation; 4) Habitat predictive modeling follows the philosophy of

Australian River Assessment System (AusRivAS) and River Invertebrate Prediction and Classification Scheme (RIVPACS); it encompasses large-scale catchment characteristics to strongly discriminate among the local-scale groups and variables; through this model, the local-scale habitat features can be predicted in the absence of degradation from catchment-scale characteristics. Thus, it is evident that no single method will meet the requirements to lotic assessment protocol. Marchant *et al.* (2006) evaluated biological methods of assessment of streams and considered using indicator species. They have arrived at a broad agreement about the criteria of biological indicators to be satisfied. They should: 1) provide early warning of a wide range of environmental stressors at appropriate time and spatial-scales; 2) indicate cause of change as well as existence of degradation; 3) be cost-effective; 4) provide easily interpretable outputs that relate to management goals. New challenges and research opportunities for preserving biodiversity and ecological services of rivers have been elegantly addressed by Arthington *et al.* (2010).

Human Impacts

Several physical and chemical perturbations exhibit direct or indirect influences on the structure as well as function of lotic ecosystem (Townsend and Riley, 1999). For instance, algae respond to changed nutrient levels, shade, turbidity and physical disturbance. It leads to changes in the stream of macroinvertebrates, as they are dependent on different food resources (*e.g.* grazer-scrapers and browsers use algal biomass and bacteria in organic matter in bed surface; collector-gatherers and collector-filters depends on fine particular organic matter; shedders rely on the coarse particular organic matter). Similarly, the hyporheic fauna are severely affected due to accumulation of fine sediments. At higher community level especially the biodiversity and food web structure, several ecosystem attributes (*e.g.* primary productivity, respiration, P: R ratio, decomposition rate and spiraling length of nutrients) will suffer due to change in land use pattern. Many pathways influence the catchment area, aquatic communities and the ecosystem functioning (some connections have been indicated in Table 1.

Table 1: Connections between catchment area, aquatic community and ecosystem functioning (modified from Townsend and Riley, 1999)

Catchment features	Aquatic community	Ecosystem functioning
Channel form	Algae	Productivity
Channel flow	Invertebrates	Respiration
Sediment	Fungi	Decomposition
Organic matter	Bacteria	Nutrient spiral
Nutrients	Higher fauna	
Turbidity		
Shade		
Temperature		
Pollutants		
Disturbance		

The widespread human impacts on streams include modification of forests, addition of pollutants and eurtrophication. It is clearly known that disturbance regimes result in perturbation in community composition and food web pattern. Invertebrate fauna (*e.g.* shredders) become depauperated in New Zealand aquatic system due to transformation of native tussock grassland landscape into exotic conifer plantations or pasture use, which resulted in increased rate of organic matter deposition in streams and declined the autochthonous production due to shading (Townsend and Riley, 1999). Shading profoundly influenced the algal biomass and the chlorophyll-a density (Townsend *et al.*, 1997b). The carbon spiraling lengths become narrow due to rapid biological oxidation of organic seston (Young and Huryn, 1997). The loading of nitrogen, usually leads to uptake by stream autotrophs and loss takes place by denitrification. Loading phosphorus (which is devoid of gaseous forms for loss) and nitrogen leads to storage in biota or may be routed through hyporheic zone (Townsend and Riley, 1999).

Refugia in streams and rivers lessen the impacts of discharge disturbances on biota. Potential value of refugia include large substratum, dead zones (shear stress on the bed is low), hyporheic zone, upstream areas, adjacent tributaries and riparian zone (Townsend and Riley, 1999). In such habitats with a wide range of refugia, the recovery of community will be rapid. However, several aspects of refugia in lotic ecosystems are yet to be invented and roles need to be understood. Townsend *et al.* (1997a) found maximum invertebrate species richness at the intermediate levels of disturbance in streams as predicted by Connell (Connell, 1978). According to Connell's hypothesis, the highest diversity is maintained at intermediate scales of disturbance and is a useful criterion to assess the impact of disturbance on biodiversity (Figure1).

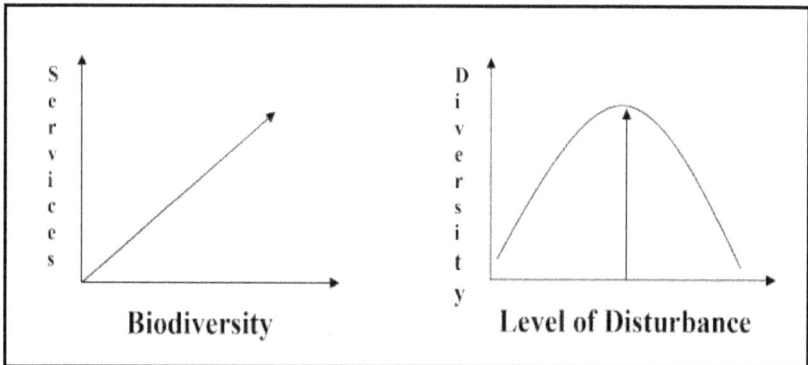

Figure:1 Biodiversity vs. ecosystem services and impact of disturbance on biodiversity

A variety of perturbations influences the lotic ecosystem (*e.g.* eutrophication, pollution and modification of riparian vegetation). In the past two centuries, water chemistry has changed dramatically especially due to atmospheric pollution, land use pattern and direct disposal of wastes into aquatic bodies. As the stream biogeochemical processes are controlled by the riparian vegetation, its modification (e.g. by agriculture, forestry and replacement by exotic species) has a dramatic effect on stream functioning (Graça *et al.*, 2002). Based on earlier studies and their own study, Lecerf and Chauvet (2008) performed a meta-analysis on above human interferences of stream ecosystem. The fungal species richness was consistently reduced in eutrophic streams. In heavily eutrophicated streams, factors other than nutrient enrichment were responsible for the loss of fungal (aquatic hyphomycetes) diversity and functions (Raviraja *et al.*, 1998). Oxygen depletion by eutrophication seems to be detrimental to aquatic hyphomycetes in streams (Raviraja *et al.*, 1998; Pascoal *et al.*, 2005; Mesquita *et al.*, 2007; Medeiros *et al.*, 2009). Mine pollution consistently reduced leaf breakdown rates, fungal diversity and fungal biomass. Alteration of riparian vegetation had significant positive effect on maximum spore production. The maximum spore production by aquatic hyphomycetes was most sensitive indicator of human impacts on streams. Overall, Lecerf and Chauvet (2008) predicted that widespread threat of stream ecosystem can be effectively monitored based on conidial production by aquatic hyphomycetes.

Indicators

Physical and chemical indicators of lotic ecosystem are mainly dependent on water qualities (*e.g.* antagonistic interaction of heavy metals with major cations and effects of pH on solubility) (ANZECC, Austalian and New Zealand Environment and Conservation Council) (Norris and Thoms, 1999). The International Union of Geological Sciences (IUGS) has given guidelines for assessment of abiotic conditions of lotic ecosystem using 27 indicators (sediment sequence and composition, soil and sediment erosion, stream flow, stream channel morphology, stream sediment storage and load, surface water quality and floodplains/wetlands structure and hydrology) (Norris and Thoms, 1999). The physical change and continuous measurements of water and sediment discharge seems to provide valuable inputs.

A variety of guidelines have been projected to assess the lotic ecosystem health using biological assessment (or 'biocriteria'), which offers cost-effective, rapid results and can be followed by non-specialists and has been termed 'thermometers' by Norris and Thoms (1999): water quality guidelines of Australian and New Zealand Environment and Conservation Council (ANZECC); The US Environmental Protection Agency (US EPA); In Australia, US and UK, Index of Biotic Integrity (IBI), Benthic-IBI (B-IBI), Australian River Assessment System (AusRivAS) and River Invertebrate Prediction and Classification Scheme (RIVPACS).

Biological indicators (*e.g.* fish and invertebrates) have been classified into many categories (including structure, community balance and functional feeding groups): species richness and composition, trophic composition, and abundance and condition (Norris and Thoms, 1999). Three types of biological indicators have been recognized by Cairns and McCormick (1992): 1) early warning indicators (signify impending decline in health); 2) compliance indicators (reveal deviations from acceptable limits); 3) diagnostic indicators (show the causes of deviations) (cyanobacterial blooms and toxin production; Hart *et al.*, 1999). Boulton (1999) indicates the possibility of incorporating all the above parameters for assessment of lotic ecosystem health. Macroinvertebrate assemblage and composition help as indicators of lotic ecosystem health. The Index of Stream Condition (ISC) developed by Ladson et al. (1999) assesses hydrology, physical form, riparian vegetation, and water quality and macroinvertebrate assemblage. If some indicators are not worldwide, region-specific biota has to be employed as indicators. It is worth looking for 'keystone species' among the region-specific biota.

Use of single-species bioassays in impact assessment of lotic ecosystems influenced by human interference was unsuccessful, while population dynamics, food web and taxonomic structure of communities have been successful (Norris and Thoms, 1999). Integrated approach of physical, chemical and biological criteria generally includes toxicological relationships by measuring the dose-response curves, which generally yield linear decline in richness of macroinvertebrates (Norris and Thoms, 1999). Based on several criteria, Norris and Thoms (1999) give a clear picture of employing physical, chemical and biological interactions in assessment of lotic ecosystem health (*e.g.* habitat structure, energy source, flow regime, water quality and biotic interactions). Several habitat variables have been considered by AusRivAS program (*e.g.* longitude, latitude, alkalinity, altitude, distance from the source, catchment area, conductivity, stream slope, riparian width, cobble, boulder, stream order, discharge, sand, macrophyte taxa, flow pattern, macrophyte cover, shading, bedrock, stream width, riffle depth, pebble, edge bank vegetation, vegetation category, range of mean annual air temperature, and silt and clay) (Norris and Thoms, 1999).

The integrated approach to assess the health of lotic ecosystem has several strengths such as test site data independent of human influence, all reference sites are used for calculation of probabilities of taxa, predicted taxa are compared with observed site, comparisons of independent characteristics for site-specific assessment and assessment of observed vs. expected variable ratio (Norris and Thoms, 1999). A variety of ecosystem-level process has been used in stream monitoring. For instance, primary production of benthic algae and macrophytes, community respiration, sediment respiration, secondary production by macroinvertebrates, nitrogen fixation, enzymatic transformations in biofilms and leaf litter decomposition (see Gessner and Chauvet, 2002). Primary production by benthic algae will be an ideal indicator when light regimes of lotic ecosystem are altered (Gessner and Chauvet, 2002).

Leaf Litter Breakdown

For rehabilitation of degraded ecosystem, the degrading processes should be understood explicitly. Leaf litter breakdown (or decomposition) plays major roles in lotic food webs and energetics (Cummins *et al.*, 1989; Suberkropp, 1998). The litter breakdown rates are higher in lotic ecosystem than terrestrial ecosystem. Relatively, studies on detritus decomposition in lotic ecosystem are scarce in tropical regions compared to temperate regions. Leaf decomposition in lotic ecosystem involves a wide range of organisms (*e.g.* bacteria, fungi, invertebrates and fishes) and human interference of leaf breakdown will be an appropriate diagnostic tool in assessing the functional integrity of lotic ecosystem (Gessner and Chauvet, 2002). Several interesting features have been emerged in using leaf litter breakdown in polluted habitats. Pollution by mine drainage strongly inhibits leaf litter breakdown process, which affects the richness of macroinvertebrates.

Significant decrease in leaf litter breakdown was seen in streams influenced by zinc input from mining activities (Niyogi *et al.*, 2001).

A wide range of factors influences the leaf litter decomposition in aquatic habitats (leaf qualities, water qualities, impact of pollutants, and biological components such as microbes and invertebrates) (Sridhar and Raviraja, 2001). In addition to leaf breakdown parameters, abundance or biomass of invertebrates, bacteria and fungi will help assessing the lotic ecosystem more effectively (Graça *et al.*, 2005). Similarly, examination of leaf parameters such as carbon, nitrogen and phosphorus alterations in leaf litter will also serve to evaluate the extent of decomposition in reference and impacted regions more precisely (see Graça *et al.*, 2005). The extent of anthropogenic influences on leaf litter breakdown and associated fauna and microbes can be evaluated using transplant experiments by switching the leaf litter bags between reference and impacted locations within or across the lotic habitats (*e.g.* Suberkropp, 1984; Rosset and Baerlocher, 1985; Sridhar *et al.*, 2005, 2009).

The leaf litter breakdown coefficient (*k*) will be a useful index for comparison of geographical regions, plant species and impact of pollutants. Speed of leaf litter breakdown is a most important standard measure to be considered (Petersen and Cummins, 1974). On the basis of coefficient of decay (*k*), Petersen and Cummins (1974) classified the substrates into fast ($k = > 0.010/day$), medium ($k = 0.005$-$0.010/day$) and slow ($k = < 0.005/day$). Firstly, time necessary to achieve 50% or 95% mass loss of leaf litter may be selected (Wallace *et al.*, 1996). Secondly, percent leaf litter mass remaining after a specific period exposed in a stream will be also useful (Jonsson *et al.*, 2001).

Exponential breakdown rate of leaf litter (*k*) serve as functional indicator and it can be calculated using temperature corrected model:

$$k = (1 \div t) \times \ln \times (Mf \div Mi)$$

Where, Mi is the initial ash-free dry mass (AFDM), Mf is the AFDM remaining after exposure and t is the cumulative temperature (in degree-days) during the exposure (Woodcock and Huryn, 2005).

The overall (*kc*) and microbial (*kf*) breakdown rates can be calculated using the coarse and fine mesh bags respectively. The ratio of *kc:kf* was used to evaluate the relative contribution of shredders and microorganisms to the leaf breakdown process (Gessner and Chauvet, 2002). The ratios of leaf breakdown coefficients in fine and coarse mesh bags will be powerful to evaluate the contribution of microorganisms vs. invertebrates (Lecerf *et al.*, 2006). Similarly, the rate of breakdown of fast and slow decomposing leaf litter will also be useful in evaluation of functional integrity of lotic ecosystem. The leaf breakdown rate (*k*) will be calculated irrespective of leaf mass remaining of leaf packs against time used (Table 2).

Table 2: Some examples of ratios leaf litter breakdown coefficients at impacted (ki) and references (kr) in streams influenced by human impact

Stress	ki:kr (%)	Plant species	Location	Reference
Mine drainage				
Copper	45-55	Acer rubrum	Virginia, USA	Schultheis *et al.* (1997)
Zinc	18-125	Salix spp.	Colorado, USA	Niyogi *et al.* (2001)
Heavy metals	60-88	Alnus glutinosa	Germany	Sridhar *et al.* (2001)
Nutrients				
Nitrate	83-90	Acer circinatum	Washington, USA	Whiles *et al.* (1993)
Phosphate	120-127	Quercus rubra	Tennessee, USA	Elwood *et al.* (1981)
Vegetation				
Eucalypt	23-59	Eucalyptus globulus	Spain	Pozo *et al.*(1998)
Willow	800	Salix babylonica	Australia	Pidgeon and Cairns (1981)
Pollutants				
Chlorine	57-105	Potamogeton crispus	Michigan, USA	Newman *et al.* (1987)
Insecticide	42-49	Acer rubrum	North Carolina, USA	Whiles *et al.* (1993)
Acid precipitation	11	Fagus sylvatica	France	Dangles and Guérold (1998)
Thermal pollution	139-307	Acer saccharum	Virginia, USA	Paul *et al.* (1978)

The Shannon's diversity index [$H' = - \Sigma (p_i \times \ln p_i)$] (Magurran, 1988) and Pielou's equitability ($J' = H' \div H'_{max}$) (Pielou, 1975) represent structural indicators to assess the lotic ecosystems. Where p_i is the relative abundance of species i and H'_{max} is the maximum value of diversity for the number of fungal species present.

Leaf-Decaying Fungi

Besides diverse flora and fauna, a variety of microbes have been adapted to freshwater ecosystem. Recent molecular approaches indicated that aquatic microbial biodiversity is higher than inferred from classical or non-molecular evidence (Dudgion *et al.*, 2006). Among microbes, several groups of fungi have adapted to freshwater habitats (*e.g.* stramenophiles on leaves and animals, ascomycetes and their anamorphs on submerged wood, and aquatic and aero-aquatic hyphomycetes on leaf litter). Allochthonous (imported) plant litter constitutes a major part of energy source in freshwater streams and its breakdown is vital importance for stream food web and ecosystem functioning (Cummins, 1988; Baerlocher, 1992). Among heterotrophic microorganisms, aquatic hyphomycetes (also known as 'Ingoldian fungi') play an important role in litter decomposition and act as intermediaries of energy transfer to the higher tropic levels in freshwater streams (Suberkropp, 2003). Most studies on freshwater fungi belong to Europe, United Kingdom and United States. Information on freshwater fungi is scattered in the literature with a few reviews attempted to bring together on their distribution and methodology (Goh and Hyde, 1996; Wong *et al.*, 1998; Gessner *et al.*, 2003; Sivichai and Jones, 2003; Shearer *et al.*, 2007). From 1990 onwards several tropical freshwater habitats have been explored for freshwater fungi (*e.g.* Sridhar *et al.*, 1992; Goh, 1997). Studies carried out in tropical habitats revealed many new species of filamentous fungi with great diversity, lifestyles and dispersal adaptations.

Plant detritus constitutes the major source of nutrition to aquatic hyphomycetes in streams (Baerlocher and Kendrick, 1974, 1981; Suberkropp and Klug, 1976) as they produce array of exoenzymes (Chamier, 1985) and such conditioned plant litter become more palatable to leaf-eating invertebrates (Suberkropp, 1992, 2003). Out of about 350 species of aquatic hyphomycetes described, only 10% have been connected to teleomorphs by conventional methods (Webster, 1992; Marvanová, 1997; Sivichai *et al.* 2002; Sivichai and Jones, 2003). Based on molecular evidences, this fungal ecological group is now considered as a polyphyletic group having affinities mainly with ascomycetes (Belliveau and Baerlocher, 2005; Baschien *et al.*, 2006). Solé *et al.* (2008) showed that aquatic hyphomycete communities as potential indicators of assessment of lotic ecosystem. Baerlocher (2009) proposed theory of cyclic patterns of events of aquatic hyphomycetes colonizing leaves in streams and designated as 'boom-bust cycle': 1) rate of arrival of conidia on leaf litter; 2) rate of release of conidia (growth rates) from leaf litter; 3) fungal biomass (respiration) on leaf litter. These cyclic (or somewhat successional) events depend on the seasonal input of leaf litter into streams and perturbation (e.g. addition of inorganic nutrients) may accelerate (expand) and compress (shorten) into a smaller period. Fungal production in streams is reported between 16 and 46 g/ m²/annum (Methvin and Suberkropp, 2003; Carter and Suberkropp, 2004) and up to 17% of leaf detritus mass consists of fungal mycelia (Gessner, 1997). Aquatic hyphomycetes are known to channel over 50% of their biomass into conidial production (Maharning and Baerlocher, 1996; Chauvet and Suberkropp, 1998). Besides possibilities of quick assessment of diversity of aquatic hyphomycetes, detritus colonizing pattern, growth, sporulation and spore germination also serve the assessment of lotic ecosystem more precisely. As they have cosmopolitan distribution and isolated from a wide geographical range from sub-arctic stream (Mueller Haeckel and Marvanová, 1979) to desert stream (Al-Riyami *et al.*, 2009), they will be ideal candidates for evaluation using of their diversity and ecological functions in pristine and human impacted freshwater lotic ecosystems.

Outlook

The loss of lotic health mainly depends on the transmission of perturbation through space and catchment hierarchy and movement of impacts through the ecological space of aquatic food webs. Overall river ecosystem health assessment depends on the data acquired from the small streams. The potential questions are: what are the warning signals of deterioration of lotic ecosystem and what are the appropriate immediate measures to be focused for restoration of health? As single-species index will not reflect the health of lotic ecosystem and different kinds of disturbances, techniques which integrate with structural and functional assessment are necessary. To meet the objectives of structural and functional assessment of lotic ecosystem more precisely, evaluation of leaf litter breakdown and associated invertebrates and microbes will be highly useful. The leaf litter breakdown is influenced by several internal and external factors, which influence the breakdown as well as associated fauna and microbes. Aquatic hyphomycetes are known for morphologically versatile conidial shapes (sigmoid and multiradiate), enormous conidial production and easily identifiable conidial features. Leaf litter decomposition and involvement of aquatic hyphomycetes in decomposition will meet the requirement of both structural and functional assessment of lotic ecosystem.   Of late, connecting rivers for sustainable use was proposed in India, which was highly debatable. What are the results of such major human interference on lotic ecosystem and will 'biocriteria' fulfill the assessment more precisely in small-scale alterations is an immediate task to be addressed. Further, evaluation of lotic ecosystem needs to satisfy the impacts of global warming and climate change.

Acknowledgement

The author is grateful to Mangalore University for grant of study leave during the academic year 2008-09.

References

Al-Riyami, M., Victor, R., Seena, S., Elshafie, A.E. and Baerlocher, F. 2009. Leaf Decomposition in a Mountain Stream in the Sultanate of Oman. *Int. Rev. Hydrobiol. 94*, 16–28.

Arthington, A.H., Naiman, R.J., McClain, M.E. and Nilsson, C. 2010. Preserving the biodiversity and ecological services of rivers: new challenges and research opportunities. *Freshwat. Biol. 55*, 1–16.

Baerlocher, F. 1992. *The Ecology of Aquatic Hyphomycetes*. Springer-Verlag, Berlin.

Baerlocher, F. 2009. Reproduction and dispersal in aquatic hyphomycetes. *Mycoscience 50*, 3–8.

Baerlocher, F. and Kendrick, B. 1974. Dynamics of the fungal populations on leaves in a stream. *Journal of Ecology 62*, 761–791.

Baerlocher, F. and Kendrick, B., 1981: Role of aquatic hyphomycetes in the trophic structure of streams. – In: *The Fungal Community: Its Organization and Role in the Ecosystem*. Wicklow, D.T. and Carroll, G.C. (Ed.), Marcel Dekker Inc., New York, 743–760.

Barbour, M.T., Gerritsen J., Snyder, B.D. and Stribling, J.B. 1999. *Rapid Bioassessment Protocols for use in Streams and Wadeable Rivers: Periphyton, Benthic Macroinvertebrates and Fish*. EPA 841-B-99-002. U.S. Environmental Protection Agency, Office of Water, Washington DC, USA.

Baschien, C., Marvanová, L. and Szewzyk, U. 2006. Phylogeny of selected aquatic hyphomycetes based on morphological and molecular data. *Nova Hedwigia 83*, 311–352.

Belliveau, M.J.R. and Baerlocher, F. 2005. Molecular evidence confirms multiple origins of aquatic hyphomycetes. *Mycol. Res. 109*, 1407–1417.

Boulton, A.J. 1999. An overview of river health assessment: philosophies, practice, problems and prognosis. *Freshwat. Biol. 41*, 469–479.

Bunn, S.E., and Davies P.M. 2000. Biological processes in running waters and their implications for the assessment of ecological integrity. *Hydrobiologia 422/423*, 61–70.

Carins, J. and McCormick, P.V. 1992. Developing an ecosystem-based capability for ecological risk assessments. *The Environmental Professional 14*, 186–196.

Carter, M.D. and Suberkropp, K. 2004. Respiration and annual fungal production associated with decomposing leaf litter in two streams. *Freshwat. Biol. 49*, 1112–1122.

Chamier, A.-C. 1985. Cell-wall-degrading enzymes of aquatic hyphomycetes: a review. *Bot. J. Linn. Soc. 91*, 67–81.

Chauvet, E. and Suberkropp, K. 1998. Temperature and sporulation of aquatic hyphomycetes. *Appl. Environ. Microbiol. 64*, 1522–1525.

Connell, J.H. 1978. Diversity in tropical rain forests and coral reefs. *Science 199*, 1302–1309.

Cummins, K.W., 1988: The study of stream ecosystems: a functional view. – In: *Concepts of Ecosystem Ecology: A Comparative View*. Pomery, L.R. and Alberts, J.J. (Ed.), Springer-Verlag, New York, 247–262.

Cummins, K.W., Wilzbach, M.A., Gates, D.M., Perry, J.B. and Taliaferro, W.B. 1989. Shredders and riparian vegetation. *BioScience 39*, 24–30.

Dangles, O. and Guérold, F. 1998. A comparative study of beech leaf breakdown, energy content, and associated fauna in acidic and non-acidic streams. *Arch. Hydrobiol. 144*, 25–39.

Dudgeon, D., Arthington, A.H., Gessner, M.O., Kawabata, Z.-I., Knowler, D.J., Lévêque, C., Naiman, R.J., Prieur-Richard, A.-H., Soto, D., Stiassny, M.L.J. and Sullivan, C.A. 2006. Freshwater biodiversity: importance, threats, status and conservation challenges. *Biol. Rev. 81*, 163–182.

Elwood, J.W., Newbold, J.D., Trimble, A.F. and Stark, R.W. 1981. The limiting role of phosphorus in a woodland stream ecosystem: effects of P enrichment on leaf decomposition and primary producers. *Ecology 62*, 146–158.

Gessner, M.O. 1997. Fungal biomass, production and sporulation associated with particulate organic matter in streams. *Limnetica 13*, 33–44.

Gessner, M.O. and Chauvet, E. 2002. A case for using litter breakdown to assess functional stream integrity. *Ecol. Appl. 12*, 498–510.

Gessner, M.O., Baerlocher, F. and Chauvet, E. 2003. Biomass, growth and sporulation of aquatic hyphomycetes. *Fungal Diversity Research Series # 10*, 127–158.

Goh, T.K. 1997. Tropical freshwater hyphomycetes. In: *Biodiversity of Tropical Microfungi*. Hyde, K.D. (Ed.), Hong Kong University Press, Hong Kong, 189–227.

Goh, T.K. and Hyde, K.D. 1996. Biodiversity of aquatic fungi. *J. Ind. Microbiol. 17,* 328–345.

Gilbert, J. and Deharveng, L. 2002. Subterranean ecosystems: a truncated functional biodiversity. *BioScience 52,* 473–481.

Gleick, P.H. 1996. Water resources. In: *Encyclopedia of Climate and Weather.* Schneider S.H. (Ed.), Oxford University Press, New York, 817–823.

Graça, M.A.S., Baerlocher, F. and Gessner, M.O. 2005. *Methods to Study Litter Decomposition - A Practical Guide.* Springer, The Netherlands.

Graça, M.A.S., Pozo, J., Canhoto, C. and Elosegui, A. 2002. Effects of *Eucalyptus* plantations on detritus, decomposers and detritivores in streams. *The Scientific World 2,* 1173–1185.

Hannah, L., Lohse, D., Hutchinson, C., Carr, J.L. and Lankerani, A. 1994. A preliminary inventory of human disturbance of world ecosystems. *Ambio 23,* 246-250.

Harding, J. S., Benfield, E.F., Bolstad, P.V., Helfman, G.S. and Jones III, E.B.D.. 1998. Stream biodiversity: The ghost of land use past. *Proc. Natl. Acad. Sci. 95,* 14843–14847.

Hart, B.T. Maher, W. and Lawrence, I. 1999. New generation water quality guidelines for ecosystem protection. *Freshwat. Biol. 41,* 347–359.

Hawksworth, D.J. and Kalin-Arroyo, M.T. 1995. *Magnitude and Distribution of Biodiversity in Global Biodiversity Assessment.* Heywood, V.H. (Ed.), Cambridge University Press, Cambridge, 107–191.

Hill, B.H., Herlihy, A.T., Kaufmann, P.R., Stevenson, R.J., McCormick, F.H. and Johnson C.B. 2000. Use of periphyton assemblage data as an index of biotic integrity. *J. N. Am. Benthol. Soc. 19,* 50–67.

Jenkins, M. 2003. Prospects for biodiversity. *Science 302,* 1175–1177.

Jonsson, M., Malmqvist, B. and Hoffsten P.-O. 2001. Leaf litter breakdown rates in boreal streams: does shredder species richness matter? *Freshwat. Biol. 46,* 161–171.

Karr, J.R. 1991. Biological integrity: a long neglected aspect of water resource management. *Ecol. Appl. 1,* 66–84.

Karr, J.R. 1996. Ecological integrity and ecological health are not same. In: *Engineering Within Ecological Constraints.* Schultze P.C. (Ed.), National Academy Press, Washington DC, 97–109.

Karr, J.R., Fausch, K.D., Angermeter, P.L., Yant, P.R. and Schlosser, I.J. 1986. *Assessing Biological Integrity in Running Waters: A Method and its Rationale.* Illinois Natural History Survey, Champaign, IL, Special Publication # 5.

Ladson, A.R., White, L.J., Doolan, J.A., Finlayson, B.L., Hart, B.T., Lake, P.S., and Tilleard, J.W. 1999. Development and testing of an Index of Stream Condition for water way management in Australia. *Freshwat. Biol. 41,* 453–468.

Lankerani, A. 1994. A preliminary inventory of human disturbance of world ecosystems. *Ambio 23,* 246–250.

Lecerf, A. and Chauvet, E. 2008. Diversity and functions of leaf-decaying fungi in human-altered streams. *Freshwat. Biol. 53*, 1658–1672.

Lecerf, A., Usseglio-Poplatera, P., Charcosset, J.-E., Lambrigot, D., Bracht, B. and Chauvet, E. 2006. Assessment of functional integrity of eutrophic streams using litter breakdown and benthic macroinvertebrates. *Arch. Hydrobiol. 165*, 105–126.

Magurran, A.E. 1988. Ecological Diversity and Its Measurement. Princeton University Press, Cambridge.

Maharning, A.R. and Baerlocher, F. 1996. Growth and reproduction in aquatic hyphomycetes. *Mycologia 88*, 80–88.

Marchant, R., Norris, R.H. and Milligan, A. 2006. Evaluation and application of methods for biological assessment of streams: summary of papers. *Hydrobiologia 572*, 1573-5117.

Marvanová, L. 1997. Freshwater hyphomycetes: a survey with remarks on tropical taxa. In: *Tropical Mycology*. Janardhanan, K.K., Rajendran, C., Natarajan, K. and Hawksworth, D.L. (Ed.), Science Publishers Inc., New Hampshire, USA, 169–226.

Medeiros, A.O., Pascoal, C. and Graça, M.A.S. 2009. Diversity and activity of aquatic fungi under low oxygen conditions. *Freshwat. Biol. 54*, 142-149.

Mesquita, A., Pascoal, D. and Cássio, F. 2007. Assessing effects of eutrophication in streams based on breakdown of eucalypt leaves. *Fundamentals Appl. Limnol. 168*, 221–230.

Methvin, B.R. and Suberkropp, K. 2003. Annual production of leaf-decaying fungi in 2 streams. *J. N. Am. Benthol. Soc. 22*, 554–564.

Minshall, G.W. 1996. Bringing biology back into water quality assessments. In: *Committee on Inland Aquatic Ecosystems - Freshwater Ecosystems: Revitalizing Educational Programs in Limnology*. Water Science and Technology Board, Commission on Geosciences, Environment and Resources, National Research Council, USA, 289–324.

Mueller Haeckel, A. and Marvanová, L. 1979. Periodicity of aquatic hyphomycetes in the subarctic. *Trans. Br. Mycol. Soc. 73*, 109–116.

Newman, R. M., J. A. Perry, E. Tam, and R. L. Crawford. 1987. Effects of chronic chlorine exposure on litter processing in outdoor experimental streams. *Freshwat. Biol. 18*, 415–428.

Niyogi, D.K., Lewis Jr., W.M. and McKnight, D.M. 2001. Litter breakdown in mountain streams affected by acid mine drainage: biotic mediation of abiotic controls. *Ecol. Appl. 11*, 506–516.

Norris, R.H. and Hawkins C.P. 2000. Monitoring river health. *Hydrobiologia 435*, 5–17.

Norris, R.H. and Norris, K.H. 1995. The need for biological assessment of water quality: Australian perspective. *Aust. J. Ecol. 20*, 1–6.

Norris, R.H. and Thoms, M.C. 1999. What is river health? *Freshwat. Biol. 41*, 197–209.

Parsons, M., Thoms, M.C. and Norris, R.H. 2004. Development of a standardised approach to river habitat assessment in Australia. *Environ. Monit. Assoc. 98*, 109-130.

Pascoal, C., Cássio, F., Marcotegui, A., Sanz, B. and Gomes, P. 2005. Role of fungi, bacteria and invertebrates in leaf litter breakdown in a polluted river. *J. N. Am. Benthol. Soc. 24*, 784–797.

Paul Jr., R.W., Benfield, E.F. and Cairns Jr., J. 1978. Effects of thermal discharge on leaf decomposition in a river ecosystem. *Verh. Int. Verein, Theor. Angew. Limnol. 20*, 1759–1766.

Petersen, R.C. and Cummins, K.W. 1974. Leaf processing in woodland stream. *Freshwat. Biol. 4*, 343–368.

Pidgeon, R.W.J., and Cairns, S.C. 1981. Decomposition and colonisation by invertebrates of native and exotic leaf material in a small stream in New England (Australia). *Hydrobiologia 77*, 113–127.

Pielou, E.C. 1975. *Ecological Diversity*. John Wiley, New York.

Pozo, J., Basaguren, A., Elo´segui, A., Molinero, J., Fabre, E. and Chauvet E. 1998. Afforestation with *Eucalyptus globulus* and leaf litter decomposition in streams of northern Spain. *Hydrobiologia 373/374*, 101–109.

Raviraja, N.S., Sridhar, K.R. and Baerlocher, F. 1998. Breakdown of *Ficus* and *Eucalyptus* leaves in an organically polluted river in India: fungal diversity and ecological functions. *Freshwat. Biol. 39*, 537–545.

Resh, V.H., Myers, M.J. and Hannaford, M.J. 1996. Macroinvertebrates as biotic indicators of environmental quality. In: *Methods in Stream Ecology*. Hauer, F.R. and Lamberti, G.A. (Ed.), Academic Press, San Diego, 647–667.

Rosenberg, D.M., McCully, P. and Pringle C.M. 2000. Global-scale environmental effects of hydrological alterations: introduction. *BioScience 50*, 746–751.

Rosset, J. and Baerlocher, F. 1985. Transplant experiments with aquatic hyphomycetes. *Verh. Internat. Verein. Limnol. 22*, 2786–2790.

Sala, O.E., Chapin, F.S., Armesto, J.J., Berlow, R., Bloomfield, J., Dirzo, R., Huber-Sanwald, E., Huenneke, L.F., Jackson, R.B., Kinzig, A., Leemans, R., Lodge, D., Mooney, H.A., Oesterheld, M., Poff, N.L., Sykes, M.T., Walker, B.H., Walker, M. and Wall, D.H. 2000. Global biodiversity scenarios for the year 2100. *Science 287*, 1770–1774.

Schultheis, A.S., Sanchez, M. and Hendricks A.C. 1997. Structural and functional responses of stream insects to copper pollution. *Hydrobiologia 346*, 85–93.

Shearer, C.A., Descals, E., Kohlmeyer, B., Kohlmeyer, J., Marvanová, L., Padgett, D., Porter, D., Raja, H.A., Schmit, J.P., Thorton, H.A. and Voglymayr, H. 2007. Fungal biodiversity in aquatic habitats. *Biodivers. Conser. 16*, 49-67.

Sivichai, S. and Jones, E.B.G. 2003. Teleomorphic-anamorphic connections of freshwater fungi. *Fungal Diversity Research Series # 10*, 259–272.

Sivichai, S., Jones, E.B.G. and Hywel-Jones, N.L. 2002. Lignicolous freshwater higher fungi with reference to their teleomorph and anamorph stages. In: *Tropical Mycology: Micromycetes.* Watling, R., Frankland, J.C., Ainsworth, A.M., Isaac, S. and Robinson, C.H. (Ed.), CAB International, UK, 41–49.

Solé, M., Fetzer, I., Wennrich, R., Sridhar, K.R., Harms, H. and Krauss, G. 2008. Aquatic hyphomycete communities as potential bioindicators for assessing anthropogenic stress. *Sci. Total Environ. 389,* 557–565.

Sridhar, K.R. and Raviraja, N.S. 2001. Aquatic hyphomycetes and leaf litter processing in unpolluted and polluted habitats. In: *Trichomycetes and Other Fungal Groups.* Misra, J.K. and Horn, B.W. (Ed.). Science Publishers, Enfield, New Hempshire, USA, 293–314.

Sridhar, K.R., Chandrashekar, K.R. and Kaveriappa, K.M. 1992. Research on the Indian subcontinent. In: *The Ecology of Aquatic Hyphomycetes.* Baerlocher, F. (Ed.), Springer-Verlag, Berlin, 182–211.

Sridhar, K.R., Krauss, G., Baerlocher, F., Raviraja, N.S., Wennrich, R., Baumann, R. and Krauss, G.-J. 2001. Decomposition of alder leaves in two heavy metal polluted streams in Central Germany. *Aq. Microb. Ecol. 26,* 73–80.

Sridhar, K.R., Baerlocher, F., Krauss, G.-J. and Krauss, G. 2005. Response of Aquatic Hyphomycete communities to changes in heavy metal exposure. *Int. Rev. Hydrobiol. 90,* 21–32.

Sridhar, K.R., Duarte, S., Cássio, F. and Pascoal, C. 2009. The role of early fungal colonizers in leaf-litter decomposition in streams impacted by agricultural runoff. *Int. Rev. Hydrobiol. 94,* 399-409.

Statzner, B., Bis, B., Dolédec, S. and Usseglio-Polatera, P. 2001. Perspectives for biomonitoring at large scales: a unified measure for the functional composition of invertebrate communities in European running waters. *Basic and Appl. Ecol. 2,* 73–85.

Suberkropp, K. 1984. Effect of temperature on seasonal occurrence of aquatic hyphomycetes. *Trans. Br. Mycol. Soc. 82,* 53–62.

Suberkropp, K. 1992. Interactions with invertebrates. In: *The Ecology of Aquatic Hyphomycetes.* Baerlocher, F. (Ed.), Springer-Verlag, Berlin, 118–134.

Suberkropp, K. 1998: Microorganisms and organic matter decomposition. In: *River Ecology and Management: Lessons from the Pacific Coastal Ecoregion.* Naiman, R.J. and Bilby, R.E. (Ed.), Springer-Verlag, New York, 120–143.

Suberkropp, K 2003. Methods for examining interactions between freshwater fungi and macroinvertebrates. In: Freshwater mycology. Tsui, C.K.M. and Hyde, K.D. (Ed.), Fungal diversity Press, Hong Kong, 159–171.

Suberkropp, K. and Klug, M.J. 1976. Fungi and bacteria associated with leaves during processing in a woodland stream. *Ecology 57,* 707–719.

Townsend, C.R. and Riley, R.H. 1999. Assessment of river health: accounting for perturbation pathways in physical and ecological space. *Freshwat. Biol. 41,* 393–405.

Townsend, C.R., Scarsbrook, M.R. and Doledec, S. 1997a. Qualifying disturbance in streams: alternative measures of disturbance in relation to macroinvertebrate species traits and species richness. *J. N. Am. Benthol. Soc. 16*, 531–544.

Townsend, C.R., Arbuckle, C.J., Crowl, T.A. and Scarsbrook, M.R. 1997b. The relationship between land use and physicochemistry, food resources and macroinvertebrte communities in tributaries of the Taieri River, New Zealand: a hierarchically scaled approach. *Freshwat. Biol. 37*, 177–191.

Vitousek, P. M., H. A. Mooney, J. Lubchenco, and J. M. Melillo. 1997. Human domination of Earth's ecosystems. *Science 277*, 494–499.

Wallace, J. B., J. W. Grubaugh, and M. R. Whiles. 1996. Biotic indices and stream ecosystem processes: results from an experimental study. *Ecol. Appl. 61*:140–151.

Webster, J. 1992. Anamorph and teleomorph relationships. In: *The Ecology of Aquatic Hyphomycetes*. Baerlocher, F. (Ed.), Springer-Verlag, Berlin, 99–117.

Whiles, M. R., J. B. Wallace, and K. Chung. 1993. The influence of *Lepidostoma* (Trichoptera: Lepidostomatidae) on recovery of leaf-litter processing in disturbed headwater streams. *Am. Midland Natur. 130*, 356–363.

Wong, M.K.M., Goh, T.K., Hodgkiss, I.J., Hye, K.D., Ranghoo, V.M., Tsui, C.K.M., Ho, W.H., Wong, W.S.W. and Yuen, T.K. 1998. Role of fungi in freshwater ecosystem. *Biodeivers. Conser. 7*, 1187–1206.

Woodcock, T.S. and Huryn, A.D. 2005. Leaf litter processing and invertebrate assemblages along a pollution gradient in a Maine (USA) headwater stream. *Environ. Pollut. 134*, 363–375.

Wright, J.F. 1995. Development and use of a system for predicting macroinvertebrates in flowing waters. *Aust. J. Ecol. 20*, 181–197.

Young, R.G. and Huryn, A.D. 1997. Longitudinal patterns of organic matter transport and turnover along a New Zealand grassland river. *Freshwat. Biol. 53*, 2199–2211.

Chapter 3

Conservation and Management of Biodiversity through Cultural Practices in Sacred Groves

A .G. Devi Prasad and Shwetha

DOS in Environmental Science, University of Mysore, Manasagangotri, Mysore-06, Karnataka

ABSTRACT

Sacred groves are a biological heritage and a system that has helped to preserve the representative genetic resources existing in the surrounding regions for generations. They represent our forefather's reverential attitude towards nature and its all life forms. They are distributed around the globe. In India, about 13,720 sacred groves have been enumerated so far from 19 states. Vast diversity among these sacred groves has been observed such as bamboo groves on the eastern coast to clumps of trees in the northwestern deserts, and from scrub jungles in the tropical south to dense Himalayan forests in the north. Many endemic species such as Cycas beddomei, Gymnakranthera canarica, Semecarpus kathalekanesis have been recorded.

Sacred groves have well developed forest ecosystems and high degree of species richness. This, however, depends on the extent of preservation of the grove. Many, rather most, of these pristine ecosystems have either vanished or are disturbed to a great extent. Of late, erosion of traditional belief system and nature worship system, conversion and fragmentation of the area due to agriculture, mining or other projects and lack of interest in youths and poor management have weakened the sanctity of these groves. Under these circumstances, preservation, conservation and management of the sacred groves are an important necessity and warrants top priority.

Key words: Sanctity, Endemic species, Deity, Taboos, Sanskritization

Introduction

Prehistorically, conservation of natural resources has been an integral part of diverse cultures in different ways. The traditional worship practices show the symbiotic relation of human beings and nature. Human, for various needs, ranging from economic to belief system has utilized natural resources, which have been gradually integrated and incorporated in the social system of human beings by means of various institutional procedures. The economically developed world as well as some of the underdeveloped economies still depend on their own and the world's forest for survival, yet in different ways and meanings (Seeland, 1997). Nevertheless, communities all over the world lived in harmony with the nature and conserved its valuable biodiversity.

Various anthropogenic activities have altered the structure and function of different ecosystems all over the world. One of the most conspicuous effects of ecosystem perturbation has been the depletion of biodiversity. Disappearance of species due to habitat alteration, overexploitation, pollution, global climate change and invasion of exotic species is so rapid that many valuable taxa may vanish even before they are identified and their scientific value is discovered. Conservation and sustainable management of natural resources therefore has become central theme for almost all societies in order to meet shortages in timber or non-timber forest produce and overcome the impact of environmental pollution. Economic development in times of increasing environmental hazards tends to create a dilemma in many societies which is bound to grow despite the often narrow limitations of declining resource base (Seeland and Schmithusen, 2000).

In this changing scenario of unabated exploitation of natural resources, many areas have been identified and declared as protected areas and various *in-situ* and *ex-situ* conservation practices have been undertaken in different parts of the world. Of these, traditional conservation practices of indigenous communities for conservation and protection of biodiversity is gaining attention and utmost importance. A good example of such traditional practices is the conservation and protection of small forest patches by dedicating them to the local deities by various indigenous communities of the world. Such forest patches are called "sacred groves" (Khan *et al.*, 2008).

Sacred groves have existed for many years all over the world, harbouring rich and unique flora and fauna. They are the tracts of virgin forests that were left untouched and protected by the local inhabitants, due to their cultural and religious beliefs and taboos that the deities reside in them. They are the unique and significant, examples of *in-situ* biodiversity conservation, ensuring the first major effort to recognize and conserve biodiversity.

Sacred groves can be defined as nature worship, where all forms of vegetation including shrubs and climbers are under the protection of the reigning deity of the grove and the removal of any material; even dead wood of twig is a taboo. This preservation of the entire vegetation in association with a deity is quite distinct phenomenon from the preservation of isolated specimens of sacred tree species such as Peepal *(Ficus religiosa)* which are often preserved and worshipped even without any association with a deity (Gadgil and Vartak, 1976:152). These sacred elements such as woods, forests, rivers, streams, rocks, mountains, peaks and trees equated to ancestral spirits or deities are found throughout the world. Embedded in traditional cultural and religious belief systems, they provide a cultural identity to each community, besides contributing environmental conservation. (Bhagwat and Rutte, 2006).

Origin of the Groves

Kosambi (1962) traces the origin of sacred groves to the pre-agrarian hunting-gathering phase of human civilization. This view is supported by Chandran (1997), who traced the ecological history of Western Ghats. Historically, when the hunter-gatherer society took up cultivation they left segments of the original forest undisturbed either out of reverence or as a mark of gratitude. Sooner or later, clear felling surpassed the restoration activities due to population pressure and threatened the very survival of forests. Unable to arrest this pernicious trend, which stemmed from avarice than prudence, they sanctified forests and trees (Rappaport 1971; Vannucci 1991). As the population swelled and societies broke down, people migrated to the plains which were also covered largely by forests. To take up agriculture, forests had to be cleared on a massive scale but the lingering ecological wisdom prompted them to leave patches of original vegetation undisturbed as an index of their reverence and declared them sacred and inviolable.

In Indian context, they represent a new dimension of eco-culture of the most pre-Brahminic agricultural societies in India, particularly of the tribe-dominated mountainous tracts. They are found to exist in the villages of homogeneous communities (tribes) as well as heterogeneous communities (caste groups) (Chandran *et al.*, 1998).

Though the link between hunter-gatherer and shifting cultivators and the origin of groves has been widely accepted, there are other reasons for their existence such as ancestor worshiping. For example, several groves on the hilly interiors of Kerala have tall monolithic memorials eulogizing the sacrifice of warriors and the service of the elders. Amorphous stones are still installed on the burial places and worshipped by the tribals of Kerala. They believe that the departed souls rest in peace there (Basha, 1998).

Distribution of Sacred Groves around the World

Sacred groves are distributed across the globe, acting as an ideal centre for biodiversity conservation and diverse cultures recognize them in different ways encoding various rules for their protection. In India as well as in parts of Asia and Africa, care and respect for nature has been influenced by religious beliefs and indigenous practices.

The existence of sacred groves has been reported in many parts of Africa, Asia, Europe, Australia and America by Hughes and Chandra (1998). In East Africa, among the Kikuyu, Frazer (1935) reports that the 'mugumu' trees are sacred. In West Africa, the aura of holiness has kept botanically rich groves from falling prey to lumber companies (Lebbie and Guries, 1995). For nearly three centuries, the community of Malshegu in Northern Ghana has preserved the sacred groves that they believe house a local spirit. There are many sacred groves associated with villages in the southern nations of Africa such as Zimbabwe and South Africa.

Sacred groves exist in most regions of Asia, and they are associated with many different traditions. In Western Asia, one of the earliest literary traditions deriving from ancient Sumeria reveals their existence. In north Asia, many Siberian people honoured sacred groves and in the sacred groves nothing could be touched, no grass or wood might be cut, no game hunted, no fish caught, nor a drop of water drunk (Frazer, 1935). In North Asia, many Siberian people honoured sacred groves and used them for the ritual practices. Many Buddhist temples both in Japan and China have carefully tended gardens, including trees that have aspect of groves. traditional Chinese religion honoured sacred mountains which were tree covered and Taoism is well-known for its respect for nature; its shrines and hermitages were often located in forests.

Several small size sacred groves were reported from Nepal by Ingles (1994). Various sacred sites associated with rich vegetation in Bangladesh were reported by Hussain (1998). The Dubla Island sacred grove in Sundarbans mangrove forest in Bangladesh harbours rich vegetation and is a place of worship for low caste Hindus, who visit it once in a year for prayer (Islam *et al.*, 1998).

Europe had thousands of sacred groves in ancient times although most of them have disappeared (Hughes, 1984). The Celts, Slavs and Germans, the Romans' northern Barbarian neighbours, customarily worshipped in groves and regarded the oak as the most divine tree (Vest, 1985). Australian aborigines found groves among the many natural spots made sacred by the experience of ancestral spirits (Molyneaux, 1995). In the Americas, a widespread view in pre-Columbian times was that earth and all her creatures are sacred. Trees were seen as having spirits and the power to help or hurt, their permission would have to be sought before any part of them was taken or used (Hughes, 1996).

Indian Scenario

India has long tradition of nature conservation as part of its culture. Sacred trees, animals, groves and landscapes are few of the living examples of this unique tradition. The groves are located in a variety of habitats ranging from resource-rich forested landscapes, such as in the Western Ghats and north eastern part of the country to extremely resource-poor desert conditions in western and central India. Though clearly a manifestation of nature worship, it is likely that the groves also performed specific functions under these different habitat conditions.

In India, the earliest documented work on sacred grove is that of the first Inspector General of Forests, Dietrich Brandis in 1897. Later, Gadgil and Vartak (1976, 1981ab) traced the historical link of sacred groves with the pre-agricultural, hunting and gathering stage, before human being had settled down to raise livestock. Most of the sacred groves reported from India are in the Western Ghats, North Eastern India and Central India (Gadgil and Vartak 1976, Burman 1992, Balasubramanyam and Induchoodan 1996, Tripathi 2001, Khumbongmayum *et al.*, 2005a).

Sacred groves have been reported in Meghalaya (Boojh and Ramakrishnan 1983, Ramakrishnan 1996, Tiwari *et al.*,. 1998a, Jamir 2002, Law 2002, Upadhaya 2002, Mishra *et al.*, 2004), Manipur (Khumbongmayum *et al.*,. 2005a), Western Ghats (Gadgil and Vartak 1976, Chandran *et al.*, 1998, Kushalappa *et al.*, 2001). Mitra and Pal (1994) also reported the occurrence of sacred groves in Meghalaya, Bihar, Rajasthan and the states along the Western Ghats. Their existence along the Himalaya, from northwest to northeast, was described by Burman (1992). Sacred mangroves, experiencing little or no damage at all, with some religious significance, were reported from Rann of Kutch, Maharashtra, Goa, Tamil Nadu and West Bengal (Untawale *et al.*, 1998).

Table:1 Sacred groves of India.

Sl No.	State	Number of Documented Sacred Groves
1.	Andhra Pradesh	750
2.	Arunachal Pradesh	58
3.	Assam	40
4.	Chhattisgarh	600
5.	Gujarat	29
6.	Haryana	248
7.	Himachal Pradesh	5,000
8.	Jharkhand	21
9.	Karnataka	1,531
10.	Kerala	2,000
11.	Maharashtra	1,600
12.	Manipur	365
13.	Meghalaya	79
14.	Odisha	322
15.	Rajasthan	9
16.	Sikkim	56
17.	Tamil Nadu	448
18.	Uttarkhand	1
19.	West Bengal	670
	Total	**13,827**

Source: Khiewtan and Ramakrishna (1989); Tripathi *et al.*, (1995); Malhotra *et al.*, (1997); Jamir and Pandey (2002).

In India, sacred groves are found mainly in tribal dominated areas and are known by different names in ethnic terms (Bhakat 1990) such as:

- *Sarna* or *Dev* in Madhya Pradesh,
- *Devrai* or *Deovani* or *Deorais* in Maharashtra,

- *Sarnas* in Bihar, *Orans* in Rajasthan,
- *Devaravana* or *Devarakadu* in Karnataka,
- *Sarpakavu* and *Kavu* in Tamil Nadu and Kerala,
- *Dev van* in Himachal Pradesh,
- *Machhiyals* in Uttharkhand,
- *Law Lyngdoh* or *Law Kyntang* etc. in Meghalaya,
- *Sarana* or *Jaherthan* in Jharkhand and *Lai umang* in Manipur.

They are protected and managed by local people on religious grounds and traditional beliefs. Wherever the sacred groves existed, the indigenous traditional societies, which have a spiritual relationship with their physical environment, sustained them.

It is very difficult to give the exact number of sacred groves in India. Malhotra (2000) in his partial enumeration of groves in India reports 13,827 sacred groves (Table 1) covering an area of about 39,063 ha and this is only an indication of the extent and magnitude of the presence of the institution of sacred groves in the country. However further inventory covering all regions of the country may lead to encounter as 1,00,000 to 15,00,000 sacred groves (Malhotra, 1998). They occur in many parts of India viz., Western Ghats, Central India, northeast India, etc. particularly where the indigenous communities live.

Diversity in Sacred Groves

There is vast diversity among India's sacred groves from bamboo groves on the eastern coast to clumps of trees in the northwestern deserts, and from scrub jungles in the tropical south to dense Himalayan forests in the north.

Types of Groves

Godbole *et al.*, (1998) recognized three types of groves, namely informal, formal and memorial groves, on the basis of their functional attributes: (i) informal or traditional groves are those where no temple or idols exist; the entire vegetation is considered sacred. Generally, such groves are at a distance from village and are not visited frequently; there is another supplementary grove near or within the village with temples and idols, where the rituals and festivals are performed regularly. (ii) the formal or temple groves are those dedicated to a village God or deity. It is also the place of important festivals like Navarathri, Holi, or Diwali and ceremonial rites and rituals are organized within. (iii) Memorial or Burial or Cremation groves have clusters of trees raised and nurtured on the site of cremation/burial ground in memory of village elders.

Number and Area of Groves

It may be recalled that D. Brandis, the first Inspector General of Forests of British India found them abundantly in hills of North and South India. Almost every village in the Sahaydri-Konkan region (North Western Ghats) has at least one Sacred Grove ranging from just a few acres to hundreds of acres. Ward and Conner (1894) observed thousands of groves, called Kavu in Malayalam, in Travancore region of Kerala like "dots on the leopard's skin" in Malabar region. About 15,000 Kavus existed prior to 1800 A.D. (Velupillai 1940) and they were innumerable in Malabar too (Logan, 1887). About 25000 groves are estimated in Rajasthan alone (Saxena *et al.,*. 1998). A grove exists in every three km2 in Uttara Kannada. Since there has been no comprehensive survey of sacred groves in India, their exact number and area is unknown. However, Malhotra *et al.,*. (2001) opine that in view of their known presence and pattern of distribution of sacred groves in Chhattisgarh, Jharkhand, Orissa, Uttaranchal, Madhya Pradesh and West Bengal, for which detailed inventories are not available, the number of sacred groves in India is likely to be between 100,000 and 150,000 (Vajpeyi, 2001).

Some groves contain only a few trees, while many others cross hundreds of acres in size. Sometimes groves overlap with larger forested areas, while others exist as islands in open plains or desert. Amirthalingam (1998) analyzed the size class of 448 groves in Tamil Nadu and found 20% of them very small (<1 ha); rarely they measure 20 ha in the plains; in the hills only 58 of them exceeded 100 ha. Even in the mountainous regions of Western Ghats 399 out of 761 groves (52.17%) were of less than 0.02 ha in size; only 41 were in 1-5 ha range and 11 exceeded 5 ha. (Induchoodan 1996). Exceptionally, a few groves like Haryali in Himachal Pradesh and Orans of Aravalli region are very large exceeding 1000 ha. (Singh ,2011). And this seems to be the trend all over India.

Belief and Taboo system

Reasonably, ancient cultures have imposed restrictions, regulations and handed down hard prescriptions mainly to arrest the attitudinal change towards the vitals of conservation such as:

- Restricted human activities within the grove with exception of festivals and prayer times.
- Taboos and folklores to restrain people from disturbing the forest: violations of restrictions might cause crop failure, epidemic strike or natural calamities.
- Prohibition of cattle grazing which was sometimes ensured by digging trenches along the periphery.
- Restriction to enter the groves for lower caste people and women during menstrual period.
- Prohibition of ploughing, sowing and erection of unauthorized structures.
- Complete ban on lopping or axing of trees: fallen twigs may be collected.
- No animal can be harmed within the grove.

These beliefs and taboos are the fundamental pillars in the system of sacred groves. The sanctity or the sacredness imposed on these sacred groves as a result of socio-cultural beliefs is the only driving force that made protection and conservation of these forest patches made possible till today.

Values of Sacred Groves

The sacred groves are not mere biophysical entities. They reflect the value system evolved and preserved through long years by the ancient communities. They are priceless treasures of great ecological, biological, cultural and historical values.

Abode of Endemic and Threatened Species

Sacred groves are considered as an asylum for many endemic and threatened, vulnerable and endangered species. More than 750 species of flowering plants have been enumerated in the sacred groves of Maharashtra. About 134 rare and endemic taxa survive in Andhra Pradesh groves. Many typical forest species like *Mesua ferrea, Vateria indica, Dysoxylum binectariferum* and *Leea guianensis* grow luxuriously in the Kans of Uttara Kannada (Table 2) (Chandran 1997). Broadly, the vegetation of Kerala kavus is of either evergreen or moist deciduous type. About 800 flowering species have been enumerated in them; 150 of them are endangered taxa and 40% of them are of medicinal value (Chandrasekara and Sankar 1998).

Ashalatha Devi *et al.,.,* (2005) recorded altogether 173 species from 4 sacred groves of Manipur State of India, out of which many species found to be disappeared from local area and exist only in the sacred groves, coming under threatened category such as: *Bombax ceiba, Anaphalis contorta, Bonnaya brachiata, Scutellaria discolor* and *Blumea hieracifolia* fall under the rare and vulnerable categories, *Dioscorea bulbifera* is the vulnerable species, *Alnus nepalensis, Schima wallichii* and *Pasania polystachya* are threatened species, *Sida rhombifolia* fall under vulnerable category and *Aphanamixis polystachya* is an endangered species.

The prevalence of some of the Critically Endangered tree species of the Western Ghats such as, *Syzygium travancoricum* and *Vateria indica* in some of the small groves studied in Siddapur Taluk, Uttara Kannda, Karanataka, which are obviously the relics of the primary forests, highlights that sacred groves can have a greater role especially in conservation of the threatened endemic biota if they are restored and managed well (Rajashry Ray, 2011). A total of 25 RET species were identified in the sacred groves of Kodagu by Kushalappa and Bhagwat (2001) some of which include; *Aegle marmelos, Canarium strictum, Cinnamomum sulphuratum, Cinnamomummacrocarpum, Dysoxylum malabaricum, Garcinia gummigutta, Myristica malabarica and Vateria indica.*

Table 2: Endemic species of sacred groves

No.	Species	Place	Reference
1.	*Boswellia ovalifoliata*	Tirupati Hills	Rao (1996)
2.	*Cinnamomum* sp.	Karnataka	Chandran *et al.,.* (1998)
3.	*Cycas beddomii*	Tirupati Hills	Raviprasada Rao and Jayaprada (1997)
4.	*Dipterocarpus indicus*	Uttara Kannada	Chandran *et al.,.* (1993)
5.	*Garcinia gummigutta*	Karnataka	Chandran *et al.,.* (1998)
6.	*Gymnakranthera canarica*	Myristica swamp, Uttara Kannada	Chandran (1997)

Condt...

Table 2: Condt...

7.	*Myristica fatua var. magnifica*	Myristica swamp, Uttara Kannada	Chandran (1997)
8.	*Myristica malabarica*	Karnataka	Chandran *et al.*,. (1998)
9.	*Pimpinella tirupatyensis*	Tirupati Hills	Raviprasada Rao and Jayaprada (1997)
10.	*Pinanga dicksonii*	Myristica swamp, Uttara Kannada	Chandran (1997)
11.	*Pterocarpus santalinus*	Tirupati Hills	Raviprasada Rao and Jayaprada (1997)
12.	*Semecarpus auriculata*	Myristica swamp, Uttara Kannada	Chandran (1997)
13.	*Shorea tumbaggaia*	Tirupati Hills	Raviprasada Rao and Jayaprada (1997)
14.	*Syzygium alternifolium*	Tirupati Hills	Raviprasada Rao and Jayaprada (1997)
15.	*Syzygium travancoricum*	Manipur valley	Balasubramaniam (1997)
16.	*Terminalia pallida*	Tirupati Hills	Raviprasada Rao (1997)
17.	*Semecarpus kathalekanesis*	Kaans, Shimoga region	Srikanth *et al.* (2012)

Source: Ramanujam *et al.*, 2002

Animal wealth of sacred groves has not received the attention it deserves since they are not regarded as wild life sanctuaries. Although, major mammalian life cannot be expected, they harbor numerous birds, butterflies, bats, apart from primates and minor mammals. The animals found in the sacred grove are of two types; those which inhabit the groves like snakes, frogs, lizards and other lower group of organisms and higher group of fauna who nests and dens there and those who visit the grove temporarily for food, shelter etc. Apart from providing transitory home for migratory birds, they serve as sanctuaries for many rare, endemic, endangered species and of animals particularly the avifauna.

From the sporadic reports available, it can be said that the zoological significance of the groves is equally impressive. The termite mounds, because of incomprehensible growth, were among the primitive cult objects. Peafowl, cobra and other animals received absolute protection since hunting is a taboo. Some groves were specifically dedicated to tigers (Hulidevaru) and snakes (Nagadevaru) in Uttara Kannada, Karnataka and Kerala's sarpakavus are legendary. Unnikrishnan (1995) has recorded the protection of hundreds of white tortoises in a pond in Melothunkavu in Kasargode district of Kerala. These Kavus are the asylum for rare snakes, nine species of frogs and a rare bird; the white - bellied sea eagle considered divine by the fisher-folk of west- coast has found asylum in some kavus only. In a recent survey, Jayarajan (2004) has recorded 117 species of butterflies, 8 spiders, 11 amphibians, 23 reptiles, 178 birds and 24 mammals from the north Kerala groves. The endangered Nilgiri Langur in Thavidisserikavu, Bonnet monkey in Aravanchalkavu, found refugia there.

Ecological Services

Biodiversity keeps the ecological processes in a balanced state, which is necessary for human survival. Therefore, the biodiversity-rich sacred groves are of immense ecological significance. They also play an important role in the conservation of flora and fauna. Besides, several rare and threatened species are found only in sacred groves, which are, perhaps, the last refuge for these vulnerable species.

A factor that amplifies the significance of sacred groves is the fact that most of them contain water in the form of springs, ponds, lakes, streams or rivers. Not only that, but the vegetative mass of the grove itself conserves water, soaking it up like a sponge during wet periods and releasing it slowly in times of drought. One of the functions of groves in preserving biodiversity is to provide a more dependable source of water for the organisms inside them and in the neighborhood. In addition, transpiration will increase the humidity in the immediate vicinity and produce a more favourable microclimate for many organisms. When forest litter is allowed to accumulate, organic material builds the soil and is returned to the biomass of the standing forest (Ramakrishna *et al.*,, 1998).

Most sacred groves of Western Ghats are associated with ponds, streams, springs or rivers. Wingate (1888) stated that the kaans of Utthara Kannada were of great economic and climatic importance. They favour the existence of springs, swamps and perennial streams and generally indicate the proximity of valuable spice gardens which derive from them both shade and moisture. The kuvus of Kerala, including the smaller serpent groves in the premises of houses, are associated with water bodies. The people believed that the groves were responsible for permanent water bodies in their habitation.

Threats to Sacred Groves

Belief and taboos are the constructive tools for conserving the sacred groves, and erosion of belief and taboos has led to deterioration of groves (Gadgil and Vartak, 1981a, Tiwari *et al.*,., 1998). It has been seen that religious beliefs and taboos that were central to the protection of sacred groves are being eroded over the years due to various reasons and thus the present status of sacred groves is rather precarious. Various anthropogenic pressures due to developmental activities, urbanization, exploitation of resources, increase in human population and sanskritisation have threatened many sacred groves of the country.

A study on the status of some sacred groves in the Himalayan region indicated that the economic forces are influencing the traditional communities to discard the community-oriented protection to these groves and they are now being exploited (Saxena *et al.*,. 1998). Conversion of sacred groves into coffee plantations and human habitation is the major threat to the conservation of groves in Kodagu district of Karnataka (Kushalappa and Bhagwat. 2001). It can be summarized based on various researchers' work that the common and major reasons for the degradation of sacred groves are:

1.   Erosion of traditional belief system and nature worship system

2.   Conversion and fragmentation of the forest area due to agriculture, mining or other developmental projects

3.   Spread of alien religions and sanskritization

4. Lack of interest among the present generation and poor management of the resources

5. Invasion of exotic weeds

6. Overgrazing, fire and over exploitation of plant resources

7. Anthropogenic activities including use and abuse of the forest area

It has been found that cultural changes among the young people are so rapid that they no longer believe in the methods their ancestors followed to maintain the fragile ecosystem. This is a global tragedy, because "with the disappearance of each indigenous group, the world loses an accumulated wealth of millennia of human experience and adaptation'. For ecologists, traditional ecological knowledge offers a means to improve research and also to improve resource management and environment impact assessment. One unfortunate matter that hinders the conservation of sacred grove is that the village people living nearby the sacred groves are poor and so they depend on the grove to meet their vital domestic necessities, such as fuel wood, vegetables, medicinal plants etc. in developing countries, the rural poor depend upon biological resources for meeting 90% of their day-to-day needs. Hence, until and unless a viable option is provided to these people for sustaining their economic condition, any step for the conservation of the sacred groves will not be successful.

Research Trends in Sacred Groves

Sacred groves are gaining importance nowadays as an area of interest in field of research. As they are sustained pristine ecosystems, the biodiversity of these sacred groves offer excellent opportunity for bioprospection, conservation and sustainable development. Scientific studies on floristic composition and structure in forests are instrumental in the sustainability of forests since they play a major role in the conservation of plant species and the management of forest ecosystems as a whole (Addo-Fordjour *et al.*, 2009). Understanding the socio-cultural beliefs related to sacred groves will strengthen the sustainable management of these pristine forest patches. Various researchers have been working on sacred groves across the globe, focusing on floral and faunal diversity, socio-cultural beliefs and its impacts, anthropogenic activities, management strategies etc.

Most of the research work on sacred groves is focused on study of floral diversity and richness of trees, lianas, shrubs and herbs distributed throughout India, which provides baseline data for management of sacred groves. Significance and uniqueness of these sacred groves are time and again proved by comparing their floral diversity with nearby protected forests.

Navendu Page *et al.*, (2009) studied the effect of degree of fragmentation and surrounding matrix on trees, lianas, shrubs and epiphytes in tropical forest fragments of Kodagu, Western Ghats, India. These fragments exist as sacred groves amidst a highly modified agricultural landscape, and have been preserved by the religious sentiments of local communities. The study revealed that 74% of the regional diversity for trees was contributed by diversity among plots, highlighting the importance of inter-patch habitat diversity in maintaining the total regional species pool. Hence it could be concluded that the trees alone cannot serve as good indicator for taking appropriate conservation measures to mitigate species loss resulting from habitat fragmentation. Rajashry Ray (2011) studied the sacred groves of Central Western Ghats, Uttara Kannada District, Karnataka. Sacred groves harboured 138 woody species among which 19 (13.7%) are endemic to Western Ghats region. Grove area and total species richness show typical species-area relationship in power form. 14 out of 19 endemics are exclusively confined to the sacred groves, the other five occurred outside forest patch as well as in some other landscape elements. The author mainly focused on four endemic species for detailed study, because of their restricted distribution and association with shaded humid forests. These were *Vateria indica, Syzygium travancoricum, Calophyllum apetalum* and *Diospyros assimillis,* of which the first two were Critically Endangered (IUCN 2011). GBH class distribution of the endemic members in the sacred grove showed typical "L" shaped curve indicating dominance of younger members. Regeneration study on all four species has revealed marked differences among them. Sixteen (16) of the groves found to be of highly disturbed category where relative disturbance score exceeded 76% (small area < 1000 sq. m). Thirty five (35) groves were ranked in moderately disturbed category (disturbance scores 51-75%) and two (2) groves that are less disturbed have their areas ranging from 2,000-10,000 sq. m.

Raviprasd Rao *et al.,* (2011) studied tree diversity, population structure and their relation to site disturbances in five replicate stands each of sacred forests and reference reserve forests in southern Eastern Ghats of Andhra Pradesh by belt method. A total of 7836 trees belonging to 158 species were inventoried in all the stands. The stands in the sacred forests were more diverse, had higher basal area, and showed fewer signs of disturbance than the reference forest stands, supporting the view that local communities afford better protection and management to sacred groves. A total of 4858 trees (mean ± SD, 971.6 ± 38.49 trees per ha; range, 929 - 1018 trees ha-1) were present in the five 1-ha plots of sacred tropical dry deciduous forest stands, which was about 37 % greater than the reference forest sites (total 3078 trees; mean ± SD, 615.6 ± 67.81 trees ha-^{1}; range, 563 - 732 trees per ha). Ramanujam and Praveen Kumar (2003) studied the plant wealth and diversity of four sacred groves – two anthropogenic stands and two natural forest patches – along the southeast coast of India adjoining Pondicherry. A total of 111species, belonging to 103 genera in 53 families, were recorded from the four sites, which together measure 15.6 ha. The presence of a stout liana of *Secamone emetica* (gbh 35cm), the robustness of *Cretaeva magna* (gbh 220cm), *Syzigium cumini* (gbh 207.45cm), *P. suberifolium*(gbh 128.7cm) and *Tamarindus indica* (gbh 250cm), and survival of evergreen species like *A.elaeagnoidea* and *Pamburus missionis* is botanically significant; *Polyalthia suberosa* is a rare taxon found only within the groves.

Cherrapunjee in Meghalaya, India, is one of the wettest places on Earth. The left of the forest area is mainly in the form of patches, most of which are treated as sacred groves by the local Khasi communities. A survey on the avifauna of these sacred groves was conducted by Firoz Ahmed (2004). The study recorded 153 species of birds from the sacred groves, including 4 threatened species (Dark-rumped Swift, Lesser Kestrel *Falco naumanni,* Grey crowned Prinia *cinereocapilla* and Swamp Francolin *Francolinus gularis),* 3 restricted range species (Dark rumped swift, Streak throated Barwing *Actinodura waldeni* and Tawny breasted Wren Babbler *Spelaeornis longicaudatus)* and 42 biome restricted species. The study helped in monitoring the population of Dark rumped Swift, and estimated this at just over 350 in three breeding sites. Mithra *et al.,* (2004) analysed the effects of anthropogenic disturbance on plant diversity and community attributes of a sacred grove (montane subtropical forest) at Swer in the East Khasi Hills district of Meghalaya in northeast India. The undisturbed, moderately disturbed and highly disturbed stands were identified within the sacred grove on the basis of canopy cover, light interception and tree (cbh $ 15 cm) density. The study revealed that the mild disturbance favoured species richness, but with increased degree of disturbance, as was the case in the highly disturbed stand, the species richness markedly decreased. The number of families of angiosperms was highest (63) in the undisturbed stand, followed by the moderately (60) and highly disturbed (46) stands. The similarity index was maximum (71%) between the undisturbed and moderately disturbed stand and minimum (33%) between the undisturbed and highly disturbed stands. The Margalef index, Shannon diversity index and evenness index exhibited a similar trend, with highest values in the moderately disturbed stand. In contrast, the Simpson dominance index was highest in the highly disturbed stand. There was a sharp decline in tree density and basal area from the undisturbed (2103 trees ha and 26.9 m ha) to the moderately disturbed (1268 trees ha and 18.6 m ha) and finally to the highly disturbed (852 trees ha and 7.1 m ha) stand indicated the complex and stable nature of the community. Ualpadhyaya *et al.,* (2002) investigated the biodiversity of woody species in Ialong and Raliang sacred groves of the Jaintia hills in Meghalaya, northeast India. These groves represent the climax subtropical broad-leaved forest of the area. A total of 738 individuals belonging to 82 species, 59 genera and 39 families was identified in a 0.5 ha plot of the Ialong sacred grove, whereas the same area in the Raliang sacred grove had 469 individuals of 80 species, 62 genera and 41 families. About 32% species were common to both groves. The majority of the species showed a contagious distribution pattern and low frequency. The basal area varied from 57.4 to 71.4 m² ha⁻¹. Species richness within the forest varied from 3 to 15 per 100 m² in Ialong and 3 to 12 per 100 m² in Raliang.

Mount Kawa Karpo of the Menri ('Medicine Mountains' in Tibetan), in the eastern Himalayas, is one of the most sacred mountains to Tibetan Buddhists. Numerous sacred sites are found between 1900 and 4000 m, and at higher elevations the area as a whole is considered a sacred landscape. Religious beliefs may affect the ecology of these sacred areas, resulting in unique ecological characteristics of importance to conservation; recent studies have demonstrated that sacred areas can often play a major role in conservation. Danica *et al.,* (2005) studied the vegetation of sacred areas in the Menri region using vegetation maps and a Geographical Information System (GIS) for remote assessment. Sacred sites were compared to random points in the landscape, in terms of: elevation, vegetation, and nearness to villages; species composition, diversity, and richness; and frequency of useful and endemic plant species. Detrended correspondence analysis (DCA) ordination revealed that sacred sites differ significantly in both useful species composition (p=0.034) and endemic species composition (p=0.045). Sacred sites are located at lower elevations, and closer to villages, than randomly selected, non-sacred sites (p< 0.0001), and have higher overall species richness (p=0.033) and diversity (p=0.042). In addition, the high-elevation (> 4000 m) areas of the mountain – a sacred landscape – are found to have significantly more endemics than low elevation areas (p<0.0001). These findings represent an initial analysis of sacred sites and suggest that sacred sites in the Menri region may be ecologically and ethno-botanically unique. Jan Salick *et al.*, (2007) studied the role of sanctity in biodiversity conservation within habitats in the Khawa Karpo region, in Eastern Himalayas, by pairing plots within the same habitats in sacred and non-sacred areas. Understory richness, diversity, cover, and number of useful species were measured; for trees, richness, diversity, cover, and density were measured. Results indicated that within habitats sanctity does not affect understory plant communities; however, within sacred areas trees are larger (p = 0.003) and forests have greater cover (p = 0.003) than non-sacred areas. Their results showed that, whereas placement of sacred areas and preservation of vegetation cover affects useful plants, biodiversity and endemism, within habitats sacred sites preserve old growth trees and forest structure. In sum, Tibetan sacred sites are ecologically unique and important for conservation on varying scales of landscape, community, and species.

The sacred groves in the Pachmarhi Biosphere Reserve (PBR) of India were studied by Chandra Prakash Kala (2011) to understand the concept of traditional, ecological and biodiversity conservation systems in the sacred groves maintained by Mawasi Gond tribal communities. Different deities were worshipped in the sacred groves and each grove was named after the deity dwelling in the respective sacred grove. Various traditional customs associated with sacred groves were in practice. The sacred groves were rich in plant genetic diversity and were composed of many ethno-botanically useful species, including wild edible fruits, medicinal plants, fodder, fuelwood and timber yielding species. Kassilly and Tsingalia (2009) studied the Tiriki sacred groves of Hamisi District, Kenya and reported that breakdown in the socio-cultural fabric of the Tiriki community due to the influence of modern religion, education and government regulations were responsible for loss of cultural values and indigenous knowledge associated with sacred groves among local people. Cultural evolution has made local people abandon sentiments and acts that at one time ensured preservation of the sacred groves. As the Tiriki people adapt to the social, cultural, political and economic miasma of a modern Kenya, their cultural links with the sacred groves have had to suffer.

Ashalatha *et al.*, (2004) carried out ethnobotanical studies in four sacred groves of Manipur, results revealed therapeutic applications of 120 plant species representing 106 genera and 57 families. Tree species contributed the maximum having 42% while herbs recorded 33% of the total medicinal plants. The study revealed that the species that are scarce locally in the forest due to various development activities, deforestation, over-exploitation etc are abundant in the sacred groves. Swamynathan and Rammoorthy (2011) carried out ethno-medical field survey Cuddalore district, Tamil Nadu and recorded that 40 medicinal plant species belonging to a total of 29 families in scared groves which are being used by the local people. Based on the habit classification of the 40 species, maximum 50 % of species were trees, 37.5 % of species were lianas, 10 % of species were herb and 2.5 % were shrub. Among the family, Capparaceae was the dominant family and plants like *Memecylon umbellatum, Garcinia spicata, Olax scandens* and *Atalantia monophylla* are frequently used for medicinal purpose and dominant species in the sacred groves.

Conclusion

The glimpse of the above studies point out that the preservation of sacred groves is of key importance in maintaining biodiversity and the comprehensive health of any landscape. They provide greater benefits than their small size would indicate. They are essential for understanding the cosmovision of any society, and consequently they have great cultural importance. At the same time, they possess biological importance which is a vital factor to the environmental conservation. Sacred site conservation has guided local environmental stewardship for millennia. It is this continued stewardship that will ultimately determine the future of these biodiversity hotspots.

References

Amirthalingam, M. 1998. Sacred groves of Tamil Nadu – Asurvey. C. P. R. Environmental Education Centre, Chennai. India. pp. 191.

Ashalatha Devi Khumbongmayum, M L Khan and R S Tripathi. 2005. Ethno-medicinal plants in the sacred groves of Manipur. *Indian journal of traditional knowledge,* Vol. **4(1)** pp. 21-35.

Bahsa, S. C. 1998. Conservation and Management of sacred groves in Kerala. In: Conserving the sacred for biodiversity management (Eds. Ramakrishnan P S., Saxena K G. and Chandra Shekara U M). Oxford IBH Publishing Co. Pvt. Ld. New Delhi. pp. 337-348.

Balasubramanyan, K. And Induchoodan, N.C. 1996. Plant Diversity In Sacred Groves of Kerala. *Evergreen* **36**: 3-4.

Bhagwat, S. A. and Rutte, C. 2006. Sacred grove: potential for biodiversity management. *Front. Ecol. Environ* **4**: 519-524.

Bhakat, R.K. 1990. Tribal Ethics of Forest Conservation. *Yojana* (March 16-31): 23-27.

Boojh, R. And Ramakrishnan, P.S. 1983. Sacred Groves And Their Role In Environmental Conservation. Strategies For Environmental Management, Souvenir Volume: **6-8**. Department of Science and Environment of Uttar Pradesh, Lucknow.

Burman, R. J. J., 1992. The institution of sacred grove. *Journal of Indian Anthropological Society* **27**: 219-238.

Chandra Prakash Kala. 2011. Traditional ecological knowledge, sacred groves and conservation of biodiversity in the Pachmarhi biosphere reserve of India. *Journal of Environmental Protection,* **2,** 967-973.

Chandran, M. D. S. 1997. On the ecological history of Western Ghats. *Current Science* 73:146-155.

Chandran, M. D. S., Gadgil, M. and Hughes, J. D. 1998. Sacred groves of Western Ghats of India. In: Conserving the sacred for biodiversity management (Eds. Ramakrishnan P S., Saxena K G. and Chandra Shekara U M). Oxford IBH Publishing Co. Pvt. Ld. New Delhi. pp. 211-232.

Chandrashekara, U. M. and Sankar, S. 1998. Structure and functions of sacred groves: Case studies in Kerala. In: *Conserving The Sacred For Biodiversity Management.* Eds. P. S. Ramakrishana, K. G. Saxena And U. M. Chandrashekhara. Oxford And Ibh Publishing Company Private Limited, pp. 323-336.

Frazer, J. G. 1935. The Golden Bough, Part I: The magic art and evolution of Kings, 2 Vols. Macmillan, New York.

Gadgil, M. And V.D Vartak. 1976. Sacred Groves Of Western Ghats Of India. *Ecological Botany* 30: 152-160.

Gadgil, M. And Vartak, V.D. 1981a. Studies On Sacred Groves Along The Western Ghats From Maharashtra And Goa; Role of Beliefs And Folklores. Pages 272-278, In: Jain. S.K. (Editor), Glimpses Of Indian Ethnobotany. Oxford University Press, Bombay.

Godbole, A. , Watve, A., Prabhu, S and Sarnaik, J. 1998. Role of sacred groves in biodiversity conservation with local people's participation: A case study from Rathnagiri district, Maharastra In: Conserving the sacred for biodiversity management (Eds. Ramakrishnan P S., Saxena K G. and Chandra Shekara U M). Oxford IBH Publishing Co. Pvt. Ld. New Delhi. pp. 233-246.

Hughes, D.J. and Chandran, S.M.D. 1998. Sacred grove around the earth: An Overview. Pages 69-86, In: Ramakrishnan, P.S., Saxena, K.G. and Chandrashekara, U.M. (Editors) Conserving the Sacred for Biodiversity Management. UNESCO and Oxford-IBH Publishing, New Delhi.

Hughes, J.D. And Chandran, M.D.S. 1998. Sacred Groves Around The Earth: An Overview. In *Conserving The Sacred For Biodiversity Management.* Eds. P. S. Ramakrishana, K. G. Saxena And U. M. Chandrashekhara. Oxford And Ibh Publishing Company Private Limited, pp. 69-86 .

Hughes, J.D., 1984, Sacred Groves: The Gods, Forest Protection, and Sustained Yield in the Ancient World. In *History of Sustained-Yield Forestry: A Symposium.* Ed. Harold K. Streen, Forest History of Society, Durham, North Carolina, pp. 331- 343.

Hughes, J.D., 1996, *North American Indian Ecology.* Texas Western Press, El Paso.

Hussain, A.B.M.E. 1998. Scared sites in Bangladesh: Country report. Pages 167, In: Ramakrishnan, P.S., Saxena, K.G. and Chandrashekara, U.M. (Editors) Conserving the Sacred for Biodiversity Management. UNESCO and Oxford-IBH Publishing, New Delhi.

Induchoodan, N. C. 1996. Ecology of sacred groves of Kerala. Ph D thesis, Pondicherry University, Pondicherry.

Islam, A.K.M.N., Islam, M. A. and Hoque, A.E. 1998. Species composition of sacred groves, their diversity and conservation in Bangladesh. Pages 163-165, In: Ramakrishnan, P.S., Saxena, K.G. and Chandrashekara, U.M. (Editors) Conserving the Sacred for Biodiversity Management. UNESCO and Oxford-IBH Publishing, New Delhi.

Jamir, S.A. And Pandey, H.N. Status Of Biodiversity In the Sacred Groves of Jaintia Hills, Meghalaya. *Indian Forester* 128(7): 738-744, 2002.

Jan Salick, Anthony Amend, Danica Anderson, Kurt Hoffmeister, Bee Gunn and Fang Zhendong. 2007. Tibetan Sacred Sites Conserve Old Growth Trees And Cover In The Eastern Himalayas. *Biodiversity And Conservation,* **16**:693–706

Jayarajan, M. 2004. Sacred groves of North Malabar. KRPLLD, Trivandrum, pp. 109.

Khan, M. L., Ashalata Dei V Khumbongmayum2 and R.S. tripathi. 2008 The Sacred Groves and Their Significance in Conserving Biodiversity - An Overview. *International Journal of Ecology and Environmental Sciences* 34 (3): 277-291

Khiewtam, R.S. and Ramakrishnan, P.S. 1989. Socio-cultural studies of the sacred groves at Cherrapunji and adjoining areas in North-Eastern India. Man in India 69 (1): 64-71.

Khumbongmayum, A.D., Khan, M.L. And Tripathi, R.S. 2005a. Sacred Groves of Manipur, Northeast India: Biodiversity Value, Status And Strategies For Their Conservation. *Biodiversity and Conservation* 14(7): 1541-1582.

Kosambi, D. D. 1962. Myth and Reality. Popular press. Bombay.

Kushalappa, C. G. and Bhagwat, S. A. 2001. Sacred groves: biodiversity, threats and conservation. In: Shankar, R., Ganeshaiah, K. N. and K. S. Bawa (eds) *Forest genetic resources: status, threats and conservation strategies.* Oxford and IBH publishing Co Pvt Ltd, pp 21-29

Law, P. 2002. Studies On Population Ecology of Keystone Species and their Role In Ecosystem Function in the two Sacred Groves of Meghalaya. Ph. D. Thesis, North- Eastern Hill University, Shillong, India. 115 Pages.

Lebbie, A. R. and Guries, R. P. 1995. Ethnobotanical value and conservation of sacred groves of the Kpaa Mende in Sierra Leone. *Economic Botany,* **49:** 297-308

Logan, W. 1887. *Manual of the Malabar district.* Govt press Madras (Reprinted 1906) pp 185-186.

Malhothra, K. C., Yoge sh Gokhale, Snajeev Srivastava and Sudipto Chatterjee. 2000. Sacred groves in India: An overview. Indira Gandhi Rastriya Manava Sangrahalaya, Bhopal. pp. 1-230.

Malhotra, K.C. 1998. Anthropological dimensions of sacred groves in India: an overview. Pages 423-438, In: Ramakrishnan, P.S., Saxena, K.G. and Chandrashekara, U.M. (Editors) Conserving the Sacred for Biodiversity Management. UNESCO and Oxford-IBH Publishing, New Delhi.

Malhotra, K.C., Gokhale, Y., Chatterjee, S. and Srivastava, S. 2001. Cultural and Ecological Dimensions of Sacred Groves in India. Indian National Science Academy, New Delhi, and Indira Gandhi Rashtriya Manav Sangrahalaya, Bhopal. 30 pages.

Malhotra, K.C., Stanley, S., Heman, N.S. and Das, K. 1997. Biodiversity conservation and ethics: sacred groves and pools. Pages 338-345, In: Fujiki, N. and Macer, R.J. (Editors) Bioethics in Asia. Eubois Ethics Institute, Kobe, Japan.

Mishra, B. P., O.P. Tripathi, R.S. Tripathi and H.N. Pandey. 2004. Effects of anthropogenic disturbance on plant diversity and community structure of a sacred grove in Meghalaya, northeast India. *Biodiversity and Conservation,* **13**: 421–436.

Mishra, B.P., Tripathi, O.P., Tripathi, R.S. And Pandey, H.N. 2004. Effect of Anthropogenic Disturbance on Plant Diversity and Community Structure of A Sacred Grove In Meghalaya, North East India. *Biodiversity and Conservation* **13**: 421-436

Molyneaux, B.L. 1995, *The Sacred Earth.* Little, Brown and Company, Delhi.

Navendu V. Page, Qamar Qureshi, Gopal S. Rawat and Cheppudira G. Kushalappa. 2009. Plant diversity in sacred forest fragments of Western Ghats: a comparative study of four life forms. *Plant Ecol.*

Rajashry Ray. 2011. Developing strategies for conservation of threatened endemic biodiversity of the sacred groves of Central Western Ghats. Project report: Centre for Ecological Sciences, Indian Institute of Science, Bangalore 560012, India.

Ramakrishna P S, K G Saxena and U M Chandrashekara. 1998. Conserving The Sacred For Biodiversity Management. Oxford and IBH Publishing Co Pvt Ltd.

Ramakrishnan, P.S. 1996. Conserving The Sacred: From Species To Landscapes. Nature And Resources; Unesco 32: 11-19. Ramakrishnan, P.S. 1998. Conserving The Sacred For Biodiversity: The Conceptual Framework. Pages 3-15, In:Ramakrishnan, P.S., Saxena, K.G. and Chandrashekara, U.M. (Editors) Conserving the Sacred for Biodiversity Management. Unesco And Oxford-IBH Publishing, New Delhi.

Rappaport, A. 1971. The sacred in Human evaluation. *Annual review of systematic and ecology* **2**:23-44.

Ravi Prasad Rao, B., M.V. Suresh Babu, M. Sridhar Reddy, A. Madhusudhana Reddy, V. Srinivasa Rao, S. Sunitha & K.N. Ganeshaiah. 2011. Sacred groves in southern eastern ghats, India: Are they better managed than forest reserves? *Tropical Ecology* **52**(1): 79-90

Saxena, K. G., Rao, K. S. and Maikhuri, R. K. 1998. Religious and cultural perspective of biodiversity conservation in India. In: *Conserving The Sacred For Biodiversity Management.* Eds. P. S. Ramakrishana, K. G. Saxena and U. M. Chandrashekhara. Oxford and IBH Publishing Company Private Limited, pp 153-162.

Singh, A. 2011. 'Orans' community led biodiversity conservation: A case study from Krapavis: National conference on conservation of sacred groves to protect local biodiversity. C. P. R. Environmental Education Centre, Chennai. India 12-14 Feb, abstract 1.

Tiwari, B. K., Barik, S. K. and Tripati, R. S. 1998. Biodiversity value, status and strategies for conservation of sacred groves of Meghalaya, India. *Ecosystem health* **4**: 20-33

Tripathi, R.S. 2001. Sacred groves: Community biodiversity conservation model in north-east India. Pages 104-107, In: Ganeshaiah, K.N., Uma Shaanker, R. and Bawa, K.S. (Editors) Tropical Ecosystems: Structure, Diversity and Human Welfare (Supplement). Proceedings of the International Conference on Tropical Ecosystems. Ashoka Trust for Research in Ecology and Environment (ATREE), Bangalore.

Unnikrishnan, E. 1995. Sacred groves of North Kerala: An eco-folklore study, Jeevareksha, Trissur, Kerala, India (in Malayalam).

Upadhaya, H.N. Pandey*, P.S. Law And R.S. Tripathi. 2002. Tree Diversity In Sacred Groves Of The Jaintia Hills In Meghalaya, Northeast India. *Biodiversity And Conservation* **12**: 583–597.

Upadhaya, K., H.N. Pandey, P.S. Law and R.S. Tripathi. 2003. K. Upadhaya, H.N. Pandey, P.S. Law And R.S. Tripathi. *Biodiversity and Conservation,* **12:** 583–597

Vajpeyi, Y. 2001. Tree of Life. Indian Express (Sunday magazine) September 3

Veluppillai, T. K. 1940. *The Travencore state manual.* Vol 1. Govt of Travencore. pp 623-628.

Vest, J.H.C., 1985, Will-of-the-land: among primal Indo-Europeans. *Environmental Review,* **9**: 323-329.

Ward and Conner. 1827. *Memoir of the survey of the Travencore and Cochin States.* Govt of Kerala.

Wingate, R. T. 1888. Settelement proposals of 16 villages of Kumata taluk, No. 210, Forst Settlement office, Karwar.

Chapter 4

Honeybee Diversity and their Role in Environment Management

S. Basavarajappa

Apidology Laboratory, DOS in Zoology, University of Mysore, Manasagangotri, Mysore-570 006, Karnataka

ABSTRACT

India is known for its salubrious climate and diversified environment amidst temperate, tropical and sub-tropical conditions. It is housed with more than 2, 00, 000 varieties of animal pollinators known around the world and considered as highly valued resource for biodiversity conservation and environment management. Of all, honeybees are considered as highly valued resources; include several hundred species distributed around the world. They are major pollinators of several plant species. Since, pollination is one of the essential biological processes, where in many flowering plants depend on this vital process for their successful reproduction, propagation and to continue their generations with safe survival at various ecosystems. By doing pollination, honeybees bring qualitative and quantitative changes significantly in the economic and biological traits by facilitating genetic enrichment of various plant species through cross-pollination. They influence the growth of local and regional vegetation, bring considerable variation and enhance the development and adaptation qualities among the plant species that help maintain wider gene pool to the changing environment. When the vegetation status improves, certainly the local environmental conditions improve. In turn this would bring change in local weather and ultimately the climate. Therefore, honeybee's pollination is essential for the maintenance of plants diversity at natural and man-made environments. They act as unseen bio-engines driving an ecosystem with their intangible pollination service. They couple plant to plant and plant to animals, spinning the verdant biological world through endless pollination service and indirectly help provide fuel (energy source i.e., nectar and pollen), fuses (reproduction and propagation) and safety valves (survival of several species) at various ecosystems. In this way, honeybees act as 'vital agents of environment management'.

Further, honeybees are social insects; thrive well by nesting single or multiples of variously sized colonies on several tree species including man-made structures at specific elevation with unique comb architecture. They live at varied environmental conditions, ready to experience hardships while availing required ecological and biological factors for their safe survival. While attending all these hardships, they never destroy or pollute the environment. Therefore, all these unique behaviors of honeybees act as *'role models'* to many animals including man. Although their existence is complex and coupled with several factors, their conservation is inevitable and it is essential for the survival of human race. Therefore, the best strategy should be devised to maintain their diversity by taking the help of biologists, wildlife biologists, environmentalists, naturalists, policy makers and economists. Further, efforts should be made to launch educative programs on these bioengineers that are unique in supporting the biological world with good environment for future generation to come.

Key words: Pollinator, Nest, Pollen, Cross pollination

Introduction

Insect's form 80% faunal component among the biodiversity groups analyzed thus so far (Kumar, 2007). Among insects, certain hymenopterans are commercially known since time immoral. Hymenoptera is one of the commercially important groups in class Insecta (Gillot, 1995) include solitary, sub-social and social bee species. The honeybees are beneficial to mankind since prehistoric times (Varadharajan, 2002). They are social insects (Gillot, 1995) include more than 20,000 identified bee species (Reddy, 2002). Honeybees provide free ecosystem services in the form of cross-pollination and propagation of several plant species at various environments by maintaining biological diversity (Verma, 1990), boosting crop productivity (Bright *et al.*, 1998) and help in the conservation of several plant species from extinction (Sihag, 2002). Therefore, honeybees are treated as 'vital components of every ecosystem' (Varadharajan, 2002) attracted much attention from the 'environmentalists' in general and 'biologists' in particular (Engel, 2002) around the globe. Further, honeybees are known for their valuable pollination service under various ecosystems. Majority of these honeybee species have the ability to manage surrounding environment by doing pollination service to local and regional flora. The honeybee species diversity promotes prospective sources for basic research on pollination biology, ecology and environment management. Further, the economic potentials of honeybees have been so far understood only in terms of their hive products (e.g. honey, beeswax) at various habitats (Reddy *et al.*, 1986; Basavarajappa, 1998; Reddy, 2002; Sattagi *et al.*, 2002; Siddappa Setty and Bawa, 2002; Basavarajappa, 2004). Moreover, the role of honeybees as one of the most effective and cheapest biological input by way of cross-pollination in increasing the density of different plant species amidst various ecosystems is remained unexplored.

Distributions of honeybee population have a direct correlation with the available flora (Krishnamurthy, 2002 and Basavarajappa, 2004) and it is depended on the availability of desired ecological conditions (Khan *et al.*, 2007). Honeybees enhance the productivity levels of agricultural, horticultural, ornamental and fodder crops through cross-pollination (Bright *et al.*, 1998) and thus provide important linkages to the botanical resources both in rural and urban environments (Nikam, 2002). Therefore, honeybee diversity and their role in environment management are discussed in this communication.

Honeybee Diversity

The biodiversity of honeybees in Asia is most researched by bee biologists since many decades. Among the social Hymenopterans, around 20,000 bee species have been recognized worldwide. Today, vast amount of scientific reports are available on honeybees that could attributed to the unique features in their distribution (Ahmed and Abbas, 1985) biology and life history (Seeley, 1995), population dynamics (Goyal, 1978; Shah and Shah, 1982), pollen foraging (Dreller and Tarpy, 2000), nectar foraging (Wells *et al.*, 1981) etc. The solitary, sub-social and social bee species are recognized at different parts of the world and are placed under 425 genera, 22 subfamilies and 6 families (Michener, 2000). The social bees (Apidae) belong to three tribes namely Bombini, Apini and Meliponini (Kerr, 1950). These three tribes are phylogenetically related to one another and have certain common features, hence united them in the subfamily 'Apinae'. Among Apinae, honeybees (Apini) and Stingless bees (Meliponini) are 'eusocial' species live in complex colonies forms a sister groups (Michener, 1990).

Further, honeybees are broadly grouped into stinging and Stingless bees (Gillot, 1995). The stinging bees belong to family Apidae and the tribe Apini consists of only one small monophyletic genus *Apis*. The genus *Apis* comprises nine species viz., *Apis laboriosa, A. dorsata, A. florae, A. andreniformis, A. mellifera, A. cerana, A. koschevnikovi, A. nigrocincta* and *A. nuluensis* which are native to Asia (Wongsiri *et al.*, 2002). *Apis* species are divided into three lineages: the cavity nesting bees (*e.g. A. nuluensis, A. koschevnikovi, A. nigrocincta, A. cerana* and *A. mellifera*), the dwarf bees (*e.g. A. florae* and *A. andreniformis*) and the giant bees (*e.g. A. laboriosa, A. dorsata, A. florae* and *A. andreniformis*). The cavity nesting bees nests in cavities/crevices, and have multiples of parallel combs. However, both dwarf and giant bees are open nesters, and build a single vertical comb. Among the cavity-nesting bees, *A. mellifera* and *A. cerana* have been 'domesticated' for a long time for commercial exploitation. However, the 'Stingless bees' are very small in size and doesn't equip with stinging apparatus (Gillot, 1995). They represent one of the most diversified components of Apoidea family (Gillot, 1995), include small to medium sized bees with vestigial stings, found in tropical and many subtropical parts of the world (Heard, 1999 and Leonhardt *et al.*, 2007). They are perennial eusocial insects (Nagamitsu *et al.*, 1999; Karunaratne and Edirisinghe, 2007), constitute the Meliponini tribe of the family Apidae. The Stingless bees are classified into five genera: *Melipona, Trigona, Meliponula, Dectylurina,* and *Lestrimelitta* (Klakasikorn *et al.*, 2005). They are grouped into two principal genera *Trigona* and *Melipona* (Schwarz, 1948), which are divided into 23 genera and 18 subgenera with 374 recognized species (Moure, 1961; Michener, 2000). The genus *Melipona* has 40 species and the genus *Trigona* consists of 10 subgenera in which about 120 species have been recognized in the tropics from Mexico to Argentina, and in the Indo-Australian region from India and Sri Lanka to Taiwan, east to the Caroline Islands, the Solomon Islands and South throughout Indonesia and New Guinea to Australia. Thus, the tribe Meliponini far exceeds the Apini in diversity and original native distribution. Moreover, Meliponini members are divided into two main groups: the more primitive short-tongued bees and the more advanced long-tongued bees. The long-tongued bees are honeybees and stingless bees, which live in complex colonies. Interestingly, the morphological distinctions between *Apis* and non-*Apis* species show clear distinctions with respect to their morphology, nesting, distribution etc. In general honeybee family includes fertile female (queen), sterile females (workers) and fertile males (drones) (Gillot, 1995). The female castes, reproductive division of labour, generation overlap and a complicated communication and recruitment related to foraging system is well developed among honeybees and the stingless bees. All these species are distributed at diversified environmental conditions and help support natural environment in various ways.

Nesting Ecology

Nests are spectacular examples of animal architecture, highly versatile visible structures in honeybees. It is a home of honeybees with specific colony profile. The individual species are recognized by their specific nest, and often their particular site with variety of attributes. Both inside and outside the nest, there are different shaped compartments arranged to provide space for brood cells, food storage cells, cells for male bees (drones) and the queen cell. The nesting site and architecture of nests made with distinctive traits which are species specific, and are cryptic (built inside the cavity or crevices) or exposed openly on/in various substrates amidst diversified environment. Usually, honeybees built two types of nests *viz.*, 'arboreal nests' and 'terrestrial nests'. The arboreal nests are open and directly exposed to climatic conditions, constructed by wild honeybees (ex. *A. laboriosa*, *A. dorsata* and *A. florea*). The terrestrial nests are of various types namely 'cavity nests', partially exposed nests or 'cryptic nests'. The stingless bees (ex. *Trigona* sp.) built cryptic nests, inside the crevices or in the hallow tree. The nest is made up of moderate sized pot shaped cells used for brood rearing, storing nectar and pollen and also for queen housing.

Further, nesting is a skilled activity, requires elite strategies. In honeybees, nesting is highly diversified and it is species specific activity. Honeybees adopt unique traits with greater variability in the mechanisms of nesting. The nests are constructed in cavities, cracks, culms that are hollow or have a soft interior, constructed in tree trunks, twigs, on the eaves of tree limbs, undersurface of man-made structures etc. Therefore, by establishing their nests on/or at different substrates amidst various ecosystems, honeybee species become part of the 'natural environment' at various geographical regions.

The trees such as *Samanea saman, Ficus religiosa, F. benghalensis, Mangifera indica, Ceiba petandra, Eucalyptus* sp. are important nest hosting trees to *A. dorsata* (Manjunath, 2008; Basavarajappa, 2012; Raghunandan, 2014). Moreover, *Bamboosa bamboo* and *Cocus nucifera* also support the giant honeybee (*e.g. A. dorsata*) population at different regions under tropical conditions (Raghunandan, 2014). Further, the dwarf honeybee, *Apis florea* is the smallest bee in the genus *Apis* (Oldroyd *et al.*, 2000), distributed in the warmer parts of Oman, Iran, Pakistan, India and Srilanka. It build small sized single vertical comb at arboreal conditions in the wild. It is frequently found in tropical forests, in scrubby, bushy vegetation and even in farming areas (Basavarajappa, 2010). The pollination behaviour of *A. florea* has been reported by Thapa and Wongsiri (1996) in Nepal. Being a migratory species, *A. florea* more often nest on small to medium sized trees, bushy vegetation and shrubs at one to eight meters elevations during migration and emigration.

Around 13 plants, which belong to 13 genera and 12 families, were selected by *A. florea* for nesting under tropical conditions of Karnataka (Basavarajappa, 2010). Of all, nine trees and four shrubs extended the nesting platform to *A. florea*. The Fabaceae family members namely *Pongamia pinnata*, *Millettia ovalifolia* followed by the members of Anacardiaceae (*Mangifera indica*), Bombaceae (*Cebex petandra*) Caesalpinaceae (*Tamarindus indica*), Poaceae (*Bambusa bambos*), Annonaceae (*Polyalthia longifolia*), Magnoliaceae (*Michelia champaca*) and Moraceae (*Artocarpus heterophyllus*) represented as tree components among the nesting plants. The shrubs belong to family Nycteginaceae (*Bougainvillea* sp.), Malvaceae (*Hibiscus rosa-sinensis*), Apocyanaceae (*Nerrium* sp.) Bignoniaceae (*Tacoma stans*) have provided the nesting platform to dwarf honeybee colonies in Karnataka (Basavarajappa, 2010; Narayanaswamy, 2013). By nesting on these plant species, honeybees (e.g. giant and dwarf honeybees) help maintain the local vegetation under natural and man-made environments (Manjunath, 2008; Narayanaswamy, 2013; Basavarajappa, 2010; Raghunandan, 2014). Neupane (2004), Reddy and Reddy, (1989), Basavarajappa (2004) have reported on the selection of tree species to construct nests by honeybees under different agro-ecosystems of Karnataka. Thus, honeybees nesting activities is very much associated with the components of natural environment.

Honeybees and Floral Source

Manjunath (2008) has reported the floral source of giant honeybees (e.g. *A. dorsata* under urban environment. Giant honeybees pollinate various plants which belong to Magnoliaceae, Nelumbonaceae, Nymphacaceae, Menispermaceae, Casurinaceae, Portulaceae, Basellaceae, Amaranthaceae, Dipterocarpaceae, Lecythidaceae, Cochlospermaceae, Santalaceae, Euphorbiaceae, Balsaminaceae, Asclepediaceae, Convolvulaceae, Oleaceae, Asteraceae, Meliaceae, Compositaceae, Punicaceae, Palmaceae, Cariaceae, Musaceae Asteraceae, Acanthaceae, Anacardaceae, Apocynaceae, Rosaceae, Solanaceae, Laminaceae, Sapotaceae, Nyctiginaceae, Sterculiaceae, Bombaceae, Rubiaceae Annonaceae, Fabaceae, Myrataceae, Combarataceae, Verbenaceae Rutaceae, Poaceae, Malvaceae Caesalpinaceae, Moraceae Mimosaceae families amidst urban environment. Moreover, honeybees offer variety of flowering plants to collect nectar and pollen in different places of Mysore (Manjunath, 2008). Further, Sheetal (2009) has reported about the foraging plants of stingless bees at Manasagangotri campus, Mysore. Total 71 plant species which belongs to 36 families were recorded as potential foraging plants for *T. iridipennis* in Manasagangotri campus. Among them, herbs (35%), shrubs (17%), trees (37%) and climbers (11%) showed ornamental, fruit yielding, vegetable and economical values and important from the apicultural point of view to stingless bees (Sheetal, 2009). The Apiaceae, Arecaceae, Asclepidaceae, Balsaminaceae, Bignoniaceae, Caricaceae, Cruciferae, Compositae, Magnoliaceae, Meliaceae, Moringaceae, Moraceae, Musaceae, Nyctanthaceae, Oleaceae, Punicaceae, Rosaceae, Santalaceae, Sapotaceae, Verbenaceae, Cucurbitaceae, Mimosaceae, Myrtaceae,

Asteraceae, Euphorbiaceae, Rutaceae, Lamiaceae, Solanaceae, Acanthaceae, Anacardiaceae, Annonaceae, Apocynaceae, Caesalpinaceae, Fabaceae, Liliaceae, Malvaceae family members have provided pollen and nectar to stingless bees amidst urban environment (Sheetal, 2009). Raghunandan (2014) have reported on the foraging source for *A. dorsata* in southern Karnataka. *A. dorsata* depended on 78 plant families for its floral source. Mimosaceae, Caesalpinaceae, Rutaceae, Cucurbitaceae, Malvaceae, Myrtaceae, Clusiaceae, Fabaceae, Anacardiaceae, Euphorbiaceae, Lamiaceae, Verbenaceae, Asteraceae, Meliaceae, Acanthaceae, Combretaceae, Papilionaceae, Bignoniaceae, Compositae, Moraceae, Rubiaceae, Amaranthaceae, Araceae, Balsaminaceae, Convolvulaceae, Elaeocarpaceae, Flacourtiaceae, Sapinadaceae, Sterculaceae, Annonaceae, Apocynaceae, Asclepidaceae, Bombaceae, Brassicaceae, Celasteraceae, Dipterocapaceae, Ebenaceae, Lythraceae, Menispermaceae, Nycteginaceae, Palmae, Periplocaceae, Portualaceae, Sapotaceae, Violaceae, Vitaceae, Alangiaceae, Apiaceae, Basellaceae, Burseraceae, Cactaceae, Caprifoliaceae, Caricaceae, Casuarinaceae, Datiscaceae, Dilleniaceae, Droseraceae, Elatinaceae, Hydrocotylaceae, Icacinaceae, Lauraceae, Lecythidaceae, Lophopetalum, Magnoliaceae, Melastomataceae, Molluginaceae, Moringaceae, Musaceae, Nyctanthaceae, Oxalidaceae, Piperaceae, Poaceae, Santalaceae, Scrophulariaceae, Smilaceae, Ulmaceae and Zingiberaceae have supported the giant honeybee (*e.g. A. dorsata*) population during different seasons at southern Karnataka (Raghunandan, 2014). Thus, different species of honeybees depend on diversified flora for their survival under natural environment. As, the plant species are partly or fully depend on certain honeybee species for their pollination and propagation, this showed perfect interdependence between animals and plants in natural environment.

Honeybees Associations with Flowers

Flowers are colorful and ultraviolet light reflected by them is very well seen by honeybees. Flowers have scent, nectar, shape and other identifying characters, which are favorable in the competitive pollination services. Flowers have also evolved structures and behavior to minimize or prevent self-fertilization. For example some plants have staminate (male) and pistillate (female) parts on separate flowers on the same plant (monoecious) or on separate plants (dioecious). Some plants increase the chances of cross-pollination by having stamens and pistils ripen at different times. Flowers provide the vital source of food for honeybees amidst diversified geographical regions. To get good source of pollen and nectar, they visit a large number of different plant species.

The bees collect readily from different flower species according to the availability of the resources and their needs, but they show 'flower constancy' associated with the abundance of a certain flower species at a fixed time of the day. *Trigona* species are diurnal in their existence, flying under very low light intensity. Few subspecies of *Trigona* visit flowers after sunset and before dawn. Certain plant species bloom shortly after sunrise (ex. *Crotalaria* sp.) i.e., starting at 10.00hrs and from *Callinandra* sp. between 15.00hrs and 16.00hrs. Visits to passion fruit flowers coincided with their open hours. *Xylocopa mordax* on St. Vincent Island and *X. sonorina* in Hawaii paid most of their visits between 11.00hrs and 15.00hrs.

Some flowers possess a long tubular corolla in which nectar is inaccessible to the honeybees. Therefore, obtain nectar honeybees select small sized flowers to access nectar easily. The range of flowers selected for nectar collection differs greatly among the honeybees. Smaller species (*e.g. A. florea, Trigona* sp.) may be able to reach the nectar sources in the legitimate way, but they have difficulty in piercing the walls of the petals or calyxes. Sugar concentration in the nectar of many flower species visited by *Trigona* sp. varies between 25 and 27%, whereas concentration of 60 to 62% is used for 'bee bread' preparation. To obtain the required concentration, the honeybees expose the collected nectar on its proboscis drop by drop until the excess water has evaporated.

This nectar ripening or dehydration has been observed principally under tropical and subtropical conditions. It occurs when the honeybee is on the plant or at the entrance of the nest. Nectar dehydration by males (drones) increases the energy/volume of the nectar in their crops, thus enabling them to stay longer in their territory. Moreover, Honeybees obtain pollen by brushing off anthers with exposed pollen. It is also obtained from anthers that are periodically dehiscent. Pollen is carded in the crop or on the hind legs. Further, Honeybees have evolved thick hairs and pollen baskets for carrying pollen grains, mouth parts and honey sac for handling nectar, and the combs for storing nectar and pollen. These are related to their association with or adaptation to flowering plants. Advantages of crop-pollination induce plants to evolve flowers for attracting cross pollinating honeybees and also flower mechanisms for preventing self-fertilization. Following are the unique features shown by honeybees at diversified environments.

- Honeybees are the most efficient pollinators of numerous cultivated and wild plants owing to their unique characteristics.

- Honeybee body is specially adapted to pick-up and handle pollen grains,

- Honeybees show flower fidelity and constancy.

- Honeybees are capable of working for a long period of time, micro-manipulating flowers and maintaining high population whenever and wherever needed.

- Honeybees are well adaptable to diverse climates (Verma, 1992).

Ecological Relationships and Pollination Ecology

Ecological relationship between honeybees and the flowering plants is well known since prehistoric times. There are reports states about the evolution of honeybees. The honeybees have evolved from wasp-like ancestors, contemporarily with the flowering plants, 60 to 100 million years ago and they have had mutual benefits and relationship developed during the course of evolution and it is persisted till date. Basically flowers provide nectar and pollen for honeybees and honeybee's inturn provide pollination service for plants. Such a mutual interdependence has resulted in various types of 'co-evolution'. There are about one-quarter million species of flowering plants on the earth. Off all, many of which have amazing complex relationships with honeybees and other pollinators including flies, beetles, moths, butterflies, birds and bats (Meeuse, 1961; Dowden

1964). Cross pollination provides greater genetic diversity in the offspring than that by self-pollination. It provides greater opportunity to produce new varieties and which can adapt to adverse environmental conditions. Therefore, honeybees are very essential for the maintenance of crop-diversity and sustainable agriculture. Further, they bring significant qualitative and quantitative changes in the economic and biological characteristics of crop plants. They facilitate genetic enrichment of native plants through cross pollination offering to their varietal development and adaptation in the changing environment. Thus, mutual relationship between honeybees and plants is well known since prehistoric times. Further, cross-pollination by honeybees provides the plant species with greater genetic variability in the offspring than that by self pollination.

The genetic variability thus created gives the plant species greater opportunity to produce new varieties enabling them to adapt to new environments and to occupy different ecological areas (Verma, 1992). Seeds from cross-pollination have greater potentials for survival of the plant species, because the genetic interchanges add to the adaptations by plant species to new and changing environments. Somatic, reproductive and adaptive heterosis or hybrid effects occur in plant progeny due to cross-pollination. Such effects, either in a single way or in different combinations, bring about significant qualitative and quantitative changes in the economic and biological characters of plants (Verma, 1992). Numerous cultivated crops are not able to set seeds or produce fruits without cross-pollination of their flowers by honeybees or other wild insects. Cross-pollination of entomophilies crops by honeybees is one of the most effective and cheap methods of increasing their yield or quality of their products.

Further, many cultivated crops do not produce fruits or seeds without cross-pollination of their flowers by pollinators such as honeybees and other insects. About one-third of our total diet comes directly or indirectly from bee pollinated crop plants (Hoopingarner and Waller, 1992). Various agronomic practices namely the use of manure, fertilizers, pesticides and method irrigation couldn't bring desired results without the use of honeybee pollination to enhance the productivity levels of different cultivated crops. Thus, both self-sterile and self-fertile flowering plant varieties require cross-pollination to produce more and better quality seeds and fruits. Moreover, genetic enrichment of plants is essential to acclimatize them to changing habitat. Hence, honeybees, as pollinators, contribute to the sustenance of biodiversity and inturn bring lot of changes in the environment. In spite of the great economic and biological importance of honeybees as pollinators of agricultural crops, unfortunately honeybees are not been made as an integral part of agriculture and forest management in the developing countries (Verma, 1992). Further, honeybees not only gather nectar and pollen from flowers for food, but while doing so they also visit flowers and pollinate numerous different kinds of plants.

Especially in developing countries, the pivotal role played by honeybees in pollination is under estimated. In fact, the main significance of honeybees and beekeeping is pollination, whereas hive products such as honey and beeswax, are of secondary value. If we consider bees' value to be based on the fruit, vegetables and seed resulting directly from pollination, we have a value that is about 150 times the value of the honey and beeswax. On this line more research is required. Furthermore, honeybees and many species of solitary bees are resource factors upon which food production, health and aesthetic aspects of our environment depend. The impact of pollination, i.e., by bees on the ecosystem as a whole is probably of great importance. Pollination of plant species that contribute to soil conservation and control erosion and those that provide valuable resources for wildlife can be of considerable importance both for now and for the future. Much could be said for the undocumented contributions made by the pollinating activities of honeybees to thousands of wild plants that are integral parts of natural ecosystems and also of undocumented value to wildlife.

It has been estimated that there are between 25 to 50 different forms of life dependent upon each plant. When key plants die out of ecosystem that system goes into a decline and dozens of other life forms in that ecosystem also disappear. Deforestation followed by overgrazing worldwide land use pattern is giving rise to increased desertification and collapsed ecosystem everywhere in the world. Here, it is intended to highlight the role and contributions of honeybees towards saving the collapsing eco-systems or reversing desertification. Thus, honeybees contribute much to the natural environment while maintaining the healthy status amidst biotic and abiotic components of this planet earth.

Honeybees and Biodiversity as Interdependent Components of Natural Environment

The pollination of crops and wild flowers by honeybees and conservation of 'biodiversity' or 'natural environment' are interdependent. Crops and wild flowers are dependent on honeybees for their pollination, honeybees are dependent on crops and wild flowers for series of forages to sustain them and the natural environment is required to provide them with nesting sites. The potential for the survival of plant species is ensured by the seeds produced. If the seed results from cross-pollination, the potential for survival of the plant species is further enhanced, because the genetic interchange involve all adaptation to new and changing environments. Furthermore, seeds and plants form parts of the food chain for seed eating and herbivorous insects, birds and mammals (Verma, 1992). As a result of cross-pollination by honeybees, 'somatic', 'reproductive' and 'adaptive heterosis' or 'hybrid effects' occur in plant progeny, either in a single way or in different combinations. Such 'hybrid and synergistic effects' bring about following significant qualitative and quantitative changes in the economic and biological characters of plants by:

- Stimulating germination of pollen on stigmas of flowers and improve selectivity in fertilization,
- Increasing viability of seeds, embryos and plants,
- Enhancing the formation of more nutritious and aromatic fruits,
- Increasing vegetative mass and stimulate faster growth of plants,
- Increasing number and size of seeds and yield of crops,

- Enhancing resistance to diseases and other adverse environmental conditions,
- Helping to increase nectar production potential,
- Helping to increase oil content potential in oil-seed crops and
- Helping to increase fruit set and reduce fruit drop (Verma, 1992).

Genetic enrichment of indigenous plants through cross-pollination activities is necessary for the adaptation of plants to accommodate a wide array of environmental variables existing on degraded lands. Moreover, genetic enrichment of plants is essential to develop varieties that can change as the ecosystem (or habitat) it evolves. Honeybees often facilitate these genetic enrichment activities.

Honeybees and their Hive Products as Monitors of the Environmental Contamination

Honeybees are good biological indicators; indicate the chemical impairment of the environment by showing signals such as mortality (pesticide poisoning) and the residues present in their bodies or in beehive products (contaminants like heavy metals and radioactive compounds in honey) detected by means of suitable laboratory analyses. Several ethological and morphological characteristics make the honeybee a reliable and easy ecological detector. Various environmental sectors such as soil, vegetation, water, and air are sampled by honeybees, providing numerous indicators (through foraging) for each season. A variety of materials that are gathered into the hive (nectar, pollen, honeydew, propolis and water) and stored according to verifiable criteria and these can be used as indicators of environmental contamination (Porrini *et al.*, 2003). Native honeybees provide following advantages for monitoring biodiversity and environmental pollution.

- Honeybees are distributed over a broad geographic area,
- They are capable of providing continuous assessment,
- They are easy to collect, assay and /or analyze,
- They are sensitive to provide early warning of change,
- They are relevant to ecologically significant phenomenon,
- They are rich in genetic diversity and
- They are sympatric in distribution for monitoring pesticides; the number of dead honeybees in front of hive is the most important variable to be considered (Verma, 1992).

Further, the toxicity of the active chemical ingredient used on blooming cultivated or spontaneous plants could be noticed by recording the presence of honeybees on the site and at the time of chemical treatment. Honeybees directly affected by an insecticide. When they come in touch with noxious chemicals, unable to return to their hive because of their poor strength against chemical ingredients and will die in the field. Few honeybees marginally struck while foraging the flowers of the treated plant species or collecting nectar and pollen from spontaneous species contaminated by chemical pesticide drift' that will ultimately die in the hive. Thus, 'honeybee acts as a direct indicator' of chemical contamination in the environment. In case of compounds that are not particularly dangerous, the 'honeybee acts as an indirect indicator', i.e. not sensitive, but exposed, and will provide us with information in form of residues. Monitoring of heavy metals such as lead, nickel or chromium can be done by analyzing the amount of contaminants accumulated inside the honeybee and deposited on the honeybee surface. Heavy metals present in the atmosphere can deposit on the hairy bodies of honeybees and brought back to the hive with pollen, or they may be absorbed together with the nectar of flowers, or through water or honeydew collected, and analyzed for the result as an indicator. Monitoring of radioactive compounds could be done by means of analyzing pollen grains collected by honeybees. The findings demonstrated that pollen was the most efficient indicator of atmospheric radionuclide contamination. Thus, honeybees help indicate the quality of natural environment.

Honeybees and Environment

The presence of honeybees in every environment helps contribute to soil conservation, control of soil erosion and provide valuable resources for wildlife. Basically, flowers provide nectar and pollen for honeybees and honeybees provide cross-pollination for plants and there one could see a kind of interdependence and that results co-evolution. It is the duty of everyone to know the potentials of honeybees in the natural environment and their contribution in food production and human nutrition. General public and policy-makers should be made aware of the contribution of pollination or of the role played by insects including honeybees while protecting and conserving innumerable plant species. Plants are the essential components of all ecosystems.

Honeybees are basic or fundamental to many plants and plant communities and most of wildlife appears to be dependent upon both. It may never be possible or even necessary to identify the many and often complex interactions in these relationships, but their importance must be taken into consideration. It is in the best of national interest, honeybee species should be conserved at natural environment. Honeybee's presence is indispensable for the maintenance of plant biodiversity, regional and local vegetation. When vegetation is rich, naturally environment would be good and suitable for all the biotic components on this earth. Considering great economic and biological significance of honeybees, their habitat protection and conservation should be seriously taken into account. Honeybees should be protected by employing various management techniques in the natural environment. Conserving honeybees need more efforts by various classes of people including policy makers and conservation biologists along with better education, increased research and government support. Collaboration among various stakeholders, including international organizations, research, trade and policy making institutes, grass-root community groups and consumers is a key to ensure integrated efforts to such endeavor.

Conclusion

It is necessary to update our knowledge about the importance of honeybees in food production, human nutrition and in turn natural environment management. It is every ones responsibility to aware about the contribution of honeybees in pollination in order to protect and restore our natural environment. Honeybees are most efficient pollinators of diversified flowering plants. Their presence is indispensable for the maintenance of biodiversity, sustainable agriculture and natural environment. Considering the pollinators economic and biological significance, pollination activities with honeybees should be introduced in crop production. Planting of different species not only protects the pollinators but also enhance the regional biodiversity.

References

Ahmed, R. and S. R. Abbas. 1985. Some observations on *Apis dorsata* F. in Andaman and Nicobar Islands, India. *Indian Bee J.* 47:46-47.

Basavarajappa, S. 1998. Status of natural colonies of *Apis dorsata* in maidan region of Karnataka. *Indian Bee J.* 60 (3): 143-146.

Basavarajappa, S. 2004. Status of Asian giant honeybee, *Apis dorsata* F. and its conservation in southern part of Deccan peninsula, Karnataka, India: In perspectives of animal ecology and reproduction (edn. V. K. Gupta and A. K. Verma). *Daya Publishing House.* Delhi. pp. 45-71.

Basavarajappa, S. 2010. Nesting plants of dwarf honeybee, *Apis florea* F. under tropical conditions of Karnataka, India. *Ani.Biol.* 60:437-447.

Basavarajappa, S. 2012. Study of biological constraints of rock bee, *Apis dorsata* Fabr. at southern Karnataka. *UGC Major Project Final Report*, New Delhi. pp. 1-100.

Bright, A. A., Chandrashekaran, M. and M. Muthuswami. 1998. Bee pollinators importance and preservation. *Kissan World.* 25(4): 61-63.

Dowden, A. O. 1964. The secret life of flowers. *The Odyssey Press,* New York. Pp.45.

Dreller, C. and D.R. Tarpy. 2000. Perception of the pollen need by foragers in a honeybee colony. *Animal Behaviour.* 59: 91- 96.

Engel, M. S. 2002. The honeybees of India, Hymenoptera: Apidae. *JBNHS.* 99(1): 3-7.

Gillot, C. 1995. Entomology, (2nd edn.), *Plenum Press.* New York. Pp. 324-345.

Goyal, N. P. 1978. Performance of *Apis mellifera* and *Apis indica* as observed in Punjab plains. *Indian Bee J.* 40: 3- 5.

Heard, T. A. 1999. The role of stingless bees in crop pollination. *CSIRO Entomology.* 44: 183-206.

Hoopingarner, R. A. and G. D. Waller.1992. Crop pollination. The hive and the honeybee. (Graham J. edn.). *Dadant and Sons.* Hamilton, Illinois. Pp. 1043-1082

Karunaratne, W. M. K. K. and J.P. Edirisinghe. 2007. Appearance and recruitment of *Trigona iridipennis* nests in a selected area in Peradeniya University Park. *Pro. Peradiniya Uni. Res. Ses.* Sri Lanka. Vol. 12. Part, 1: 74-76.

Kerr, E. W. 1950. Evolution of the mechanism of caste determination in the genus Melipona, *Evolution.* 4: 7-13.

Khan, S.M., Kaushik, H.D. and H. R. Rohilla. 2007. Nesting behaviour of rock bee, *Apis dorsata,* II-Height and directional preferences for comb building. *Indian Bee J.* 69: 8-12.

Klakasikorn, N.A. Wongsiri, S., Deowanish, S. and O. Duangphakdee. 2005. New record of stingless bee (Meliponini: Trigona) in Thailand. *The Natural J. Chulalongkorn* Uni. Thailand. pp. 5(1):1-7.

Krishnamurthy, D. V. 2002. Status of giant honeybee *Apis dorsata* F. colonies in B. R Project Area. *M.Sc., Project Report,* Kuvempu Uni. Shankaraghatta, India. pp. 1-80.

Kumar, P. M. 2007. Ecology of Butterflies of family Papilionidae in mid Western Ghats of Shimogga District, Karnataka. *Ph. D. Thesis.* Kuvempu Uni. Shankaraghatta. India. pp. 1-100.

Leonhardt, S. D., Dworschak, K., Eltz. T. and N. Bluthgen. 2007. Foraging loads of stingless bees and utilization of stored nectar for pollen harvesting, *Apidologie.* 38: 125-135.

Manjunath, M. G. 2008. Studies on nesting behavior of giant honeybee, *Apis dorsata* F. in Mysore, Karnataka. *M. Phil., dissertation,* Uni. of Mysore, Mysore. pp. 1-103.

Meeuse, B. J. D. 1961. The story of pollination, *Ronald Press.* New York . Pp. 243.

Michner, C. D. 1990. Classification of the Apidae (Hymenoptera). *Univ. Kans. Sci. Bull.* 54: 75-164.

Michner, C. D. 2000. The bees of the World. Johns Hopkins Uni. Press. Baltimore. Maryland, USA. Pp. 1-500.

Moure, J.S. 1961. A preliminary supra-specific classification of the old world Meliponine bees (Hymenoptera: Apoidae). *Studia Entomologica.* 4:181-242.

Nagamitsu, T., Momose, K., Inoue, T. and D.W. Roubik. 1999. Preference in flower visits and partitioning in pollen diets of stingless bees in an Asian tropical rain forest. *Res. Popul. Ecol.* 41: 195-202.

Narayanaswamy, R. 2013. Study on dwarf honeybee, Apis flora F. (Hymenoptera: Apidae) in Manasagangotri campus, Mysore.*M.Phil, Dissertation,* Uni. of Mysore, India. pp. 8-67.

Neupane, K. R. 2004. Nesting behavior of giant honeybees, *Apis dorsata. J. Forestry Nepal.* 16(49): 351-357.

Nikam, T. B. 2002. A need to conserve honeybees for sustainable Agro-Horticultural and Forestry Development. *6th AAA. Inter. Conf. & World Apiexpo. Bangalore. India.* 24th Feb.–1st March. pp. 211.

Oldroyd, B. P., Osborne, E. and M. Mardan. 2000. Colony relatedness in aggregations of *Apis dorsata* Fabricius (Hymenoptera: Apidae). *Insect Soc.* 47: 94-95.

Porrini, C., Sabatini, A. G., Girotti, S., Ghini, S., Medrzycki, P., Grillenzoni, F.,

Bortolotti, L., Gattavecchia, E. And G. Celli. 2003. Honeybees and bee products as monitors of the environmental contamination. *Apiacta.* 38 (2003) 63-70.

Raghunandan, K.S. 2014. Bio-ecology of *Apis dorsata* Fabr. in few areas of south-western Karnataka. *Ph.D. Thesis,* Uni. of Mysore, India. pp. 89-124.

Reddy, C. C., Reddy, S. and S.P. Rajesh. 1986. Studies on the colony population of *Apis dorsata* F. *Indian Bee J.* 48: 57- 58.

Reddy, M.S. and C.C. Reddy. 1989. Height dependents nest site selection in *Apis dorsata* F. *Indian Bee J.* 51(3): 105-106.

Reddy, M. S. 2002. Prospects of *Apis mellifera* beekeeping in South India. *6th AAA. Inter. Conf. & World APIEXPO*, Bangalore, India. 24th Feb.–1st March. Pp. 152.

Sattagi, H. N., Kulkarni, K. A., Rajasekhar, D. W. And N. Kambrekar. 2002. Techniques to remove honey comb from rock bee *Apis dorsata* F. colonies for honey extraction. *6th AAA. Inter. Conf. and World APIEXPO,* Bangalore, India. 24th Feb.–1st March. pp. 54.

Schwarz, H.F. 1948. Stingless bees (Meliponidae) of the western hemisphere. *Bull. Am. Mus. Nat. Hist.* 90:1-546.

Seeley, T. D. 1995. The wisdom of the hive. *Belknap Press of Harvard University,* Cambridge. UK. pp. 1-200.

Shah, F. A. and T.A. Shah. 1982. The role of Kashmir bee in exploiting bee keeping potential in India. *Indian Bee J.* 44: 37- 42.

Sheetal, V. K. 2009. Nesting ecology and flora of stingless bee, *Trigona iridipennis* Smith. in Manasagangotri campus, Mysore. *M.Phil, Dissertation,* Uni. Of Mysore, India. pp. 11-81.

Siddappa Setty, R. and K.S. Bawa. 2002. Characteristics of honey resources in a tropical forest: Productivity and extractions of *Apis dorsata* honey in Biligiri Rangayan Hills Wildlife Sanctuary, India. *Proc. 6th AAA Inter. Con. & World APIEXPO- 2002,* Bangalore, India. 24th Feb.–1st March. pp. 54.

Sihag, R. C. 2002. Conservation of bee diversity challenges in the millennium. *Proc. 6th AAA Inter. Conf. & World APIEXO,* B'lore. pp. 190.

Thapa, R and S. Wongsiri. 1996. Toxicity of azadirachtin derivatives and synthetic pesticides on oil seed rape to *Apis cerana* (Hymenoptera: Apidae) biopesticides toxicity, safety, development and proper use. In: *Proc. Inter. Symp. On Biopesticides Phitsanulok,* Thailand, pp.82-86.

Varadharajan, M. 2002. Foraging strategies of the Indian honeybee, *Apis cerana indica,* with reference to pollen preferences and nectar collection under central place foraging and risk sensitive foraging in artificial nectaries. *Ph.D., Thesis.* Bharathidasan Uni. Mannampandal. Tamil nadu. pp. 1-300.

Verma, L. R. 1990. Beekeeping-an integrated mountain development. *Oxford and IBH Publishing Co, Ltd.,* New Delhi. pp. 57- 80.

Verma, L. R. 1992. Beekeeping and pollination ecology of mountain crops. (Verma, L. R.edn.) Honeybees in mountain agriculture. *ICIMOD, Kathmandu, Nepal.*

Wells, H., Wells, P. H. and D.M. Smith. 1981. Honeybee response to reward size and color in an artificial flower patch. *J. Apic. Res.* 20: 172- 179.

Wongsiri, S. Koeniger, G. and N. Koeniger. 2002. Bee biology in the new millennium. *6th AAA. Intr. Conf. & World APIEXPO*, Bangalore, India. 24th Feb.–1st March. pp: 5.

Chapter 5

Regeneration Studies – A Tool for Forest Management

A.G. Devi Prasad and Sapna Rao

*DOS in Environmental Science, University of Mysore, Manasagangotri,
Mysore-06, Karnataka*
Correspondence Email Id: agdprasad@yahoo.com

ABSTRACT

This paper describes the important processes of regeneration by which a forest is naturally regenerated. It embraces all methods employed by plants in producing their juvenile stages. Forest management has become a focal issue today. To understand the process of natural regeneration in forest as well as formulate methods to promote regeneration artificially is gaining attention. The study of regeneration helps to understand and maintain a stable population structure of species in a community. The understanding of various processes involved in regeneration such as seed production, germination, establishment etc. will help in giving emphasis to the tree species which have been greatly exploited and not regenerating in that particular area. This study would give valuable information which would help in maintaining the forests and trees providing the essential forest products at the same time assuring multipurpose functions such as recreation and preservation of habitat of wildlife.

Key words: Germination, Logging, Disturbance, Coppice, Saplings, Predation

Introduction

Forests cover 40 per cent of the earth's land surface and are home to more than 70 per cent of plants and animals living on land. Forest resources provide a wide range of ecological, social and economical benefits. For example, Tropical forests, in particular, have high productive and protective natural value. An estimate of 10-30 million species is found in tropical forests alone comprising of very complex ecological communities of different species that have unique ecological importance (Sandalow, 2000). But destruction of these natural forests of the world in the last two decades has been rampant due to the illegal encroachment by agricultural activities and urbanization. Uncontrolled exploitation has reduced the abundance and diversity of trees in the forest reserves.

All populations in a forest are under the flux of two vital and opposite processes which are growth and death. Regeneration is an intermediate process which leads to an increase in population number (Krebs, 1972). Forests depend on adequate regeneration of tree species to be healthy and sustainable. The presence of young trees in the forest understory is necessary to sustain forest canopy development after timber harvest or natural disturbances such as windstorms, or individual tree mortality, which creates forest canopy gaps. The density of regeneration is expected to vary spatially due to forest type and physical site conditions (Liang and Seagle 2002; Ward *et al.*, 2006) and temporally due to changes in seed production (Boerner and Brinkman 1996). Biological factors such as insects, disease, herbivory, and competing vegetation also influence regeneration (Ward *et al.* 2006).

Adequate seed supply, effective dispersal, good viability and longevity of seeds, successful establishment of seedlings and good conversion to mature trees are all unavoidable for a sustainable forest management. Therefore, the population structure at each of these life stages, viz., adult trees, flower, fruit, seed, seedling, sapling, pole, etc. determine the structure of mature tree populations of the future in a forest. Hence successful management and conservation of natural forests require reliable data on all aspects of above mentioned regeneration trends of the tree species in the forest and also on the factors affecting their regeneration.

The term regeneration means the process by which a forest is naturally renewed. This process embraces all methods employed by plants in producing their juvenile stages.

In a wider sense the term regeneration applies to all life stages of the plant beginning with flowering and ending with formation of adult trees, passing through the stages of fruits, seeds, seedlings, saplings and poles (Nair, 1961).

Glimpse of Research on Regeneration

Various environmental factors such as disturbance, fire, natural calamities, anthropogenic activities have a profound influence on the regeneration of plants in their wild.

Chesser and Brewer (2011) studied Factors influencing seedling recruitment in a critically endangered pitcher plant, *Sarracenia rubra ssp. Alabamensis*, which is endemic to Alabama (USA). Despite their various efforts, seedling recruitment remained very low within the largest population of *S. rubra ssp. alabamensis*. They considered the possibility that other factors such as germinability, seed removal (by predators or dispersal agents), and inadequate soil moisture were reducing or preventing seedling recruitment. They reported the results of laboratory and field experiments investigating factors limiting seedling recruitment and establishment in the 2 largest remaining populations, which were observed at sites that differed in soil moisture, fire management, and the abundance of sphagnum moss. Seedling recruitment was higher at the wet unmanaged site than at the dry, fire-managed site. The major contributing factors to site differences in recruitment were a greater abundance of safe sites for germination (specifically, patches of sphagnum) and lower seedling mortality at the wet unmanaged site as a result of higher soil moisture. In conclusion they recommend that land managers shift some of their attention to this and other wet sites and seek permission to increase population density and restore fire at these sites.

A study carried out by Guzman *et al.*, (2011) predicted that seeds deposited below parent trees after fruiting fall has finished, is advantageous to minimize seed predators and should show higher survival rates. Four Amazonian plant species, *Dicranostyles ampla, Oenocarpus bataua, Guatteria atabapensis* and *Ocotea floribunda*, were tested for seed survival probabilities in two periods: during fruiting and 10-21 days after fruiting. Experiments were carried out in two biological stations located in the Colombian Amazon (Caparú and Zafire Biological Stations). Seed predation was high and mainly caused by non-vertebrates. Out of the four plant species tested, for *Guatteria atabapensis*, seed predation by vertebrates increased after the fruiting period (from 4.1% to 9.2%) while seed predation by nonvertebrates decreased (from 54.0% to 40.2%). In contrast, seed predation by vertebrates and by non-vertebrates after the fruiting period in *D. ampla* increased (from 7.9% to 22.8% and from 40.4% to 50.6%, respectively), suggesting predator satiation. Results suggest that for some species dispersal in time could be advantageous to avoid some type of seed predators. Escape in time could be an additional dimension in which seeds may reach adequate sites for recruitment. Thus, future studies should be addressed to better understand the survival advantages given by an endozoochory time-dispersal process.

Rocky and Milligo (2011) studied the Pugu forest reserve in Tanzania, which is highly degraded due to exploitation. To conserve the forest, part of natural forest was cleared and exotic plantation was established to provide forest resource needs for the nearby residents. However, the regeneration pattern of indigenous trees in exotic plantation was not monitored. Thus, their study was aimed at assessing diversity, population structure, and size class distribution and natural regeneration pattern of indigenous trees in both exotic tree plantation and the natural forest. Significant differences were found on the abundance of indigenous plant species between those in exotic tree plantation and the natural forest. Conclusively, establishing exotic plantation through clearing natural forests required monitoring as a management strategy because indigenous woody plants recovered through natural regeneration outcompeted exotic trees in the plantation.

Pokhriyal *et al* (2010) reported regeneration status of trees in two watersheds namely Phakot and Pathri Rao in Uttarakhand. Seedling, sapling and tree density were greater in Phakot watershed forest than those in Pathri Rao watershed forest. In general, both forests were regenerating, although seedling and sapling population was higher in Phakot watershed forest. As far as the regeneration status is concerned, maximum tree species were found with fair regeneration in the forests of both the watersheds. In Phakot watershed, three species (*Accacia nilotica, Engelhardtia spicata* and *Olea glandulifera*) and in Pathri Rao watershed seven species (*Acacia nilotica, Anogeissus latifolius, Casearia elliptica, Cassia fistula, Holarrhena pubescens, Mallotus phillippensis* and *Ougeinia oojeinensis*) were found not regenerating. In Phakot watershed, general densities– diameters class distribution showed decline in density from small diameter class to higher diameter class whereas in Pathri Rao watershed no trend was evident.

Seed production, dispersal, fate of seeds, seedling recruitment, mortality and growth behavior of *Schima wallichii* (DC.) Korth. were investigated by Sahoo and Lalfakawma (2010) during 2004-05 in an undisturbed and a disturbed tropical forest stand of North-East India. Five fruiting trees, each with different diameter at breast height, growing 100 m apart, were selected and concentric circles of 2.5m radial increments were marked in both stands to study seed production, fate of seed population and seedling recruitment. Significant (P<0.01) differences were observed in mean seed yield per year between the stands and dbh class. More seeds disappeared during post-seed-fall period than during seed-fall period. Germination was lowest in the innermost circle as distance had remarkable effect (P<0.01). Seedling recruitment and relative shoot growth were higher (P<0.05) in the disturbed than the undisturbed stand in the first year. Controlled disturbance could be a favourable management tool for increased recruitment, survival and growth of *Schima wallichii* seedlings.

Reddy and Ugle (2008) investigated the regeneration status in tropical dry and moist deciduous forests of Mudumalai Wildlife Sanctuary, Western Ghats, India. A total of 124 tree species were recorded in tropical deciduous forest system. Out of the 104 species (young and mature trees) recorded, 28.8% showed good regeneration, 5.8% represented fair, 33.7% poor, 29.8% showed no regeneration and 6 (5.8%) were considered as new arrivals in moist deciduous forest. In the case of dry deciduous forest out of 86 (young and mature trees) 33.7 % showed good regeneration, 3.5% fair, 16.3% poor, 17.4% showed no regeneration and 9 species (10.5%) were considered as new arrivals. Absence of Younger type of most of the species inferred impact of anthropogenic disturbances such as recurrent forest fires, cattle grazing and biological invasion of exotic weeds on natural regeneration. The basic analysis may be considered here to be driven by two criteria: Species endemism and degree of threat, and therefore survival threat to the flora of the Mudumalai wildlife sanctuary was studied.

In 2008 Chauhan *et al.* studied regeneration and tree diversity in natural and planted forests in a Terai - Bhabhar forest in Katarniaghat Wildlife Sanctuary, India. They compared regeneration, tree diversity and floristic diversity of natural and planted tropical deciduous forests (dominated by *Shorea robusta* and *Tectona grandis*; *Acacia catechu* and *Syzygium cumini*, respectively) in western Uttar Pradesh, India. Species diversity (70 species in natural and 59 species in planted forests) as well as species evenness was higher in natural forests than in planted forests. Natural forest sites also had higher mature tree, pole, sapling and seedling densities compared to planted forests. In spite of differences in diversity, natural and planted forests did not differ strongly in species composition, fifty-six species occurred in both sites. This may reflect similar soil types but differences in soil moisture, organic carbon, and available nitrogen, and phosphorus, potassium and soil pH in natural and planted forests. Of the 126 species found in both sites, 32.5% showed good regeneration, 19.8% fair, 24.6% poor and 11.1% lacked regeneration. The remaining 11.9% of species were present as seedlings but not as adult individuals. Good quality timber species are not regenerating, with the exception of *Shorea robusta*, although mortality at seedling stage of this species is high. In all, the results suggest that species richness and diversity differed between natural forest and planted forest and regeneration of some important tree species also varied in natural and planted forests because of variation in their microclimate and edaphic characteristics. Moreover, these conditions indicate succession pattern and a potential for forestry plantations in dry forests. This study will help in the formulation of effective forest management and conservation strategies.

Dhaulkhandi *et al.,* (2008) reported the regeneration potential and community structure of natural forest site in Gangotri, Uttarakhand. A total of seven tree species were recorded from the site. Among the trees, *Picea smithiana* was the dominant and *Cedrus deodara* was found co-dominant species. However, the highest (240 trees/ha) density was reported for *Pinus wallichiana* while least number of individuals (30 trees/ha) were recorded for *Acer caesium* and *Pinus wallichiana*. In tree layer the most of the species (65.16%) were distributed contagiously and few (34.84%) were distributed randomly. However, none of the species should regular distribution pattern. *Artemesia gamillinea* and *Cotoneaster gilgitansis* were the most and least dominant shrub species respectively. All species of shrub layer were distributed contagiously (100%). In the seedling stage, maximum number was observed for *Pinus wallichiana* (1080 seedling/ha) followed by *Picea smithiana* (1040 seedling/ha) which was recorded just after in sapling stage, because it shows more survival rate of *Picea smithiana* (600 sapling/ha) as compared to *Pinus wallichiana* (520 sapling/ha). As far as regeneration status was concerned, 71.4% species showed good regeneration, 14.3% species were facing the problem of poor regeneration whereas; only 14.3% species were not regenerating.

Tropical dry forests are the most threatened tropical terrestrial ecosystem. However, few studies have been conducted on the natural regeneration necessary to restore these forests. Vieira and Scariot (2006) reviewed the ecology of regeneration of tropical dry forests as a tool to restore disturbed lands. Dry forests are characterized by a relatively high number of tree species with small, dry, wind-dispersed seeds. Over small scales, wind-dispersed seeds are better able to colonize degraded areas than vertebrate-dispersed plants. Small seeds and those with low water content are less susceptible to desiccation, which is a major barrier for establishment in open areas. Seeds are available in the soil in the early rainy season to maximize the time to grow. However, highly variable precipitation and frequent dry spells are important sources of mortality in seeds and seedlings. Collecting seeds at the end of the dry season and planting them when soil has sufficient moisture may increase seedling establishment and reduce the time they are exposed to seed predators. Germination and early establishment in the field are favored in shaded sites, which have milder environment and moister soil than open sites during low rainfall periods. Growth of established seedlings, however, is favored in open areas. Therefore, clipping plants around established seedlings may be a good management option to improve growth and survival. Although dry forests have species either resistant to fire or that benefit from it, frequent fires simplify community species composition. Resprouting ability is a noticeable mechanism of regeneration in dry forests and must be considered for restoration. The approach to dry-forest restoration should be tailored to this ecosystem instead of merely following approaches developed for moister forests.

Eilu and Obua (2005) studied tree condition and natural regeneration in disturbed sites of Bwindi Impenetrable forest national Park, southwestern Uganda. Sampling was done in anthropogenically undisturbed, lightly disturbed, heavily disturbed and completely disturbed forest types. Nested plots measuring 25 x 30 m were established to assess tree size classes from seedlings (diameter < 2 cm and height ≤ 150 cm) to large trees (DBH > 15 cm). Higher stem densities were found in undisturbed and lightly disturbed forest types compared to heavily or completely disturbed types. Tree species richness and diversity were highest in lightly disturbed forest. *Acacia mearnsii*, an introduced tree species in Uganda, was recorded in completely disturbed forest. Regeneration from vegetative sprouts was highest in disturbed sites while regeneration from seeds was highest in undisturbed sites. High intensity human disturbance was associated with fewer signs of mammal damage. Damage to trees by physical agents and climber abundance increased with intensity of disturbance except in completely disturbed forest. High intensity human disturbance adversely affected tree species abundance, diversity and regeneration and increased the incidence of damage to trees. Regeneration from vegetative sprouts was most important in heavily disturbed sites. Intensity of disturbance and slope influenced the distribution of regenerating tree species.

Khan *et al.* (2005) Tropical wet evergreen forests of Arunachal Pradesh, in northeast India are being modified and degraded due to increased anthropogenic pressure. Natural populations of *Elaeocarpus ganitrus Roxb.* (Elaeocarpaceae) are threatened due to household and industrial uses. Natural regeneration is scarce, partly due to extensive harvesting of the tubercled nuts for use in jewellery. Effects of disturbance on natural regeneration are poorly understood. They studied the impact of disturbance on *E. ganitrus* demography by monitoring flowering, fruiting and seed dispersal during three consecutive years and at four sites that varied in the degree of disturbance. Pollination rate in reproductive individuals varied significantly between years, but increased with tree diameter. Fruit production differed significantly among sites, and was highest in the moderately disturbed forest and lowest in the undisturbed forest. Mean fruit weight decreased significantly with increasing disturbance. About 40–70% of the total fruits produced by *E. ganitrus* were removed during the fruit-fall period, mainly by arboreal frugivores and seed-hoarding rodents. After the fruit-fall period, most (55–99%) of the remaining fruits also disappeared, while rates increased with disturbance index. No seeds were found to be germinated. In general, about 80% of the fruits produced by *E. ganitrus* disappeared from the forest floor or was damaged severely by insects, especially ants and termites. The proportion of fruits disappearing decreased significantly with increasing distance from the parent tree. Ripe fruits with intact exocarp were removed more frequently than both unripe fruits with exocarp, and nuts. The nut bank on the soil surface decreased with increasing disturbance index. More than 85% of the nuts of the nut bank was predated. The findings on fruit set and dispersal of *E. ganitrus* may have great implications for regeneration of the species.

Omeja *et al.*, (2004) A study was carried out between 1999 and 2001 in Degeya, Lufuka and Mpanga forests in central Uganda to determine the regeneration, density and size class distribution of trees used for making drums. Thirty sample plots measuring 20 · 20 m were established at 250 m intervals along transects laid in the northeast direction across swamps, mid-slopes and hilltops. Diameter at breast height (DBH) of trees and number and species of seedlings, saplings and poles of six tree species were determined. The tree species were: *Antiaris toxicaria, Erythrina excelsa, Ficus mucuso, Ficus exasperata, Funtumia africana* and *Polyscias*

fulva. Antiaris toxicaria was the most abundant and *P. fulva* was the least abundant in the forests. The intensity of use and concentration on a limited number of tree species has resulted in localized exploitation with potential knock-on effects on the forests' health. On-farm tree planting by local communities should be encouraged to supply logs for making drums and reduce pressure on the forests.

Hooper *et al.*, (2004) tested alternative hypotheses concerning factors affecting early forest succession and community composition in deforested and abandoned areas invaded by an exotic grass, *Saccharum spontaneum*, in Panama. They hypothesized three barriers to natural regeneration: (1) Saccharum competition, (2) seed dispersal limitations, and (3) fire. They also measured natural tree and shrub regeneration in a factorial experiment combining distances from adjacent forest, mowing treatments of the Saccharum, and a prescribed burn. To determine the applicability of the general model of neotropical succession and the nucleation model of succession to species composition of forest regeneration in these anthropogenic grasslands in Panama the effect of time since fire and distance to remnant vegetation (isolated trees, shrubs, and large monocots) was measured. Fire significantly affected species composition and decreased species richness because most species had either their resprouting ability or seed germination inhibited by fire; the few species that had regeneration enhanced by fire dominated early successional communities. Sites differed in time since fire, ranging from 1 to 4 yr; the interaction of site and distance from the forest significantly affected community composition and the prevalence of species with different dispersal mechanisms and shade tolerance. At recently burned sites, light-dependent wind-dispersed species predominated; most were found near the forest edge. As time since fire increased, significantly more shade-tolerant, larger-animal-dispersed species were recorded, and the proximity to and species identity of remnant vegetationbecame more important in affecting species composition of the natural regeneration; no significant effect of distance from the forest was found at sites that were unburned for three or more years. Our results support both successional models; the temporal sequence of species composition corresponded to later stages of the general model, while the spatial distribution of species followed the nucleation model. Their results highlight the importance of effective seed dispersal in structuring successional species composition and distribution and in regaining lost diversity resulting from frequent fires in the Saccharum.

Mersha Gebrehiwot (2003) studied the Wondo-Wesha Catchment Awassa Watershed Sourthern Ethiopia. Preliminary survey found that Ethiopia did not have sufficient data to make decisions regarding the management of the natural forests. Thus they applied statistical and geostatistical methods to analyze the regeneration diversity and the spatial distribution of the regeneration of tree species in the natural forests of Hayena Valley and Solomon valley at Wondo-Wesha Catchment and generate information about the status of the forests. The statistical method focused on analyzing the relationships between the different stem groups (seedlings, saplings and trees) with stand structure parameters (crown cover, shrub cover, basal area and stand density), seed trees and environmental factors (slope, elevation and aspect), within and between the forests. Considering the status of regeneration of the forest, both the statistics and geostatistical analysis confirmed that the relatively disturbed forest has higher regeneration diversity and higher variation in diversity distribution in the study area.

Matts Karlson (2001) studied the natural regeneration of broadleaved tree species in Southern Sweden. The objective of the present thesis was to examine the effects of silvicultural treatments and seed dispersal from surrounding stands on the establishment of natural regeneration of broadleaved tree species in southern Sweden. Most of the broadleaved tree species that occur naturally in forests in southern Sweden were studied but birch (Betulapendula Ehrh.1 B. pi~besceizsR oth) was the most common species and present in equal numbers in all studies. The wind dispe.rsa1 of see.ds of se.ven species was studied and great variations were found. This could mainly be explained by differences in seed morphology. The effect of soil scarification was examined in all five studies and was generally found to be positive for the establishment of the studied broadleaved species. However, in some cases the scarification was not positive for the establishment. The reason for tlis was hypothesised to be that the seed supply was limited, or an effect of large and/or animal-dispersed seeds. The effect of shelterwood was examined in three studies and was found to be positive for animal-dispersed species but negative for shade-intolerant species, although a sparse shelterwood can be used to rcgcncratc birch. Slash removal was included in onc study and found to be positive for thc cstnblishrncnt of birch. This thcsis showccl that regeneration treatment can be used to incrcasc the e.stablishnent of naturally regene.rated broadleaved tree seedlings, but the stand structure and species composition must be regulated with pre-commercial thinning. However, the effect of variations in seed production and seed dispersal must be closely examined from a time and a space perspective prior to any forecasts regarding the effects of regeneration treatment.

The degraded sal forests of north-eastern Uttar Pradesh were observed by Pandey and Shukla (2001) for the regeneration strategy of constituent woody perennials and the status of resultant plant diversity. The species showing poor sprouting were much greater in number at low disturbance. Conversely, the species showing rich sprouting and ramet formation were much more at high disturbance. The diversity index (H) was always greater when a genet complex was treated as a single individual than in case when each ramet, distinct at soil surface, was treated as a separate individual. The value of H, however, was lower at low disturbance. It is a moot point whether the diversity index should be based on the number of genets (biotypes) or superficially distinct shoots (including ramets) irrespective of their genetic status. The species like *Clerodendron infortunatum, Croton oblongifolius, Mallotus philippensis* and *Flacourtia indica* increased their ramet production with increase in disturbance level, but recurrent disturbance of high intensity affected ramet proliferation quite adversely. *Bridelia retusa, Casearia tomentosa* and *Holarrhena antidysenterica* produced comparatively much lesser number of ramets per genet. The inter-ramet distances or spacers on root-stock as well as the number of ramets per genet showed significant differences with respect to the level of disturbance. The age structure and spatial pattern of ramet population were also correlated with the level of disturbance. In a forest environment which is too harsh to allow regeneration through seed, a non-seed regeneration of a group of woody perennials may help maintain the minimal vegetation cover and considerable plant diversity. The non-seed regeneration strategy of prolific ramet producers, therefore, shows a promise to the quick recovery of forest ecosystems ravaged by anthropogenic perturbations.

Barik *et al.*, (1996) carried out a study on tree regeneration in a subtropical humid forest: effect of cultural disturbance on seed production, disturbance and germination. The subtropical humid forest of Meghalaya, India has been exposed to various kinds of cultural disturbances. In order to analyse the effect of disturbance on regeneration, they studied seed production, dispersal and germination of a few dominant and commercially important species in three stands, differing in degree of disturbance, for a period of four years. The study revealed that seed production in *L. dealbatus* and *S. khasiana* significantly varied between the three stands and increased with increasing disturbance. It increased with increase in d.b.h. of the trees in all tree species. *Lithocarpus* and *Quercus* species produced heavier seeds in the disturbed stands than in undisturbed stands.

The number of seeds dispersed from the trees decreased with the distance from the parent tree. Seed predation decreased and germination increased with distance from the parent tree in all three stands, suggesting that distance – related seed predation was not influenced by disturbance. An analysis of the fate of seed populations of the oak species revealed that loss of seeds caused through consumption of rodents and insects and transportation by various agents accounted for more than 98% of the seeds, while less than 1% of them germinated. The findings of the study are discussed in relation to their potential application in management of the disturbed subtropical forests.

Today, the concept of conservation in resource management is spreading widely. Many international forums find their efforts to develop suitable methods for practicing the concept of sustained yield in forestry (UNESCO, 1975). Therefore, the subject of natural regeneration is receiving greater attention. Different kinds of organisms have different kinds of regenerative strategies (Grime, 1979). Of these, forest trees by and large have seed based regenerative strategies (i e, by genets) although some species also show a certain degree of vegetative regeneration (i.e., by ramets).

Natural Regeneration From Seeds

Regeneration of plants in the wild mainly depend on:

a. **Seed production:** The most important consideration for natural regeneration is the production of adequate amount of fertile seeds by the trees of that area.

b. **Seed dispersal:** Seed dispersal is transport of produced seeds away from the parent plant. The place of production of seeds does not have the carrying capacity to grow and sustain all of them (Gadgil, 1971). Thus, competition is avoided by dispersing seeds even at the danger of casualties. Plants have limited mobility thus they rely on biotic and abiotic vectors for their dispersal. The mechanism of dispersal involves wind, water, frugivorous birds and animals (Ridley, 1930, Pijl, 1969). In Wet Forests seeds of more than 60 percent of the trees are dispersed by sarcochorous means (eaten by animals) (Danserau and Lems, 1957); while, the Dry Forests show a greater percentage of wind dispersal (Baker *et al.*, 1983). The seed dispersal mechanism is species specific too. For example: seeds of *Dalbergia, Acacia, Terminalia, Bombax* species are dispersed by wind; seeds of Mangrove, teak species are dispersed by water; seeds of *Diospyros* are dispersed by birds and seeds of *Acacia, Prosopis, Zizyphus* species dispersed by animals. Seed dispersal allows plants to reach specific habitats that are favourable for survival.

c. **Seed germination:** Germination is the growth of an embryonic plant contained within a seed; it results in the formation of the seedling. After dispersal, insects, birds and rodents destroy many seeds. The others which survive germinate, provided they are deposited on suitable area or soil.

d. **Seed establishment:** Even if germination is good, the process of natural regeneration may be affected by factors such as weeds, grazing, fire, which may kill the seedlings, thus, preventing them from establishing themselves in a suitable environment.

If the established seedlings of tree species in a forest community have successfully survived all three phases mentioned above and recruited in their prevailing environmental conditions, then the regeneration process is said to be successful. Thus according to Good and Good 1972 the three major components which cause successful regeneration of tree species are:

1. **Ability to initiate new seedlings;**

2. **Ability of seedlings and saplings to survive;**

3. **Ability of seedlings and saplings to grow.**

Natural Regeneration from Vegetative Parts (Coppice)

Coppice arising from the stool or a living stump of a tree is called stool coppice. In this method, regeneration is obtained from the shoots arising from the adventitious buds of the stump of felled trees. The coppice shoots generally arise either from near the base of the stump or from its top. The shoots arising from near the top of the stump are liable to be damaged by the rotting of the upper portion of the stump as well as by wind. When regeneration obtained by coppice develops into a forest, it is called coppice forest to differentiate it from the high forest.

The selection of the areas for the regeneration studies is based on levels of disturbances in that area. Then Vegetation analysis is carried out by randomly placing quadrats of specific sizes along transect. The quadrats are laid at 200 to 250 m intervals on alternate sides of transect in undisturbed and highly disturbed areas. For each species, seedlings (diameter < 5 cm), Poles (DBH 10 – 19.9 cm) and mature trees (DBH > 20 cm) are identified and counted within each quadrat and the results are subjected to statistical analysis. From the results obtained the regeneration can be said:

1. Good, if the number of seedlings > saplings > tree

2. Fair, if the number of seedlings > saplings < tree

3. Poor regeneration, if species survive only in sapling stage

A species is considered as not abundant, if the species has no representatives, but only saplings and/or seedlings. And if the species are found only in adult form it is considered as not regenerating.

Some of the Important Factors Affecting the Natural Regeneration Process

Seed production is one of the most important determinants of successful natural regeneration (Shelton and Cain 2001). It varies among populations because of differences in both the number of fruiting individuals and number of seeds produced per reproductive individual. Seed production of trees depends on various factors such as:

1. **Type of species:** Some species like *Acacia, Tectona* and *Dalbergia* produce seeds annually whereas species of *Cedrus, Abies and Picea* species produce seeds at an interval of several years, thus quantity of seeds produced depends on the type of species.

2. **Age of trees:** The age of trees affects the production of fertile seeds. Adequate amount of fertile seeds are produced from middle aged trees.

3. **Size of crown:** The size of the crown tree influences seed production. As a general trend, the bigger the crown the larger is the seed production.

4. **Climate:** Warmer climate favors larger seed production. Hot dry climate is generally followed by heavy seed years on account of increased photosynthesis. Heavy rainstorm at the time of pollen dissemination reduces chances of pollination and good seed production.

5. **Other external factors**: Injury by fire and insect attack reduces seed production by damaging the crown. If damages are only concentrated on the barks then it stimulates seed production by transporting carbohydrates to the seeds and not to the roots.

6. **Disturbance**: Seed production tends to increase with increased intensity of disturbance, since, for successful regeneration of species, it is necessary that the seeds be dispersed to a specific site where they will establish and germinate.

Seed dispersal is an essential part of the regeneration or recruitment of plants (e.g. Darwin, 1859; Harper, 1977; Willson, 1992; Ribbens *et. al* 1994; Clark et al, 1999). The Janzen-Connell hypothesis proposes that seeds have higher survival probabilities when they escape distance-and density-dependent predation below parental trees, affecting plant recruitment and diversity (Janzen 1970, Connell 1971). Thus, seed dispersal away from parent trees become important for plant fitness and survival. Some predictions of the Janzen- Connell hypothesis have been tested in various plant species and habitats, generally providing positive support (Hammond & Brown 1998, Wright 2002, Hyatt et al. 2003, Peterman et al. 2008, Bagchi et al. 2010, Matthesius et al. 2010). Seeds can be dispersed by various mechanisms and seeds of specific species will have a specific dispersion mechanism. Not all fruits that are removed and ingested by the vectors result in effective seed dispersal. The effectiveness of seed dispersal depends on quantity of seeds removed and quality of dispersal. The main factors that affect the quality of dispersal are the distance moved from the parental tree, and the particular microsite the seed is deposited in. Habitat alterations could affect seed dispersal indirectly by changing fruit and seed characteristics (eg: ripeness, size and nutritional content) that determines their attractiveness as food for disperser. Dispersal especially by frugivorous animals and net scatter-hoarding by granivorous animals helps seeds escape high mortality near the parent tree, reduce sibling competition by scattering seeds and enable the species to colonize new areas. However most consumers/ dispersers like rodents, hornbills, bats, flying squirrels, monkeys, deer, wild pigs, buffaloes, mithuns and elephants are decreasing in numbers due to deforestation and habitat alteration and poaching, which is affecting the process of regeneration. The seed dispersal depends on factors such as species-specific morphological adaptations of seeds to dispersal mechanisms, available dispersal agents like animal populations, wind conditions and tree height.

Predation is an important factor controlling the viable seed population. Seed predation intensity is highly dependent on seed species (Osunkoya 1994; Holl and Lullow 1997). It has been argued that seed predation decreases with distance from the parent tree and the 'escape hypothesis' seems to hold good in many plant community (Janzen, 1971; Howe, 1982). There are instances of up to 40 % seed predation by rodents (Synnott, 1973). In *Shorea ovalis* greater than 90 % seed predation due to insects has been recorded (UNESCO, 1978 b). Generally predation decreases with distance from seed tree or with poor seed density. Janzen (1971) suggested a 'predator escape hypothesis' according to which plants escape predation by satiating them (Howe and Smallwood, 1982). In the Dipterocarp

Forests of Malaya the seeds escape predator threat by immediately germinating and building up a seedling bank (P. N. Nair, 1961; Grime, 1979). Post-dispersal predation of seeds alters the amount of germinable seeds in an area. Rapid disappearance of seeds from the forest floor could be due to transportation by mice, voles and other rodents, which consume and/or scatter and hoard seeds beneath deep litter layer or below the ground, which could lead to poor recruitment of the seedlings despite large seed production. Medium sized seeds are generally more susceptible to rodents because they are easier to find than very small seeds and are easier to manipulate than larger seeds. Small seeds are mainly eaten by ants. Seed predation is also affected by landuse. For example, less rodent abundance in pastures may result in less medium sized seed predation than in forests, whereas more abundance of ants results in higher predation of small seeds in pastures. Predation is normally higher near the tree crown and decreases with the distance from the tree crown (Vieira and Scariot, 2006).

Seed germination plays an important role in recruitment of a tree. The probability of germination increases if the seeds are transported far from the parent tree which helps in escaping the higher chances of predation near parent tree. Poor germination of seeds of tree species is inhibited during winter; when the forest floor is usually dry due to lack of soil moisture and low temperature. In a protected area where lopping is prohibited, the forest will have a good canopy cover (preventing light penetration) which might affect the germination of seeds. Thus light plays a very important role for proper germination of seeds. However shade tolerant species will still be able to grow in a high canopy cover area. Disturbance increases light availability and nutrient percolation through the soil and facilitates nutrient uptake. In general, seedlings always require partial canopy harvesting in order to increase insulation of the forest floor for survival, better germination and growth. Dispersed seeds generally show a period of rest termed 'dormancy' (Harper, 1977). Seeds of trees of mature phase in Wet Forests are generally not dormant (Tang and Tamari, 1973) while those of other species extend from two weeks to 3 years (Mensbrugi, 1966). Most species of Semi-evergreen Forests lack seed dormancy (Hoi, 1972). Dormancy is by far the chief factor determining the time of germination. Even in forests where there are two peak seasons of seed dispersal there is only one peak season for seed germination (Garwood, 1983 a), the peak being within, the first two months of the rainy (wet) season. In Tropical Seasonal Forests canopy species, lianas and the pioneer species show a unimodal pattern of germination. On the contrary in the case of understorey and shade tolerant species germination is throughout the rainy season, without a peak in any of the months (Garwood, 1983 b). Seedling emergence in canopy gaps peak 1-6 weeks prior to that in shaded understories (Garwood, 1983 a). The conditions for germination and establishment of mature phase tree species are very much specialized (Gomez-Pompa *et al.*, 1972). In the life history of a plant highest mortality rates operate between flowering and seedling establishment (Wyatt-Smith, 1963). Mortality due to vagaries in rain fall, intense drought, herbivore predation and self thinning are recorded (UNESCO, 1978 b).

Grazing by cattle in forest has been reported to cause tree damage through trampling and browsing and loss of species richness and diversity. In Switzerland, several cantons have enacted forest laws to discourage the practice of forest grazing (Mayer et al. 2006). But there are also studies been conducted which argue that forest grazing can be sustainable if grazing intensity is controlled (Krzic et al 2001; Pollock et al 2005; Mayer and Huovinen 2007). Forest grazing can enhance tree growth by reducing the biomass of grasses and sedges that otherwise outcompete tree seedlings (Belsky and Blumenthal 1997; Gratzer et al 1999; Darabant et al 2007). Grazing has also been reported to promote biodiversity (Mitchell and Kirby 1990; Mountford and Peterken 2003). In a study conducted in Bhutan (Bill Buffum, 2009) during a 5 year period concluded that the number of cattle grazing inside the forest significantly decreased and the number of naturally regenerated tree seedlings and saplings significantly increased. It also concluded that moderate intensities of forest grazing and timber harvesting can be combined in a forest without negative impacts on forest regeneration.

Fire is a major barrier to natural regeneration of native tree species because it decreases species diversity, favoring the few species that are able to survive it, and killing the rest. Fire hampers the seed germination. But there is another school of thought which proves that fire is the mechanism by which the forest id continually regenerated. Fires consume dead, decaying vegetation accumulating on the forest floor, thereby clearing the growth for new growth. Some species such as the jack pine, even rely on fire to spread their seeds. The jack pine produces "seratonous" (resin filled cones) that are very durable. The cones remain dormant until a fire occurs and melts the resin (Hall, 1999).

Forest regeneration occurs in gaps formed by selective logging. The size of the gap plays a significant role in species diversity, composition and regeneration of natural forests. Gaps in forest landscapes assume a wide range of sizes from the openings created by the deaths of single branches or trees (Spies and Franklin, 1989). The immediate and perhaps greatest effect of canopy opening is an increase in duration and intensity of direct sunlight to lower strata of the forest. Natural disturbance to forest canopies create broad varieties of opportunities for the growth of nearby plants and establishment of new ones, largely by increasing the amount of light penetrating in the forest interior (Lawton, 1990). Gap size ranges from the tiniest gaps formed by the natural death of trees in a natural forest to formation of large gaps created through intensive tree felling.

Repeated logging practice damages the structure and composition of natural forest, leading to declining of forest and agricultural field. Johnson and Cabarle, (1997) have estimated the damage to tropical forest as a result of logging, to the remaining trees, including species of economic importance, which ranged from 26 to 75 percent of its cover. It can remove progressively more of primary forest trees, reducing the number of seed sources for eventual succession and increase the damage to the forest floor and its populations of juvenile trees. Light demanding trees and herbs become increasingly prevalent. Ultimately a low, waste forest is formed, without internal structure and with only scattered big trees (Whitmore and Burnham, 1984). Another effect of logging on forest succession is from the differential removal of one species and the leaving of another, thus changing the composition of the forest.

Silvicultural practices in forests after logging and other disturbances is becoming an important method of regeneration for restoring degraded forest ecosystems. Controlled logging in a natural forest can be used as a silvicultural management system to maintain regeneration. Therefore human assistance is needed to recover forest structure, species composition, and species interaction on forests, which will in turn help in conserving the biodiversity of the area.

Keeping in View the Importance of Forest Management, Dynamics of Natural Regeneration After Exploitation has Received Particular Attention, Because:

1. Natural regeneration of forest in forest ecosystems is fundamental for evolution.

2. It is an important process to rely on for restoring forest over a large area after severe disturbances.

3. The study helps to understand and maintain a stable population structure of species in a community.

4. Research in this field contributes to planning, conservation and decision making in forest resource management programmes.

5. This study can be carried out to give emphasis to the tree species which are not regenerating in the area.

6. It is a valuable means to maintain the forests that are more efficient in providing necessary forest products and at the same time, in assuring multipurpose forest functions like recreation, site protection, water conservation, and preservation of habitat of endangered species of plants and wildlife.

7. It helps to understand as to what is the potential of the forest for the successive growth and to maintain the forest cover even for the future generation.

8. In addition such a study may help in prescribing appropriate silvicultural practices for regeneration of disturbed forest ecosystems.

9. Better performance of certain tree species in highly disturbed areas can be used as an indication to make a better choice for raising plantations on such sites.

10. The elaboration of sustainable management criteria relying on natural regeneration is considered as an important effort at the international level.

Conclusion

Assessment of regeneration in any forest is instrumental in determining the species structure. It offers scope for *in situ* conservation of ecologically and economically potential tree species for sustainable utilization and management.

References

Bagchi, R., T. Swinfield, R.E. Gallery, O.T. Lewis, S. Gripenberg, L. Narayan & R.P. Freckleton. 2010. Testing the Janzen-Connell mechanism: pathogens cause overcompensating density dependence in a tropical tree. *Ecol. Lett.* **13**: 1262-1269.

Baker, H. G., Bawa, K. S., Frankie, G. W. and Opler, P. A. 1983. Reproductive biology of plants in tropical forests. Pages **183- 215**. In: Golley, F. G. and Lieth, H. (eds.). Tropical rain forest ecosystems. Elsevier Scientific Publ.

Barik, S.K., Tripathi, R.S., Pandey, H.N., and Rao, P., 1996. Tree regeneration in a subtropical humid forest: effect of cultural disturbance on seed production , dispersal and germination. *Journal of Applied Ecology*. **33**: 1551-1560.

Belsky AJ, Blumenthal DM. 1997. Effects of livestock grazing on stand dynamics and soils in upland forests of the interior west. *Conservation Biology* 11:315–327.

Boerner, R. E., and J. A. Brinkman. 1996. Ten Years of Tree Seedling Establishment and Mortality in an Ohio Deciduous Forest Complex. Bulletin of the Torrey Botanical Club **123**:309-317.

Buffum B, Gratzer G, Tenzin Y. 2008. The sustainability of selection cutting in a late successional broadleaved community forest in Bhutan. *Forest Ecology and Management* **256**:2084–2091. http//dx.doi.org/10.1016/j.foreco.2008.07.031.

Chauhan, D.S., Dhanai, C.S., Singh, B., Chauhan, S., Todaria, N.P., Khalid, M.A., 2008. Regeneration and tree diversity in natural and planted forests in a Terai-Bhabhar forest in Katarniaghat wildlife sanctuary, India. *J. Tropical Ecology*. **49(1)**: 53-67.

Chesser, J.D. and Brewer, J.S., 2011. Factors influencing seedling recruitment in a critically endangered pitcher plant, *Sarracenia rubra* ssp. alabamensis. *Inter-research 2011, www.int-res.com.*

Clark, J.S. Beckage. B. Camill. P. Cle.ve.land,B . HilleRisLaibers, J. Lichter. J. McLachlan.

Connell, J.H. 1971. On the role of natural enemies in preventing competitive exclusion in some marine animals and rain forests trees, p. 298-312. In P.J. den Boer & G.R Gradwell (eds.). Dynamics of population. Center for Agricultural Publication and Documentation, Wageningen, The Netherlands.

Dansereau, P. and Lems, K. 1957. The grading of dispersal types in plant communities and their ecological significance. Contr. Inst. Bot.

Darabant A, Rai PB, Kenzin K, Roder W, Gratzer G. 2007. Cattle grazing facilitates tree regeneration in a conifer forest with palatable bamboo understory. *Forest Ecology and Management* 252:73–83.

Darwin, C. 1859: On the origin of species by means of natural selection. (Facsimile edition).

Dhaulkhandi, M., Dobhal, A., Bhatt, S., Kumar, M., 2008. Community structure and regeneration potential of natural forest site in Gangotri, India. *J. basic and applied sciences*. 4(1): 49-52.

Eilu, G and Obua, J., 2005. Tree condition and natural regeneration in disturbed sites of Bwindi Impenetrable Forest National Park, Southwestern Uganda. *J.Tropical Ecology.* 46(1): 99-111.

Gadgil, M. 197 1. Dispersal: population consequences and evolution. *Ecology* 52: 253-261.

Garwood, N. C. 1983 a. Seasonal rhythm of seed germination in a semideciduous tropical forest. Pages 173-185. In: Leigh, E. G. Jr., Rand, A. S ., and Windor, D. M. (eds.). The ecology of a tropical forest: seasonal rhythms and longterm changes. Smithsonian Inst., Washington.

Garwood, N. C. 1983 b. Seed germination in a seasonal tropical forest in Panama: a community study. *Ecol. Monogr.* 53: 159- 181.

Gebrehiwot, M., 2003. Assessment of natural regeneration diversity and distribution of forest tree species. Dissertation submitted to the forestry for sustainable development for the degree of master of science in geo-information science and earth science with specialization in forestry for sustainable development.

Gomez-Pompa, A., Vazquez-Yanes, C. and Guevara, S . 1972. The tropical rain forest: a non-renewable resource. *Science* 177: 762-765.

Good NF, and Good RE (1972). Populationdynamics of tree seedlings and saplings inmature Eastern hardwood forest. *Bull TorreyBot.* Club. 99

Gratzer G, Rai PB, Glatzel G. 1999. The influence of the bamboo Yushania microphylla on regeneration of Abies densa in central Bhutan. Canadian Journal of Forest Research—Revue Canadienne De Recherche Forestiere 29:1518– 1527.

Grime, J. P. 1979. Regenerative strategies. Pages 79-122. In: Grime, J. P. (ed.). Plant strategies and vegetation process. John Wiley and Sons, New York.

Grime, J. P. 1979. Regenerative strategies. Pages 79-122. In: Grime, J. P. (ed.). Plant strategies and vegetation process. John Wiley and Sons, New York.

Guzman, A. and Stevenson, P.R., 2011. A new hypothesis for the importance of seed dispersal time. *Int. J. Trop. Biol.* 59(4): 1795-1803.

Hammond, D.S. & V.K. Brown. 1998. Disturbance, phenology and life-history characteristics: factors influencing distance/density- dependent attack on tropical seeds and seedlings, p. 51-78. In D.M. Newbery, H.H. Prins & N. Brown (eds.). Dynamics of tropical communities. Blackwell Science, London, England.

Harper, J. L. 1977. Population biology of plants. Academic Press, London.

Hoi, L. V. 1972. Forest calendar and forestry seed saigon. Inst. Rech. For. 1972: 1-100.

Holl, K. D., and M. E. Lulow. 1997. Effects of species, habitat, and distance from edge on post-dispersal seed predation in a tropical rainforest. *Biotropica* 29:459–468.

Hooper, E. R., Legendre, P., Condit, R., 2004. Factors affecting community composition of forest regeneration in deforested, abandoned land in panama. *J. Ecology.* 85(12): 3313-3326.

Howe, H. F. and Smallwood, J. 1982. Ecology of seed dispersal. *Ann. Rev. Ecol. Syst.* 13: 201-228.

Howe, H.F. & Smallwood, J. (1982). Ecology of seed dispersal. Annual Review of Ecology and Systematics, 13, 201–228.

Hyatt, L.A., M.S. Rosenberg, T.G. Howard, G. Bole, W. Fang, J. Anastasia, K. Brown, R. Grella, K. Hinman, J.P. Kurdziel & J. Gurevitch. 2003. The distance dependence prediction of the Janzen-Connell hypothesis: a meta-analysis. Oikos 103: 590-602.

J. Mohan, J. & Wyckoff, P. 1999 Interpreting rwruitment limitation in forests. *American Journal of Botany* 86(1):1-16

Janzen, D.H. (1971). Seed predation by animals. Annual Review of Ecology and Systematics, 2, 465–492.

Janzen, D.H. 1970. Herbivores and the number of tree species in tropical forests. *Am. Nat.* 104: 501-528.

Karlsson, M., natural regeneration of broadleaved tree species in southern Sweden – effects of silvicultural treatments and seed dispersal from surrounding stands. Doctoral thesis Swedish University of Agricultural Sciences Alnarp 2001.

Khan, M.L., Bhuyan, P., Tripathi, S., 2005. Effects of forest disturbance on fruit set, seed dispersal and predation of Rudraksh (Elaeocarpus ganitrus Roxb.) in northeast India. *J. Current Science.* 88(1): 133 – 142.

Krebs, C. J. 1972. Ecology: the experimental analysis of distribution and abundance. Harper and Row, New York.

Krzic M, Broersma K, Newman RF, Ballard TM, Bomke AA. 2001. Soil quality of harvested and grazed forest cutblocks in southern British Columbia. Journal of *Soil and Water Conservation* 56:192–197.

LAWTON, R. O. 1990. Canopy gaps and light penetration into a wind-exposed tropical lower montane forest. *Can. J. For.* 20: 659–667.

Liang, S. Y., and S. W. Seagle. 2002. Browsing and microhabitat effects on riparian forest woody seedling demography. *Ecology.* **83**:212-227.

Matthesius, A., H. Chapman & D. Kelly. 2010. Testing for Janzen-Connell effects in a West African montane forest. *Biotropica* 43: 77-83.

Mayer AC, Huovinen C. 2007. Silvopastoralism in the Alps: Native plant species selection under different grazing pressure. *Ecological Engineering* 29: 372–381.

Mayer AC, Sto¨ckli V, Konold W, Kreuzer M. 2006. Influence of cattle stocking rate on browsing of Norway spruce in subalpine wood pastures. *Agroforestry Systems* 66:143–149.

Mensbruge, C. de la. 1966. La germination et les plantules des essences arborees de la foret dense humide de la cote d' ivoire Nogent- Sur-Marne. Centre Technique Forestier Tropical (CTFT) Publ. no. 389.

Mitchell FJG, Kirby KJ. 1990. The impact of large herbivores on the conservation of semi-natural woods in the British uplands. *Forestry* 63:333– 353.

Models to predict patterns of tree seedling dispersion. *Ecology* 75(6): 1794- 1806.

Mountford EP, Peterken GF. 2003. Long-term change and implications for the management of wood-pastures: Experience over 40 years from Denny Wood, New Forest. *Forestry* 76:19–43.

Nair, P. N. 1961. Regeneration of moist tropical forests with special reference to Kerala. *DAIFC Dissertation, Indian For. Coll., Dehra Dun.*

Omeja, P., Obua, J., Cunningham, A.B., 2004. Regeneration, density and size class distribution of tree species used for drum making in central Uganda. *African journal of ecology.* 42: 129 – 136.

Osunkoya, O. O. 1994. Postdispersal survivorship of North Queensland rainforest seeds and fruits: effects of forest, habitat and species. *Australian Journal of Ecology* 19:52–64.

Pandey, S.K. and Shukla, R.P. 2001. Regeneration strategy and plant diversity status in degraded sal forests. *J. Current Science.* 88(1): 95-102.

Peterman, J.S., A.J.S. Fergus, L.A. Turnbull & B. Schmid. 2008. Janzen-Connell effects are widespread and strong enough to maintain diversity in grasslands. *Ecology* 89: 2399-2406.

Pijl, van der, 1969. Principles of dispersal in higher plants. Springer- Verlag, Berlin.

Pokhriyal, P., Uniyal, P., Chauhan, D.S., Todaria, N.P. 2010. Regeneration of tree species in forest of Phakot and Pathri Rao watersheds in Garhwal Himalaya. *J. Current Science.* 98(2): 171-175.

Pollock ML, Milner JM, Waterhouse A, Holland JP, Legg CJ. 2005. Impacts of livestock in regenerating upland birch woodlands in Scotland. *Biological Conservation* 123(4):443–452.

Reddy, S and Ugle, P., 2008. Survival threat to the flora of Mudumalai wildlife sanctuary, india: an assessment based on regeneration status. *J. Nature and Science.* 6(4): 42-54.

Ribbens, E., Silander. J.A. cPr Pacala, S.W. 1994: Seedling recruitment. in forests: Calibiating.

Ridley, H. N. 1930. The dispersal of plants throughout the world. L. Reeve, London.

Rocky, J. and Mligo, C., 2011. Regeneration and size class distribution of indigenous woody species in exotic plantation in Pugu Forest Reserve, Tanzania. *International Journal of Biodiversity Conservation.* 4(1): 1-14.

Sahoo, U.K. and Lalfakawma., 2010. Population dynamics of *Schima wallichii* in an undisturbed vs disturbed tropical forest stand of north-east india. *International Journal of Ecology and Environmental Sciences.* 36(2-3): 157-165.

Shelton, M.G. and Cain, M.D. 2001. Dispersal and viability of seeds from cones in tops of harvested loblolly pines. *Canadian Journal of Forest Research* 31: 357-362.

Synnott, T. J. 1973. Seed problems. In: IUFRO Int. Symp. on Seed Processing, Rergen.

Tang, H. T. and Tamari, C. 1973. Seed description and storage tests of some Dipterocarps. Malay. For. 36: 38-53.

Unesco, 1975. Regional meeting on integrated ecological research and training needs in tropical deciduous and semideciduous forest ecosystems of South Asia. MAB Report ser. no. 35. Hamburg-Rein.

Unesco, 1978 b. Chapter 8. The natural forest: plant biology, regeneration and tree growth. In: Tropical forest ecosystems: 180-215. UNESCO-UNEP, France.

Vieira, D.L.M., Scariot, A., 2006. Principles of natural regeneration of tropical dry forests for restoration. *J. restoration ecology.* 14(1): 11-20.

Viera, D. and Scariot, A. 2006. Principles of natural regeneration of tropical dry forests for restoration. MARCH 2006 *Restoration Ecology.* Vol. 14, No. 1, pp. 11–20.

Ward, J.S.,T.E. Worthley, P.J. Smallidge, and K.P. Bennett. 2006. Northeast forest regeneration handbook: A guide for forest owners, harvesting practitioners, and public officials. USDA Forest Service, Newton Square, PA. Available online at: http://www.na.fs.fed.us/stewardship/pubs/forest_regn_hndbk06.pdf

Willson, M.F. 1942: The ecology olseed dispersal. In :Seeds -*The Eculogy of Regeneration.*

Wright, S.J. 2002. Plant diversity in tropical forests: a review of mechanisms of species coexistence. *Oecologia.* 130: 1-14.

Wyatt-Smith, J. 1963. Manual of Malayan silviculture for inland forests. Malay. For. Rec. no. 23.

Chapter 6

Consolidated Bioprocessing Technology for Efficient Production of Ethanol from Different Carbohydrate Sources

M.P. Raghavendra and A.G. Devi Prasad

P G Department of Microbiology, Maharani's Science College for Women, JLB Road, Mysore – 570 005, Karnataka,
DOS in Environmental Science, University of Mysore, Manasagangotri, Mysore-06, Karnataka

ABSTRACT

Consolidated bioprocessing (CBP) technology is considered as an efficient method for conversion of biomass into bioethanol from different carbohydrate sources including waste. Ethanol produced via microbial enzymes is considered as renewable, sustainable, efficient, cost effective and safe alternative fuel against petrochemicals. Microorganisms are considered to be efficient in degradation of many carbohydrates. Since biodegradation involves conversion of complex polysaccharides into simpler sugars which are later converted to ethanol, the combination of physicochemical reactions and enzyme degradation is making the bioethanol production commercially not viable. In this connection development of whole –cell biocatalyst, enzyme engineering, engineering of efficient cellulose degrading bacteria such as Caldicellulosiruptor bescii, development of recombinant strain of S. cerevisiae and screening of cellulolytic thermophilic microorganisms and its engineering are the promising field to achieve high titre of ethanol from different carbohydrates including lignocelluloses. CBP is considered to be cheaper because enzymes production, saccharification and fermentation required for ethanol production are carried out as a single process by engineered microorganisms. Several fungi such as Aspergillus sp., Rhizopus sp., *Monilia* sp., *Neurospora* sp., *Fusarium* sp., *Trichoderma* sp., *Rhizopus oryzae, Fusarium oxysporum, Clostridium thermocellum and Mucor, bacteria such as Clostridium cellulolyticum, Thermo-anaerobacterium saccharalyticum, Trametes hirsute and even thermophilic yeast are considered to be potential candidates of CBP technology. The present chapter deals with the several techniques and applicability of CBP to enhance ethanol production from biomass.*

Key words: *CBP; bioethanol; microorganism; biomass conversion*

Introduction

Increased use of biofuels would contribute to sustainable development by reducing greenhouse-gas emissions and the use of nonrenewable resources. Instead of traditional feed stocks (starch crops) lignocellulosic biomass, including agricultural and forestry residues, could prove to be an ideally inexpensive and abundantly available source of sugar for fermentation into transportation fuels. Lignocellulosics are composed of heterogeneous complex of carbohydrate polymers (cellulose, hemicelluloses and lignin). Even though it is available plenty in nature, cellulose crystallinity, accessible surface area, protection by lignin, and sheathing by hemicellulose all contribute to the resistance of cellulose in biomass to hydrolysis.

The biomass pretreatment and the intrinsic structure of the biomass itself are primarily responsible for its subsequent hydrolysis. The conditions employed in the chosen pretreatment method will affect various substrate characteristics, which, in turn, govern the susceptibility of the substrate to hydrolysis and the subsequent fermentation of the released sugars. Therefore, pretreatment of biomass is an extremely important step in the synthesis of biofuels from lignocellulosic biomasses.

Biological pretreatment, such as fungal, is milder in its operational conditions than physical or chemical pretreatment. The oxidative biodegradation of lignin by white-rot fungi has been widely studied in the past (Taniguchi *et al.*, 2005). The main advantages of biological pretreatment are low energy input, no chemical requirement, mild environmental conditions and environmentally friendly working manner (Lin *et al.*, 2010). However, the biological pretreatment processes usually need a long retention time and have a low yield. They are also sensitive to the process conditions such as temperature and pH. Enzymatic pretreatment of biomass has been comprehensively studied in the past and will be discussed in a separate section. In this process, enzyme is used for decomposing polysaccharides into monosaccharides.

By the use of enzymatic hydrolysis, pure cellulose can be degraded to soluble sugars which can be fermented to form ethanol, butanol, acetone, single cell protein, methane and many other products. Enzymatic saccharification remains one of the most costly steps in conversion of cellulosic biomass to ethanol and cellulase preparations dedicated for bioethanol industry are hardly available. It has been estimated that the greatest returns in cost savings will be realized by improving conversions of biomass to sugars, increasing hydrolysis yields, reducing enzyme loadings, eliminating or reducing pretreatment (Lynd *et al.*, 2008). A broader suite of enzymes is required for hydrolysis of cellulose and hemicelluloses to fermentable sugars (McMillan *et al.*, 2011). Thus enzymatic hydrolysis can be effectively carried out if a mixture of different cellulolytic and accessory enzymes is used.

Different strategies are also available for enzymatic hydrolysis and fermentation of lignocelluloses such as separate enzymatic hydrolysis and fermentation, simultaneous saccharification and fermentation, non-isothermal saccharification and fermentation, simultaneous saccarification and co-fermentation and Consolidated Bioprocessing (CBP) (Taherzadeh and Karimi, 2007). Among different strategies CBP is considered to posses outstanding potential, ethanol together with all of the required enzymes is produced by single microbial communities is also referred to as direct microbial conversion (DMC). It may involve single or multiple cultures of microorganisms to directly convert cellulose to ethanol in a single bioreactor (Wyman, 2007).

Enzymatic hydrolysis of biomass with cellulases after pretreatment has received much attention in academia and industry. Cellulases are a mixture of endoglucanases, exoglucanases and cellobiohydrolases, and catalyze the degradation of cellulose to oligomers according to the following proposed mechanism: (1) endoglucanase cleaves β-1,4 glycosidic bonds in the long-chain polymer, and free ends are created; (2) exoglucanases, such as β-glucosidases, act on the reducing and non-reducing ends to liberate oligosaccharides; and (3) cellobiohydrolases cleave the polymer from the reducing ends to liberate cellobiose. Catalyzed by the above three cellulases, cellulose hydrolyzes to glucose (Eriksson *et al.*, 2002; Hosseini and Shah 2011a; Hosseini and Shah, 2011b).

Consolidated Bioprocessing Technology (CBP)

Industrial conversion of plant biomass to fuels currently relies on thermal and chemical treatment of biomass to remove hemicellulose and lignin, followed by enzymatic hydrolysis to solubilize the plant cell walls to generate a fermentable substrate for fuel-producing organisms as explained earlier (Zhang *et al.*, 2007; Negro *et al.*, 2003; Wyman 2007). However, these methods add cost, produce hydrolysates that are toxic to microorganisms (Pauly and Scheller 2000; Klinke *et al.*, 2004; Kim and Dale 2004; Zhang *et al.*, 2010) and are destructive to the sugars in the biomass (Barakat *et al.*, 2012). An alternative approach is to use CBP, in which the fermentative organism is also responsible for production of the biomass-solubilizing enzymes (Lynd *et al.*, 2002; Hasunuma and Kondo, 2012).

Among the different wild type cellulolytic microorganisms available, high temperature CBP is highly preferred, because cellulolytic optimum enzyme performances occur at elevated temperature of around 50C. While on the other hand, optimum microbial fermentation occur best at temperature between 28°C - 37°C. This necessitates the search for more thermo tolerant micro-organisms as suitable candidates for CBP. Hence the best candidate for CBP is the members of the genus *Caldicellulosiruptor* sp., which are able to ferment all primary C5 and C6 sugars from plant biomass and are the most thermophilic cellulolytic bacteria known, with growth temperature optima between 78°C ~ 80°C (Hamilton-Brehm *et al.,* 2010). They can also grow on and degrade biomass containing high lignin content as well as highly crystalline cellulose without conventional pretreatment (Blumer-Schuette *et al.,* 2008; Yang *et al.,* 2009), raising the possibility of further economic improvement of biofuel production from plant biomass by reducing or eliminating the pretreatment step.

One of the peculiar features of CBP is that it requires highly engineered microbial strains that would be compatible with process parameters such as high temperature and simultaneously hydrolyzing biomass with enzymes on its own with high ethanol titre. Recent studies have shown that several organisms have been tested in this regard. Thermophilic yeast such as *Kluveromyces marxianus* has been engineered to explore its ability to perform CBP at elevated temperature for bioethanol production (Limayem and Ricke 2012) and extreme anaerobic thermophilic bacteria such as *Thermoanaerobacter saccharalyticum, Thermo-anaerobacter ethanolicus* and *Clostridium thermocellum* have been genetically modified to suit perform for CBP (Kumar *et al.,* 2009; Limayem and Ricke, 2012).

Fungi which are known to degrade several biomolecules due to its heterotrophic nature are also considered to be the best candidates for CBP. The cellulase system in fungi is considered to comprise three hydrolytic enzymes: (i) the endo-(1,4)-β-D-glucanase (synonyms: endoglucanase, endocellulase, carboxymethyl cellulose [EC 3.2.1.4]), which cleaves b-linkages at random, commonly in the amorphous parts of cellulose; (ii) the exo-(1,4)-β-Dglucanase (synonyms: cellobiohydrolase, exocellulase, microcrystalline cellulase, Avicelase [EC3.2.1.91]), which releases cellobiose from either the non reducing or the reducing end, generally from the crystalline parts of cellulose; and (iii) the β-glucosidase (synonym: cellobiase [EC 3.2.1.21]), which releases glucose from cellobiose and short-chain cellooligosaccharides. Although β-glucosidase has no direct action on cellulose, it is regarded as a component of cellulase system because it stimulates cellulose hydrolysis (Maheshwari *et al.,* 2000). Even *Fusarium oxysporum* (Xiros *et al.,* 2009), white rot fungi *Trametes hirsute* (Okamoto *et al.,* 2011), *Rhizopus oryzae* (Zhang and Yang 2012), *Aspergillus, Rhizopus, Monilia, Neurospora, Fusarium, Trichoderma,* and *Mucor* (Hasunuma *et al.,* 2013) are few fungi which have been explored for the improved production of ethanol from biomass.

Screening of Cellulolytic Microorganisms

In view of the urgent need to covert lignocellulosic biomass to ethanol, several microorganisms were screened for their efficiency to produce enzymes which can breakdown different carbohydrates including lignocelulose. The most potent thermophilic cellulolytic isolates of *Bacillus subtilis* was identified by Acharya *et al.,* (2012). In an effort isolate efficient microbial strain even faecal matter of herbivores was screened. Salunke (2012) successfully isolated cellulose-degrading *Kleibsiella* sp., *Enterobacter* sp., *Bacillus* sp. from herbivores excreata.

Thermoactinomycetes sp., *Pseudomonas* spp. and *Penicillium* sp. were isolated from soil samples (Gomashe *et al.,* 2013). 111 thermophilic microorganisms (91 bacteria and 20 yeasts) were isolated from 10 western Algerian sources (thermal and non-thermal) and tested for the production of cellulase. The results revealed the presence of 19 thermophilic cellulolytic isolates out of which 16 are found to be thermophilic bacteria and 3 are thermophilic yeasts. These isolates were tested for the degradation of cellulosic biomass (printable paper, filter paper and cotton) for 14 days of incubation at 60°C. The obtained results showed a great potential of these thermophilic cellulolytic microorganisms to produce thermostable cellulolytic enzymes, and was proved effective in recycling of cellulosic biomass for bioenergy production after optimization studies in the future (Khelil and Cheba, 2014).

Klebsiella sp. PRW-1 isolated from the soil sample was found to utilize pure cellulosic substrates (carboxymethylcellulose (CMC) and avicel) and different agricultural wastes like sugarcane bagasse, sugarcane barbojo, sorghum husks, grass powder, corn straw and paddy straw by producing a large amount of endoglucanase, exoglucanase, β-glucosidase, filter paperase (FPU), xylanase and glucoamylase. The reducing sugar production was found higher in the presence of grass powder and sugarcane barbojo (Waghmare *et al.*, 2014).

Four different substrates like *Acacia arabica* pod, *Bauhinia forficata* pod, *Cassia surattensis* pod and *Peltophorum pterocarpum* pods (as cellulose substrate) were used in the submerged production medium to screen cellulolytic bacteria from soil. Maximum enzyme activity were observed in *Bacillus cereus* (0.440 IU/ml/min and 0.410 IU/ml/min), followed by *Bacillus subtilius* (0.357 IU/ml/min) and *Bacillus thuringiensis* (0.334 IU/ml/min) (Patagundi *et al.*, 2014).

Crude cellulolytic enzymatic complex of *Aspergillus niger* and *Mucor recemosus* isolated from Brazilian Cerrado (Savanna) was found to produce ethanol from sugarcane bagasse. It was found that after 48h of fermentation using crude enzymatic extract produced by *Aspergillus niger* produced 11.5 g/L of ethanol and *M. racemosus* 7.2 g/L of ethanol (Fischer *et al.*, 2014).

The majority of cellulolytic bacteria in Chinese white pine beetle *Dendroctonus armandi* larva gut were identified as *Serratia* and accounted for 49.5%, followed by *Pseudomonas*, which accounted for 22%. In addition, members of *Bacillus*, *Brevundimonas*, *Paenibacillus*, *Pseudoxanthomonas*, *Methylobacterium* and *Sphingomonas* were found in the *D. armandi* larva gut. *Brevundimonas kwangchunensis*, *Brevundimonas vesicularis*, *Methylobacterium populi* and *Pseudoxanthomonas mexicana* were reported to be cellulolytic for the first time by the study conducted by (Hu *et al.*, 2014).

Even the screening of metagenomic clone libraries from diverse environmental sources has previously yielded numerous biomolecules including novel proteases, amylases, cellulases, and antibiotics (Brady and Clardy, 2000; Rondon *et al.*, 2000; Daniel, 2005; Warnecke *et al.*, 2007). Sommer *et al.*, (2010) have shown that metagenomic functional selections can successfully discover functional genetic elements encoding chemical tolerance relevant to biomass conversion. The same platform can be applied to select for microbial usage and production of specific biomass chemicals. The repertoire of biomass substrates that can be used by a microbial biocatalyst has been expanded by transfer of specific genetic machinery for substrate metabolism from other microbes with these properties (Jin *et al.*, 2005).

The search for wild type strains from different sources with more efficiency to break down cellulose will help in identifying a novel strain or candidate for CBP. Enzymes of these strains can be effectively engineered to increase its efficiency to convert biomass to ethanol.

Enzyme engineering

Enzyme description

Cooperative action of several enzymes such as endocellulases (EC 3.2.1.4), exocellulases (cellobiohydrolases, CBH, EC 3.2.1.91; glucanohydrolases, EC 3.2.1.74), and betaglucosidases (EC 3.2.1.21) are involved in degradation of cellulose to glucose. Endocellulases hydrolyze internal glycosidic linkages in a random fashion, which results in a rapid decrease in polymer length and a gradual increase in the reducing sugar concentration. Exocellulases hydrolyze cellulose chains by removing mainly cellobiose either from the reducing or the nonreducing ends, which lead to a rapid release of reducing sugars but little change in polymer length. Endocellulases and exocellulases act synergistically on cellulose to produce cello oligosaccharides and cellobiose, which are then cleaved by β-glucosidase to glucose (Kumar *et al.*, 2008).

In particular the overall structure of the glycoside hydrolase family GH-12 protein: Cel12 is available. It is a β-jellyroll fold with a six stranded antiparallel β sheet packing on the outside of a nine-stranded mostly antiparallel β sheet curved around an active site cleft The cel12 family hydrolyzes the β -1,4-glycosidic bond in cellulose in a catalytic mechanism involves two glutamic acids, one serving as a nucleophile and the other as a proton donor, in a double displacement reaction with a glycosyl enzyme intermediate that results in retention of configuration in the product(Vasella *et al.*, 2002; McCarter *et al.*, 1994). A number of crystal structures have been solved in this family, and their protein data bank (Berman *et al.*, 2000) entries. The key features in these enzymes, in addition to the catalytic glutamates, are a series of conserved aromatic residues interacting with the sugar rings on the cellulose substrate, along with specific hydrogen-bonding interactions from the backbone and polar side chains. The standard notation for these enzymes lists the interactions with each sugar as a sub site, numbered sequential from +3 to -3 from reducing end to non-reducing end, with cleavage occurring between +1,-1. At the reducing end of the active site there is a conserved Pro-X-Gly motif, termed the "cord", which causes the substrate to distort to aid in hydrolysis.

Understanding the catalytic site and its activity can further help in generating reasonable three dimensional models of the enzymes for specific cellulose degradation by homology modeling. It is also called comparative modeling where in three dimensional models of a protein structure from its amino acid sequence can be predicted. Because tertiary structures of proteins are far more conserved than their primary amino-acid sequences, many sequences will share the same overall fold, even with less than 20% identity in amino acid sequence. Thus it is possible to generate a reasonable three-dimensional model of protein from a crystal structure of a protein homologous to the query structure. The structure used is referred to as the 'template' for building the model. As a rule of thumb, usually if the template and query have > 30% identity a good quality model can be produced. With a sequence identity in the 20-30% range is considered a region where it is possible to generate a model, but may be less accurate, and below 20% is usually not considered sufficient for generating a model.

Directed Evolution

It is one of the major technologies that enzyme engineers use today to advance biofuel production. Natural evolution occurs when mutations are created during DNA replication and the sequence of genetic codons is altered. The new genes lead to the production of novel proteins, which can be advantageous or harmful to the new organism. Natural selection will then favor useful mutations while the deleterious ones are removed. The directed evolution technique mimics this process in the laboratory in which firstly mutations must be induced in the gene that codes for the enzyme of interest and library of these gene mutants can be created. These genes are then cloned in to an expression vector allowing its expression in bacterial cells that produce the desired enzyme. Those mutants which record significant increase in enzyme production can be selected. Its efficiency in different environmental parameters can also be tested to suit fermentation conditions for industrial applications.

Another method, known as site-directed saturation-mutagenesis, uses structural information to selectively mutate at a specific location in the enzyme Mutations are integrated within the primers that anneal the polymerases to the gene of interest for PCR. Using this method, researchers are able to concentrate mutations at the active site of the enzyme where they are likely to be the most useful.

Significant advances in enzyme functionality can also be achieved using another mutagenesis process called DNA shuffling. In this method *in vitro* DNA recombination is used to exchange the functional domains of two homologous endoglucanase enzymes. In this process genes for endoglucanases are fragmented at several random locations. These fragments then recombined to create new mutant chimeras that contain large stretches of genetic material from both enzymes. The mutant genes are then cloned into bacteria and advantageous properties are selected for in a high-throughput screening process. The major benefit of using this method is that enzyme activity can be substantially improved. By recombining large portions of multiple enzymes, mutant chimeras can potentially contain more than one active site, which can greatly enhance the activity of the enzyme.

Along with the different method explained above, the application of synthetic biology (Baker *et al.*, 2006; Ro *et al.*, 2006) to engineer biocatalysts that produce biofuels from diverse lignocellulosic materials including waste and low agricultural intensity biomass holds promise to deliver one such sustainable alternative (Farrell *et al.*, 2006; Tilman *et al.*, 2006; Fargione *et al.*, 2008; Searchinger *et al.*, 2008).

Metabolic Engineering for Utilization of Xylose for Ethanol Production

Genetic engineering can also be used to improve the fermentation step in biofuel production. Synthetic biology involves the design, synthesis and introduction of new genetic programming to organisms for new biological functions. This strategy is used to reengineer the metabolic pathways of the fermentation microbes that convert biomass sugar into products. Codexis has used their genetic engineering technologies to introduce a non-native pathway for the fermentation xylose in yeast. The primary sugar resulting from the deconstruction of cellulose and hemicellulose is glucose, but xylose is also produced in significant amounts. Currently, xylose is not converted to ethanol by the yeast used in today's first generation ethanol production. Thus, the conversion efficiency of a biochemical process using cellulose and hemicellulose feedstock can be substantially increased by also utilizing the xylose sugars. Metabolic engineering can also be used to design organisms that will secrete novel chemicals of interest. This marked the first time these chemicals had been produced directly from glucose in *E. coli*. Using genetic engineering to design microbes that produce useful secretions has the potential to sustainably produce chemical feedstock, not only for energy but also other chemical industries.

It is now obvious that the production of bioethanol from lignocellulose hydrolysates requires a robust, D-xylose-fermenting and inhibitor-tolerant microorganism as catalyst. To develop a efficient strain having both the characteristics Demeke *et al.*, (2013) used an expression cassette containing 13 genes including *Clostridium phytofermentans* XylA, encoding D-xylose isomerase (XI), and enzymes of the pentose phosphate pathway was inserted in two copies in the genome of Ethanol Red. Subsequent EMS mutagenesis, genome shuffling and selection in D-xylose-enriched lignocelluloses hydrolysate, followed by multiple rounds of evolutionary engineering in complex medium with D-xylose, gradually established efficient D-xylose fermentation.

The best-performing strain, GS1.11-26, was proved to possess a maximum specific D-xylose consumption rate of 1.1 g/g DW/h in synthetic medium, with complete attenuation of 35 g/L D-xylose in about 17 h. In separate hydrolysis and fermentation of lignocellulose hydrolysates of *Arundo donax* (giant reed), spruce and a wheat straw/hay mixture, the maximum specific D-xylose consumption rate observed by them was 0.36, 0.23 and 1.1 g/g DW inoculum/h, and the final ethanol titer was 4.2, 3.9 and 5.8% (v/v), respectively. In simultaneous saccharification and fermentation of Arundo hydrolysate, GS1.11-26 produced 32% more ethanol than the parent strain Ethanol Red, due to efficient D-xylose utilization. The high D-xylose fermentation capacity was stable after extended growth in glucose. Cell extracts of strain GS1.11-26 displayed 17-fold higher XI activity compared to the parent strain, but overexpression of XI alone was not enough to establish D-xylose fermentation. The high D-xylose consumption rate was due to synergistic interaction between the high XI activity and one or more mutations in the genome. The GS1.11-26 had a partial respiratory defect causing a reduced aerobic growth rate.

Development of Recombinant Strain of *S. Cerevisiae*

Saccharomyces cerevisiae which is involved in converstion of hydrolysed sugars to ethanol is another target to achieve more production of ethanol. Wen *et al.,* (2010) engineered a recombinant *Saccharomyces cerevisiae* strain displaying a trifunctional minicellulosome that could directly ferment phosphoric acid-swollen cellulose to ethanol with a titer of 1.8 g/liter. Later efforts were made to develop *Saccharomyces cerevisiae* strains as whole-cell biocatalysts capable of combining hemicellulase production, xylan hydrolysis, and hydrolysate fermentation into a single step by Sun *et al.,* (2012). These strains were made to display a series of uni-, bi-, and trifunctional minihemicellulosomes that consisted of a miniscaffoldin (CipA3/ CipA1) and up to three chimeric enzymes. The miniscaffoldin derived from *Clostridium thermocellum* contained one or three cohesin modules and was tethered to the cell surface through the *S. cerevisiae* a-agglutinin adhesion receptor. Up to three types of hemicellulases, an endoxylanase (XynII), an arabinofuranosidase (AbfB), and a β-xylosidase (XlnD), each bearing a C-terminal dockerin, were assembled onto the miniscaffoldin by high-affinity cohesin-dockerin interactions. Compared to uni- and bifunctional minihemicellulosomes, the resulting quaternary trifunctional complexes recorded an enhanced rate of hydrolysis of arabinoxylan. Furthermore, with an integrated d-xylose-utilizing pathway, the recombinant yeast displaying the bifunctional minihemicellulosome CipA3-XynII-XlnD was also found to hydrolyze and ferment birch wood xylan to ethanol with a yield of 0.31 g per g of sugar consumed.

Hemicellulose, including xylan, is also a major component of cellulosic biomass, Furthermore, ethanol production from hemicellulose is important for full utilization of cellulosic biomass. Xylan is hydrolyzed to xylo-oligosaccharides by endo-β-xylanase, and the produced xylo-oligosaccharides are hydrolyzed to D-xylose by β-xylosidase. *Saccharomyces cerevisiae* co-displaying xylanase II (XYN II) from *T. reesei* and β-xylosidase (XylA) from *A. oryzae* was constructed using the β-agglutinin-based display system, and shown to hydrolyze xylan to xylose (Katahira *et al.,* 2004). In a further development for ethanol production from xylan, xylose reductase (XR) and xylitol dehydrogenase (XDH) from *Pichia stipitis* and xylulokinase (XK) from *S. cerevisiae* were produced in the XYN II- and XylA-co-displaying yeast.

Whole Cell Biocatalysis

Industrial synthetic chemistry mainly depends on biocatalysis which means use of different enzymes or whole cells as biocatalysts. Biocatalysis is helping in industrial production of alcohol, cheese and many other industrial products from several hundred years. Microorganisms in general are gifted with array of operons which regulate and produce several enzymes capable of degrading different substrates. But these naturally occurring enzymes sometimes are not optimal for specific industrial applications due to its less stability and enzyme activity. These enzymes are known to be involved in catalyzing naturally occurring substrates than non natural components to produce desired end products. The exhaustive growth in molecular biology and understanding gene expression patterns, it is now possible to develop novel biocatalysis with specific industrial operating parameters through enzyme engineering.

Development of whole-cell biocatalysts to produce biofuel from biomass such as grain or cellulosic biomass has attracted attention as a means of creating a sustainable society based on biomass resources. Especially, CBP of lignocellulose to ethanol is an ideal system combining all processes such as enzyme production, hydrolytic degradation, and fermentation of sugar (Lynd *et al.*, 2005). The arming technology has been applied to yeasts for the construction of whole-cell biocatalysts that can perform saccharification and fermentation. Among different carbohydrate sources, starch and cellulose are major components in grain and cellulosic biomass, respectively. Therefore, the cell surface display of enzymes for hydrolytic degradation of these components is considered to be highly effective in order to achieve CBP (Kuroda and Ueda, 2013).

Several reports are available in this regard on different carbohydrate substrates. Murai *et al.*, (1997) displayed starch-degrading enzyme, an exotype glucoamylase from *Rhizopus oryzae* on *S. cerevisiae* using the β-agglutinin-based display system, which allowed the direct production of ethanol from starch through the saccharification of starch on the cell surface and the subsequent fermentation of released glucose. Later in addition to glucoamylase, an endotype β-amylase from *Bacillus stearothermophilus* was co-displayed with glucoamylase, leading to an improved production efficiency of ethanol from starch (Murai *et al.*, 1999). In the same way, β-amylase from *Streptococcus bovis* was also displayed using the Flo1p-based display system on glucoamylase-displaying yeast constructed by the β-agglutinin-based display system (Shigechi *et al.*, 2004). The arming yeasts described above were shown improved ethanol production and faster growth in the medium including starch as the sole carbon source.

As explained earlier degradation of cellulose requires various enzymes. Therefore, the potential of co-display of these enzymes on the yeast cell surface is an important feature, which is suitable for complete conversion of the cellulose to ethanol through multiple reactions carried out by multiple enzymes displayed on the cell surface. Actually, the co-display of carboxymethylcellulase (CMCase) and β-glucosidase (BGL1) from *Aspergillus aculeatus* on *S. cerevisiae* enabled assimilation of cellobiose or oligosaccharide and growth in medium containing these materials as the sole carbon source (Murai *et al.*, 1998). Endoglucanases (EG) acts randomly against the amorphous region of the cellulose chain to produce reducing and nonreducing ends, and cellobiohydrolase (CBH) releases cellobiose from both ends. Therefore, efficient degradation of cellulose to cellobiose and cellooligosaccharides is achieved by the endo-exo synergism of EG and CBH. Finally, β-glucosidase hydrolyzes the generated cello-oligosaccharides to glucose. The arming yeast constructed by co-display of EG II and CBH II from *Trichoderma reesei* and BGL1 from *A. aculeatus* can produce ethanol directly from phosphoric-acid-swollen cellulose due to the combined activities of these three cell surface enzymes (Fujita *et al.*, 2004). In sake yeast, ethanol production from β-glucan can be achieved by cell surface display of EG and BGL from *A. oryzae* (Kotaka *et al.*, 2008). Furthermore, in arming yeast co-displaying EG II, CBH II, and BGL1, four non-conserved amino acids in the Carbohydrate-Binding Module (CBM) of EG II were comprehensively mutated. In the case of co-displaying multiple kinds of enzymes on the same cell surface, the control or design of the display ratio of enzymes is a next challenge in arming technology. Recently, some approaches to this challenge were performed for optimized synergistic effects of displayed enzymes (Ito *et al.*, 2009; Yamada *et al.*, 2010). The optimal combination of yeasts displaying EG II with mutated CBMs was determined by inoculating a yeast library into selection liquid medium that included newspaper as the sole carbon source. The selected yeast mixture showed the improved ethanol production from newspaper, suggesting that this strategy is useful for the selection of the optimal combination of CBMs for each type of biomass (Nakanishi *et al.*, 2012).

The engineered scaffolding protein consisting of three cohesin domains derived from *Clostridium thermocellum*, *Clostridium cellulolyticum*, and *Ruminococcus flavefaciens* was displayed on the *S. cerevisiae* cell surface using the a-agglutinin-based display system. The incubation of the scaffolding protein-displaying yeast with three recombinant cellulases (EG, CBH, and BGL) fused with a dockerin domain produced by *Escherichia coli* led to the synergistic hydrolytic degradation of cellulose (Tsai *et al.*, 2009). Furthermore, a yeast consortium system has been developed, in which four kinds of engineered yeasts are co-cultivated (Goyal *et al.*, 2011). In this system, construction of the mini-cellulosome was achieved by co-cultivation of yeast displaying scaffolding protein and yeasts secreting three kinds of dockerin-fused enzymes.

The arming yeast catalyzes simultaneous saccharification and fermentation of xylan (Katahira *et al.*, 2004; Katahira *et al.*, 2006). In addition to the two-step isomerization of xylose into xylulose, xylose isomerase (XI) is an alternative enzyme that shows great promise, as it is not associated with a cofactor imbalance. Recently, XI from *Clostridium cellulovorans* was successfully displayed and retained activity on *S. cerevisiae*. The constructed XI-displaying yeast could grow in medium containing xylose as the sole carbon source and directly produce ethanol from xylose (Ota *et al.*, 2013).

Lignin is also a major component of cellulosic biomass, and inhibits cellulose degradation by cellulases, because of its physiological recalcitrance and its masking the cellulose fibers. Therefore, the removal of lignin from cellulosic biomass is required for efficient degradation. Laccase from white-rot fungus, which participates in several biological pathways including lignin degradation, was displayed on *S. cerevisiae* via the β-agglutinin-based display system. By pretreatment of hydrothermally processed rice straw with laccase-displaying yeast, ethanol production by yeast co-displaying EG II, CBH II, and BGL1 was improved (Nakanishi *et al.*, 2012b).

According to the new research results of the VTT Technical Research Centre of Finland, lignocellulosic biomass can be used in the production of high-quality biofuels for the price of less than one euro per litre. A new technology developed in Finland allows the transfer of more than half the energy of wood raw materials to the end-product. The technology is considered ready for the construction of a commercial-scale production plant in Europe (Technical Research Centre of

Finland, 2013). Considering all these technological developments lignocelluloses can be used as a better easily available resource for enhanced production of ethanol to address the problems of biofuel production in agrarian countries.

References

Acharya A., Joshi D.R., Shrestha K. and Bhatta D.R. 2012. Isolation and screening of thermophilic cellulolytic bacteria from compost piles. *Scientific World*, 10(10): 43-46.

Baker D., Church G., Collins J., Endy D., Jacobson J., Keasling J., Modrich P., Smolke C. and Weiss R. 2006. Engineering life: building a fab for biology. *Sci. Am.* 294: 44–51.

Barakat A., Monlau F., Steyer J.P. and Carrere H. 2012. Effect of lignin-derived and furan compounds found in lignocellulosic hydrolysates on biomethane production. *Bioresour. Technol.* 104: 90–99.

Helen M. Berman H.M., Wastbrook J., Feng Z., Gilliland G., Bhat T.N., Weissig H., Shindylov I.N. and Bourne P.E. The protein data bank. *Nucleic Acids Res*, 2000. 28(1): 235-242.

Blumer-Schuette S.E., Kataeva I., Westpheling J., Adams M.W. and Kelly R.M. 2008. Extremely thermophilic microorganisms for biomass conversion: status and prospects. *Curr. Opin. Biotechnol.* 19: 210–217.

Brady S.F. and Clardy J. 2000. Long-chain N-acyl amino acid antibiotics isolated from heterologously expressed environmental DNA. *J. Am. Chem. Soc.* 122: 12903–12904.

Daniel R. 2005. The metagenomics of soil. *Nat. Rev. Microbiol.* 3: 470–478.

Demeke M.M., Dietz H., Li Y., Foulquié-Moreno M.R., Mutturi S., Deprez S., Abt T.D., Beatriz M Bonini B.M., Liden G., Dumortier F., Verplaetse A., Boles E. and Thevelein J.M. 2013. Development of a D-xylose fermenting and inhibitor tolerant industrial *Saccharomyces cerevisiae* strain with high performance in lignocellulose hydrolysates using metabolic and evolutionary engineering. *Biotechnol. Biofuels* 6: 89-113.

Eriksson T., Börjesson J. and Tjerneld F. 2002. Mechanism of surfactant effect in enzymatic hydrolysis of lignocellulose. *Enzym. Microb. Technol.* 31: 353-364.

Fargione J., Hill J., Tilman D., Polasky S. and Hawthorne P. 2008. Land clearing and the biofuel carbon debt. *Science* 319: 1235–1238.

Farrell A.E., Plevin R.J., Turner B.T., Jones A.D., O'Hare M. and Kammen D.M. 2006. Ethanol can contribute to energy and environmental goals. *Science* 311: 506–508.

Fischer J., Lopes V., Santos E.F.Q., Filho U.C. and Cardoso V.L. 2014, Second generation ethanol production using crude enzyme complex produced by fungi collected in Brazilian cerrado (brazilian savanna). *Chem. Eng. Transact.* 38: 487-492.

Fujita Y., Ito J., Ueda M., Fukuda H. and Kondo A. 2004. Synergistic saccharification, and direct fermentation to ethanol, of amorphous cellulose by use of an engineered yeast strain codisplaying three types of cellulolytic enzyme. *Appl. Environ. Microbiol. 70*: 1207–1212.

Gomashe A.V., Gulhane P.A. and Bezalwar P.M. 2013. Isolation and screening of cellulose degrading microbes from Nagpur region soil. *Int. J. Life Sci.* **1(4)**: 291-293.

Goyal G., Tsai S.L., Madan B., DaSilva N.A. and Chen W. 2011. Simultaneous cell growth and ethanol production from cellulose by an engineered yeast consortium displaying a functional mini-cellulosome. *Microb. Cell Fact.* 2011, **10**: 89-97.

Hamilton-Brehm S.D., Mosher J.J., Vishnivetskaya T., Podar M., Carroll S., Allman S., Phelps T.J., Keller M. and Elkins J.G. 2010. *Caldicellulosiruptor obsidiansis* sp. nov., an anaerobic, extremely thermophilic, cellulolytic bacterium isolated from Obsidian Pool, Yellowstone National Park. *Appl. Environ. Microbiol.* **76**: 1014–1020.

Hasunuma T. and Kondo A. 2012. Consolidated bioprocessing and simultaneous saccharification andfermentation of lignocellulose to ethanolwith thermotolerant yeast strains. *Process Biochemistry* **47(9)**: 1287-1294.

Hasunuma T., Okazaki F., Okai N., Hara K.Y., Ishii J. and Kondo A. 2013. A review of enzymes and microbes for lignocellulosic biorefinery and the possibility of their application to consolidated bioprocessing technology. *Bioresour. Technol.* **135**: 513-522.

Hosseini S.A. and Shah N. 2011a. Enzymatic hydrolysis of cellulose part II: Population balance modelling of hydrolysis by exoglucanase and universal kinetic model. *Biomass Bioenerg.* **35**: 3830-3840.

Hosseini S.A. and Shah N. 2011 b. Modelling enzymatic hydrolysis of cellulose part I: population balance modelling of hydrolysis by endoglucanase. *Biomass Bioenerg.* **35**: 3841-3848.

Hu X., Yu J., Wang C. and Chen H. 2014. Cellulolytic bacteria associated with the gut of *Dendroctonus armandi* larvae (Coleoptera: Curculionidae: Scolytinae). *Forests* **5**: 455-465.

Ito J., Kosugi A., Tanaka T., Kuroda K., Shibasaki S., Ogino C., Ueda M., Fukuda H., Doi R.H. and Kondo A. 2009. Regulation of the display ratio of enzymes on the *Saccharomyces cerevisiae* cell surface by the immunoglobulin G and cellulosomal enzyme binding domains. *Appl. Environ. Microbiol.* **75**: 4149–4154.

Jin Y.S., Alper H., Yang Y.T. and Stephanopoulos G. 2005. Improvement of xylose uptake and ethanol production in recombinant Saccharomyces cerevisiae through an inverse metabolic engineering approach. *Appl. Environ. Microbiol.* **71**: 8249–8256.

Katahira S., Fujita Y., Mizuike A., Fukuda H. and Kondo A. 2004. Construction of a xylan-fermenting yeast strain through codisplay of xylanolytic enzymes on the surface of xylose-utilizing *Saccharomyces cerevisiae* cells. *Appl. Environ. Microbiol.* **70**: 5407–5414.

Katahira S., Mizuike A., Fukuda H. and Kondo A. 2006. Ethanol fermentation from lignocellulosic hydrolysate by a recombinant xylose- and cellooligosaccharide-assimilating yeast strain. *Appl. Microbiol. Biotechnol.* **72**: 1136–1143.

Khelil O. and Cheba B. 2014. Thermophilic cellulolytic microorganisms from western Algerian sources: promising isolates for cellulosic biomass recycling. *Procedia Technol.* **12**: 519 – 528.

Kim S. and Dale B.E. 2004. Global potential bioethanol production from wasted crops and crop residues. *Biomass Bioenerg.* **26**: 361-375.

Klinke H.B., Thomsen A.B. and Ahring B.K. 2004. Inhibition of ethanol-producing yeast and bacteria by degradation products produced during pre-treatment of biomass. *Appl. Microbiol. Biotechnol.* **66**: 10–26.

Kotaka A., Bando H., Kaya M., Kato-Murai M., Kuroda K., Sahara H., Hata Y., Kondo A. and Ueda M. 2008. Direct ethanol production from barley β-glucan by sake yeast displaying *Aspergillus oryzae* β-glucosidase and endoglucanase. *J. Biosci. Bioeng.* **105**: 622–627.

Kumar R., Singh S. and Singh. O.V. 2008. Bioconversion of lignocellulosic biomass: Biochemical and molecular perspective. *J. Ind. Microbiol. Biotechnol.* **35**: 377–391.

Kuroda K. and Ueda M. 2013. Arming Technology in Yeast—Novel Strategy for Whole-cell Biocatalyst and Protein Engineering. *Biomolecules* **3**: 632-650.

Limayem A. and Ricke S.C. 2012. Lignocellulosic biomass for bioethanol production: Current perspectives, potential issues and future prospects. *Prog. Energ. Combust. Sci.* **38(4)**: 449-467.

Lin C-W., Tran D-T., Lai C-Y., I Y-P. and Wu C-H. 2010. Response surface optimization for ethanol production from *Pennisetum alopecoider* by *Klebsiella oxytoca* THLC0409. *Biomass Bioenergy* **34**: 1922-1929.

Lynd L.R., Laser M.S., Bransby D., Dale B.E., Davidson B., Hamilton R., Himmel M.E., Keller M., McMillan J.D. and Sheehan J. 2008. How biotech can transform biofuels. *Nat. Biotechnol.* **26**: 169–172.

Lynd L.R., Weimer P.J., van Zyl W.H. and Pretorius I.S. 2002. Microbial cellulose utilization: fundamentals and biotechnology. *Microbiol. Mol. Biol. Rev.* **66**: 506–577.

Maheshwari R., Bharadwaj G. and Bhat M.K. 2000. Thermophilic fungi: Their physiology and enzymes. *Microbiol. Mol. Biol. Rev.* **64(3)**: 461–488.

McCarter J.D. and Withers G.S. 1994. Mechanisms of enzymatic glycoside hydrolysis. *Curr. Opin. Struct. Biol.* **4(6)**: 885-892.

McMillan J.D., Jenning E.W., Mohagheghi A. and Zuccarello M. 2011. Comparative performance of precommercial cellulases hydrolyzing pretreated corn stover. *Biotech. Biofuels.* **4**: 29-46.

Murai T., Ueda M., Kawaguchi T., Arai M. and Tanaka A. 1998. Assimilation of cellooligosaccharides by a cell surface-engineered yeast expressing β-glucosidase and carboxymethylcellulase from *Aspergillus aculeatus*. *Appl. Environ. Microbiol.* **64**: 4857–4861.

Murai T., Ueda M., Shibasaki Y., Kamasawa N., Osumi M., Imanaka T. and Tanaka A. 1999. Development of an arming yeast strain for efficient utilization of starch by co-display of sequential amylolytic enzymes on the cell surface. *Appl. Microbiol. Biotechnol.* **51**: 65–70.

Murai T., Ueda M., Yamamura M., Atomi H., Shibasaki Y., Kamasawa N., Osumi M., Amachi T. and Tanaka A. 1997. Construction of starch-utilizing yeast by cell surface engineering. *Appl. Environ. Microbiol.* **63**: 1362–1366.

Nakanishi A., Bae J.G., Fukai K., Tokumoto N., Kuroda K., Ogawa J., Nakatani M., Shimizu S. and Ueda M. 2012a. Effect of pretreatment of hydrothermally processed rice straw with laccase-displaying yeast on ethanol fermentation. *Appl. Microbiol. Biotechnol.* **94**: 939–948.

Nakanishi A., Bae J., Kuroda K. and Ueda M. 2012b. Construction of a novel selection system for endoglucanases exhibiting carbohydrate-binding modules optimized for biomass using yeast cell-surface engineering. *AMB Express* **2**: 56-66.

Negro M.J., Manzanares P., Ballesteros I., Oliva J.M., Cabanas A. and Ballesteros M. 2003. Hydrothermal pretreatment conditions to enhance ethanol production from poplar biomass. Appl. Biochem. Biotechnol. **105–108**:87–100.

Okamoto, K., Nitta Y., Maekawa N. and Yanase H. 2011. Direct ethanol production from starch, wheat bran and ricestraw by the white rotfungus *Trametes hirsuta*. *Enzyme Microb. Tech.* **48(3)**: 273-277.

Ota M., Sakuragi H., Morisaka H., Kuroda K., Miyake H., Tamaru Y. and Ueda M. 2013. Display of *Clostridium cellulovorans* xylose isomerase on the cell surface of *Saccharomyces cerevisiae* and its direct application to xylose fermentation. *Biotechnol. Prog.* **29**: 346–351.

Patagundi B.I., Shivasharan C.T.and Kaliwal B.B. 2014. Isolation and characterization of Cellulase producing bacteria from soil. *Int. J. Curr. Microbiol. App. Sci.* **3(5)**: 59-69.

Pauly M. and Scheller H.V. 2000. O-acetylation of plant cell wall polysaccharides: identification and partial characterization of a rhamnogalacturonan O-acetyl-transferase from potato suspension-cultured cells. *Planta* **210(4)**: 659-67.

Ro D.K., Paradise E.M., Ouellet M., Fisher K.J., Newman K.L., Ndungu J.M., Ho K.A., Eachus R.A., Ham T.S., Kirby J., Chang M.C., Withers S.T., Shiba Y., Sarpong R. and Keasling J.D. 2006. Production of the antimalarial drug precursor artemisinic acid in engineered yeast. *Nature* **440**: 940–943.

Rondon M.R., August P.R., Bettermann A.D., Brady S.F., Grossman T.H., Liles M.R., Loiacono K.A., Lynch B.A., MacNeil I.A., Minor C., Tiong C.L., Gilman M., Osburne M.S., Clardy J., Handelsman J. and Goodman R.M. 2000. Cloning the soil metagenome: a strategy for accessing the genetic and functional diversity of uncultured microorganisms. *Appl. Environ. Microbiol.* **66**: 2541–2547.

Salunke M. 2012. Isolation and screening of cellulose degrading microorganisms from fecal matter of herbivores. *Int. J. Sci. Eng. Res.* **3(10)**: 1-5.

Searchinger T., Heimlich R., Houghton R.A., Dong F., Elobeid A., Fabiosa J., Tokgoz S., Hayes D. and Yu T.H. 2008. Use of US croplands for biofuels increases greenhouse gases through emissions from land-use change. *Science* **319**: 1238–1240.

Shigechi H., Koh J., Fujita Y., Matsumoto T., Bito Y., Ueda M., Satoh E., Fukuda H. and Kondo A. 2004. Direct production of ethanol from raw corn starch via fermentation by use of a novel surface-engineered yeast strain codisplaying glucoamylase and β-amylase. *Appl. Environ. Microbiol.* **70**: 5037–5040.

Sommer M.O.A., George M.C. and Dantas G. 2010. Functional metagenomic approach for expanding the synthetic biology toolbox for biomass conversion. Molecular Systems Biology 6:360-367.

Sun J., Wen F., Si T., Xu J.H. and Zhao H. 2012. Direct conversion of xylan to ethanol by recombinant Saccharomyces cerevisiae strains displaying an engineered minihemicellulosome. *Appl Environ Microbiol.* **78(11)**: 3837-3845.

Taherzadeh M.J. and Karimi, K. 2007. Enzyme-based hydrolysis processes for ethanol from lignocellulosic materials: A review. *BioResources* **2(4)**: 707-773.

Taniguchi M., Suzuki H., Watanabe D., Sakai K., Hoshino K. and Tanak T. 2005. Evaluation of pretreatment with *Pleurotus Ostreatus* for enzymatic hydrolysis of rice Straw. *J. Biosci. Bioenergy* **100(6)**: 637-643.

Technical Research Centre of Finland (VTT). New gasification method turns forest residues to biofuel with less than a Euro per litre. ScienceDaily, 3 July 2013. ‹www.sceincedaily.com/releases/2013/07/130703101018.htm›.

Tilman D., Hill J. and Lehman C. 2006. Carbon-negative biofuels from low-input high-diversity grassland biomass. *Science* **314**: 1598–1600.

Tsai S.L., Oh J., Singh S., Chen R. and Chen W. 2009. Functional assembly of minicellulosomes on the *Saccharomyces cerevisiae* cell surface for cellulose hydrolysis and ethanol production. *Appl. Environ. Microbiol.* **75**: 6087–6093.

Vasella A., Davies G.J. and Bohm M. 2002. Glycosidase mechanisms. *Curr. Opin. Chem. Biol.* 6(5): 619-29.

Waghmare P.R., Kshirsagar S.D., Saratale R.G., Govindwar S.P. and Saratale G.D. 2014. Production and characterization of cellulolytic enzymes by isolated *Klebsiella* sp. PRW-1 using agricultural waste biomass. *Emir. J. Food Agric.* **26 (1)**: 44-59.

Warnecke F., Luginbuhl P., Ivanova N., Ghassemian M., Richardson T.H., Stege J.T., Cayouette M., McHardy A.C., Djordjevic G., Aboushadi N., Sorek R., Tringe S.G., Podar M., Martin H.G., Kunin V., Dalevi D., Madejska J., Kirton E., Platt D., Szeto E., Salamov A., Barry K., Mikhailova N., Kyrpides N.C., Matson E.G., Ottesen E.A., Zhang X., Hernández M., Murillo C., Acosta L.G., Rigoutsos I., Tamayo G., Green B.D., Chang C., Rubin E.M., Mathur E.J., Robertson D.E., Hugenholtz P. and Leadbetter J.R. 2007. Metagenomic and functional analysis of hindgut microbiota of a wood-feeding higher termite. *Nature* **450**: 560–565.

Wen F., Sun J. and Zhao H. 2010. Yeast surface display of trifunctional minicellulosomes for simultaneous saccharification and fermentation of cellulose to ethanol. *Appl. Environ. Microbiol.* **76**: 1251–1260.

Wyman C.E. 2007. What is (and is not) vital to advancing cellulosic ethanol. *Trends Biotechnol.* **25**: 153–157.

Xiros, C., Moukouli M., Topakas E. and Christakopoulos P. 2009. Factors affecting ferulic acid releasefrom Brewer's spent grain by *Fusarium oxysporum* enzymatic system. *Bioresour. Technol.* **100(23)**: 5917-5921.

Yamada R., Taniguchi N., Tanaka T., Ogino C., Fukuda H. and Kondo A. 2010. Cocktail β-integration: A novel method to construct cellulolytic enzyme expression ratio-optimized yeast strains. *Microb. Cell Fact.* **9**: 32-40.

Yang S.J., Kataeva I., Hamilton-Brehm S.D., Engle N.L., Tschaplinski T.J., Doeppke C., Davis M., Westpheling J. and Adams M.W. 2009. Efficient degradation of lignocellulosic plant biomass, without pretreatment, by the thermophilic anaerobe *Anaerocellum thermophilum* DSM 6725. *Appl. Environ. Microbiol.* **75**: 4762–4769.

Zhang M.J., Su R.G., Qi W. and He Z.M. 2010. Enhanced enzymatic hydrolysis of lignocelluloses by optimizing enzyme complexes. *Appl. Biochem. Biotech.* **160**: 1407-1414.

Zhang Y.H., Ding S.Y., Mielenz J.R., Cui J.B., Elander R.T., Laser M., Himmel M.E., McMillan J.R. and Lynd L.R. 2007. Fractionating recalcitrant lignocellulose at modest reaction conditions. *Biotechnol. Bioeng.* **97**: 214–223.

Zhang B. and Yang S.T. 2012. Metabolic engineering of *Rhizopus oryzae*: Effects of over expressing fumR gene on cell growth and fumaric acid biosynthesis from glucose. *Process Biochem.* **47(12)**: 2159-2165.

Chapter 7

Energy Resource Management for Sustainable Utilization

A.G. Devi Prasad and H.P. Komala

DOS in Environmental Science, University of Mysore, Manasagangotri,
Mysore-06, Karnataka

ABSTRACT

Energy is one of the most important inputs for economic growth and human development, this paper describes the different types of energy and their consumption and utilization trend in India. India is world's 7th largest energy producer and 5th largest energy consumer. Due to the immense demand for energy within the country dissemination of sustainable energy becomes the need of the hour. Despite the financial, political and technical constraints, utilization of sustainable energy has to be encouraged as it provides enormous benefits from ameliorating adverse impacts on environment to improve livelihoods of peasant's. Biomass, an important energy source, contributes in substituting the fossil fuels, especially in developing countries. To cater the biomass demand, more land area has to bring under cover, which would help the rural population. Prioritization of development of renewable energy sources help in self-sustenance of energy sources.

Key words: *Biomass, Renewable, Consumption, Management, Electricity*

Introduction

Energy is one of the most important inputs for economic growth and human development in raising the standard of living and quality of life. It has an influencing role in the development of key sectors of economic importance such as industry, transport and agriculture. This has motivated many researchers to focus their research on energy management (Singh, 1999; Baruah and Bora, 2008). It is a kind of strategic resource and an important substantial basis for economic increase and social development (Arevalo *et al.*, 2007; Shao and Chu, 2008).

Per capita energy consumption is an index of growth of any nation in all forms of inputs. Different forms of energy like fuel for cooking, motive power for transport and electricity for modern communication are very important factors for growth of present- day- civilization (Singh, 2002). The state of economic development of any region can be assessed from the pattern and consumption quality of its energy. It is considered as a prime agent in the generation of wealth and also a significant factor in economic development (Demirbas, 2003). The relation between use of energy and economic growth has been a subject of greater inquiry as energy is considered to be one of the important driving forces of economic growth in all economies (Pokharel, 2006).

The consumption of energy has increased steadily over the last century as the world population has grown and more countries have become industrialized. There is a strong two-way relationship between economic development and energy consumption. On one hand, growth of an economy, with its global competitiveness, depends on the availability of cost-effective and environmentally benign energy sources, and on the other hand, the level of economic development has been observed to be dependent on the energy demand (EIA, 2006).

Around 400 million Indians don't have electricity access including 10,000 unelectrified villages. India's coal reserves are projected to run out in four decades. Due to production of more than three quarters of electricity produced by burning coal and natural gas, increases carbon emission is around 1.6 billion tons which is highest in the world.

India domestically meets up 30% of its crude oil requirement and the rest is being imported from the oil producing nations. Indian transport sector is the principal consumer of petrol and diesel followed by big and small industrial units. Similarly, electricity consumption share too is the largest by this sector (GOI, 2005). The world's energy demand in 2006 amounted to about 490 Exa Joule(EJ)(11,703 MT TOE) and was made up of about 81% fossil fuels (oil, gas, and coal) about 10% biomass , about 6% nuclear and about 2.2 and 0.5% hydropower and other energy respectively (IEA, 2008).

Classification of Energy

Energy can be classified into three types:

1. Primary and secondary energy

The energy sources which are either found or stored in nature are called primary energy, *e.g.*, Coal, oil, natural gas, biomass, nuclear energy from radioactive substances, thermal energy stored in earth's interior and energy sources are converted in industrial utilities into secondary energy sources. For example- coal, oil, or gas converted into stream and electricity.

2. Commercial and non-commercial energy

The energy sources that are available in the market for a definite price are known as commercial energy. The important forms of commercial energy are electricity, coal and natural gas, lignite, refined petroleum products. These energy forms the basis of industrial, agricultural, transport and commercial development in the modern work. The energy which is not available in the commercial market for a price is called as non-commercial energy. These energy sources include fuels such as firewood, cattle dung and agricultural wastes, which are traditionally gathered and not bought at a price used especially in rural households. These are also called traditional fuels.

3. Renewable and non-renewable energy

Renewable energy is energy obtained from sources that essentially in exhaustible. These energy resources include wind power, geothermal energy, tidal power and hydroelectric power. Non – renewable energy is the converted fossil fuels such as coal, oil and gas, which are likely to deplete with time.

Sources of Energy

We use many different energy sources to do work for us. Energy sources are classified into two groups- renewable and non-renewable. Most of our energy comes from non- renewable energy sources, e.g. coal, petroleum, natural gas, propane and uranium. These energy sources are called non-renewable because their supplies are limited. Petroleum for example, was formed millions of years ago from the remains ancient sea plants and animals. It will exhaust in future days. A renewable resource is always there and will never run out as it keeps renewing itself. Renewable energy is the energy generated from natural resources such as sunlight, wind, hydro, tides and geothermal heat which are naturally replenished renewable.

Overall Status of Energy Sources In India

India is both a major energy producer and a consumer. India currently ranks as the world's seventh largest energy producer, accounting for about 2.49% of the world's total annual energy production. It is also the world's fifth largest energy consumer, accounting for about 4.1% of the world's total annual energy consumption. The current per capita energy consumption of India is 0.5 Tonnes of Oil Equivalent (TOE) as compared to the world average of 1.9 TOE, and this indicates a high potential for energy consumption.

Non-Renewable Energy Sources

Petroleum (Oil)

India presently ranks as the 25th greatest producer of crude oil, accounting for about 1% of the world's annual crude oil production. The production of crude petroleum increased from 6.82MTs during 1970-71 to 37.71MTs during 2011. In the year 2010-2011, the production of petroleum products in the country was 190.36 MTs as against 179.77 MTs during 2009-2010, an increase of 5.9%. Out of the total domestic production of 190.36 MTs of all types of petroleum products, high speed diesel oil accounted for the maximum share (41%), followed by Motor Gasoline (13.73%), Fuel Oil (10.78%), Naphtha (9.2%), Kerosene (4%) and Aviation Turbine Fuel (5%). About 30% of India's energy needs are met by oil, and more than 60% of that oil is imported (EIA, 2006). India's proved oil reserves are currently estimated at about 757 million tonnes (as on 2011). Geographical distribution of crude oil indicates that, the maximum reserves are in the Western offshore (43%) followed by Assam (22%) (Energy Statistics, 2012).

Natural Gas

Natural gas accounts for about 8.9% of energy consumption in the country. It accounts for about 1.6% of the world's natural gas production. Natural gas has experienced the fastest rate of increase of any fuel in India's primary energy supply; demand is growing at about 4.8% per year and is forecasted to rise to 1.6% Trillion Cubic Feet (TCF) per year by 2015. India's natural gas reserves are currently estimated at about 1241 billion cubic meters (as on 2011). The maximum reserves are in the Eastern Offshore (35%) followed by Western Offshore (33%).

Coal

Coal was the key energy source for the industrial revolution, which has provided amenities that most of us take for granted today—including electricity, new materials (steel, plastics, cement and fertilizers), fast transportation, and advanced communications. Coal replaced wood combustion because of coal's abundance, its higher volumetric energy density. As Nicolls (1915) writes, "With Coal, we have light, strength, power, wealth, and civilization; without Coal we have darkness, weakness, poverty, and barbarism." (as quoted in Freese, 2003). The energy density of coal is 32 MJ/kg and 42 GJ/m^3, by weight and volume, respectively; similarly, energy density for dry wood is15 MJ/kg and 10 GJ/m^3 (Sorensen, 1984). These numbers will vary depending on coal rank and wood quality.

Indian coal is primarily bituminous and sub-bituminous; there are nearly 36 gigatons (GT) of lignite resources in Tamil Nadu, Gujarat, Rajasthan, Jammu and Kashmir (Ministry of Coal, 2006a).

India has a good reserve of coal. As on 2011, the estimated reserves of coal was around 286 billion tones. These deposits are mainly found in eastern and south central parts of the country. There has been an increase of 3.1% in the estimated coal reserves during the year 2010-11 with Madhya Pradesh accounting for the maximum increase of 5%. A typical coal power plant requires 2023m^2 of land area per Mega Watt (MW) for plant installation.

India not only has 10% of the worlds' coal reserves but imports some as well to meet the rising demand from sectors like steel. According to the World Bank, 53% of India's commercial energy demand is met by coal. Coal will remain the dominant fuel in India's energy mix through 2030. Demand is projected to grow from 407 Mt in 2006 to 758 Mt in 2030, at an average rate of growth of 2.4% per year. The power sector will be the chief driver of Indian demand. Currently, 71% of India's electricity is generated from coal.

Electricity

India is the sixth greatest electricity generating country and accounts for about 4% of the world's total annual electricity generation. Electricity consumption in

India has more than doubled in the last decade. The average per capita consumption of electricity in India is estimated to be 704 Kilo Watt Hour (kWh) during 2008-09.

In Hyderabad, Electricity represents over one fourth of the total energy consumed in the household sector. The rich people consume higher quantity of electrical energy than the poor. The average monthly per household and per capita consumption of electricity is only 90 and 15 kWh respectively. The consumption level of the rich and the poor households varies widely. Monthly consumption ranges from 180 and 41kWh per household and per capita in the highest income group to 57 and 7kWh, respectively in the lowest income group (Alam *et. al.,*).

Karnataka is dependent mainly on conventional energy sources such as coal, diesel, gas, and hydro energy. Only 24% of the total installed plant capacity is based on renewable energy sources in Karnataka. Though the state get very good solar insolation, solar energy utilization is not remarkable (Ganesh and Ramachandra, 2012). It is one of the most electrical energy consuming states with an annual consumption of 36975.2 Million KWh (2010-11). Per capita consumption is around 604 KWh (Ramachanadra, 2012). Though state has total installed plant capacity of 13490.63 MW, state is an electrical energy deficit state, so solar energy harvesting could lead to the solution.

Renewable Energy sources

The country has made a remarkable growth in last 2-3 years in the field of Renewable energy power generation. The past few years saw a record addition of 2332MW of renewable energy sources *i.e.* solar, wind, bio-mass, geothermal & hydro etc., which could make important contributions to sustainable development. The bulk addition is in wind generation at 1565MW, small hydro power segment recorded an addition of 305MW, cogeneration 295MW and bio-mass 153MW. The lowest additions were in the solar at 8MW and waste-to-energy segments at 4.7MW (Raghavaiah *et. al.,* 2011).

In India, there is high potential for generation of renewable energy from various source e.g.,-wind, solar, biomass, small hydro and bagasse. The total potential for renewable power generation in the country as on 2011 is estimated at 89760 MW. This includes an estimated wind power potential of 55%, small –hydro power potential of 17%, biomass power potential of 20%, and 6% from bagasse-based cogeneration in sugar mills.

The resource potential and energy demand has been estimated in Jaunpur block of Uttaranchal state of India by Akella *et.al.,* (2007). Study shows that, total load is 808 MWh/yr and total available resources are 807 MWh/yr, whereas % age contribution of each resources are Micro Hyder Power (MHP) 15.88% (128166), solar 2.77% (22363), wind 1.89% (15251) and biomass energy 79.46% (641384) kWh/yr.

The State wise estimated potential of renewable power in India is about 89760MegaWatt (as on 20011). Karnataka accounts for about 12%, Maharashtra-11%, Gujarat-14%. Tamil Nadu-9%, Rajastha-7%, Uttar Pradesh-4%, Jammu and Kashmir-7%, Andhra Pradesh-8% and Others shares about 24% (Energy Statistics-2012).

Small Hydro power

It has been one of the earliest known renewable energy sources, in existence in the country. Hydropower is a clean source of energy. No fuel is burned, so the air is not polluted. It is the cheapest source of electricity because the water is free to use. Estimates place the small hydro potential in India at 15,000 MW (TERI 2000). 1,748 MW of installed small hydropower (SHP) operated in India in 2006. Karnataka and Maharashtra accounted for 17 and 11 percent of the total, respectively. The potential of this sector is dependent on available water resources, which are abundant in the majority of states. (Source: Ministry of Non-conventional Energy Sources, Annual Report 2005/06 (New Delhi: MNES, Government of India, 2006).

Himachal Pradesh has vast hydel potential of approx. 21,000 MW (approx. 750 MW under Small Hydro Sector) in the five river basins. Of this potential, 6037 MW has been harnessed so far. (Commission's order on Small Hydro Power Projects tariff and other issues December 18, 2007, Himachal Pradesh Electricity Regulatory Commission, Shimla).

Wind Power

The most important form of renewable energy India depends on as alternate is Wind Power. Wind energy is one of the clean, renewable energy sources that hold out the promise of meeting energy demand in developing countries like India. The development of wind power industry in India took place in 1990s and India is the fifth largest installed wind power in the world. (Gujarat Energy Development Agency portal, Government of Gujarat, http://geda.gujarat.gov.in/). Estimates place the economical wind energy potential in India at 45,000MW.

There are 39 wind potential stations in Tamil Nadu, 36 in Gujarat, 30 in Andhra Pradesh, 27 in Maharashtra, 26 in Karnataka, 16 in Kerala, 8 in Lakshadweep, 8 Rajasthan, 7 in Madhya Pradesh, 7 in Orissa, 2 in West Bengal, 1 in Andaman Nicobar and 1 in Uttar Pradesh. Out of 208 suitable stations 7 stations have shown wind power density more than 500 Watts/ m^2. (Bureau of Energy Efficiency).

Karnataka is one of the wind-rich states in India and has a potential of around 14 GW. The preferential tariff framework has enabled Karnataka to be among the top five wind-energy generating states with an installed capacity of 1,932 MW (as on March 2012).C-WET Centre for wind energy technology (CII Karnataka Conference on Power 2012, Sustainable Power through Renewable Sources).

More importantly, using wind to generate electricity emits much less harmful greenhouse gases than during the combustion of fossil fuels that are generously used to generate electricity. It is estimated that there is a saving of 300- 500 tonnes of CO_2 emission from a wind farm of 4MWh electricity generation capacity in India (C-WET).

Biomass

Biomass, consisting of wood, crop residues and animal dung continues to dominate energy supply in rural and traditional sectors. Wood wastes of all types make excellent biomass fuels (Hurst Boiler, 2012). Since millions of years are requires to form fossil fuels in the earth, their reserve are finite and subject to depletion as they are consumed (Klass, 2004). With fossil fuel prices rising daily and the adverse impact of it usage on the environment, the only substitute for fossil fuels is biomass (Balat and Ayar, 2003). It is used to meet a variety of energy needs, including generating electricity, heating homes, fueling vehicles and providing process heat for industrial facilities. It can be converted into useful forms of energy using a number of different processes (Mckendry, 2001, 2002a).

In India, biomass fuels dominate the rural energy consumption patterns, accounting for over 80% of total energy consumed (Ravindranath *et al.*, 1995). It is having about one- third share in the total primary energy consumption in the country. Energy from biomass holds promise in the Indian conditions as it encourages self- reliance through efficient use of indigenous resource and employment of simpler technologies consistent with minimal possible hazards. It meets about 75% of rural Indians energy needs, while in Karnataka, non commercial energy sources like firewood, agricultural residues, charcoal and cow dung account for 53.2%.

Biogas Energy

Biogas energy is the energy, generated from organic materials under anaerobic conditions. Feedstock's for biogas generation include cow dung, poultry droppings, kitchen waste, grass faecal matters and algae. It is one of the options, which can meet the growing energy demand of rural areas in developing countries.

It may be noted that there is a very large potential of biogas utilization at household level in India. The cumulative number of total biogas plants installed as per the latest annual report of Ministry of Non-Conventional Energy Sources (MNES) was 2.93 million in December 1999 which is about 10% of the total estimated potential in realistic scenario (based on 1991 data on bovine population).

Solar Energy

Solar energy is one of the important renewable energy. This energy is obtained from the sun and can easily absorb and converted into electrical energy for usage. Various studies have shown that solar energy utilization is economic, environmental friendly and profitable also. Conversion of solar energy into electrical energy is the better way to use the solar radiation effectively since electrical energy is one of the basic needs and easily convertible to other forms of energy (Ramachandra 2000; 2003; 2011).

Among various renewable energy resources, India possesses a very large solar energy resource which is seen as having the highest potential for the future. At present solar photovoltaic (SPV) contributes around two and a half percent of the power generation based on renewable energy technology in India. Solar thermal technologies have a very high potential for applications in solar water heating systems for industrial and domestic applications and for solar cooking in the domestic sector.

Solar Thermal Power Generation potential in India is about 35 MW per Sq. Km estimates indicate 800 MW per year potential for solar thermal based power generation in India during the period 2010 to 2015. The estimate shows that 0.97% of total land area or 3.1% of the total uncultivable land area of India would be required to generate 3400 TWh/yr from solar energy power systems in conjunction with other renewable energy sources (Mitavachan *et. al.*, 2012).

Several designs and capacities of Solar Water Heating System (SWHS) are available to suit the user's needs. A system of Solar Photovoltaic pump which produces electricity form sunlight and operates the pump to lift water from wells. A 900 watts pump can deliver about 50,000 litres of water per day over a total head of 30 feet. It can irrigate one to two acres (http://www.tn.gov.in/spc/tenthplan/CH_11_2.PDF). Over the past few decades the efficiency of the photovoltaic cells has increased from 8% in 1976 to around 43% in 2011 (NREL, 2011).

Rajasthan has about 2,08,110 Km^2 of desert land, which is 60% of the total area of the state. Interestingly, Rajasthan receives solar radiation of 6.0-7.0 kWh/ m^2. As the area has low rainfall, about 325 days have good sunshine in a year (Barbalace, 2011), and in western areas in Thar desert it may extend up to 345-355 days as rains occur only for 10.4-20.5 days in a year (Baruah and Bora, 2008).

The average intensity of solar radiation received over India is 200 MW/km square (megawatt per kilometre square) with 250–325 sunny days in a year. Solar energy intensity varies geographically in India, but the highest annual global radiation (≥ 2400 kWh/m^2) is received in Rajasthan and northern Gujarat (CEA, GOI). India receives the solar energy equivalent of more than 5000 trillion kWh/year. Depending on the location, the daily incidence ranges from 4 to 7 kWh/m^2, with the hours of sunshine ranging from 2300 to 3200 per year (C-WET; Demirbas, 2003). Recent research has shown that India has a vast potential for solar power generation since about 58% of the total land area (1.89 million km2) receives annual average Global insolation above 5 kWh/m^2/day. Indeed, at present efficiency levels, 1% of land area is sufficient to meet electricity needs of India till 2031 (Fthenakis and Kim, 2009; GOI, 2001).

Geothermal Energy

Geothermal Energy is a non-conventional energy sources and it is the vast reservoir of heat energy in the earth's interior, whose surface manifestations are the volcanoes, fumaroles, geysers, steaming grounds and hot springs. About 300 thermal springs are known to occur in India, falling in orogenic(Himalayan) as well as non-orogenic (peninsular) province(Razdan *et al.*,2008).

All the geothermal provinces of India are located in areas with high heat flow and geothermal gradients. The heat flow and thermal gradient values vary from 75-468 mW/m^2 and 59-234^{0}C respectively. These provinces are capable of generating 10,600 MW of power (Kumar *et al.*, 2008). At present the total estimated electric power production from geothermal energy is in the range of 8771 MWe all over the globe (Lund, 2004).

Thus a sound geochemical database and geological and tectonic setting of the thermal provinces is now available for future use (Chandrasekharam, 2000; Minissale *et al.*, 2000, 2003;). But for direct utilization, India is yet to attract independent power producers to utilize this source for power generation. Direct utilization of this energy source is in practice at Manikaran, Vasist in Himachal Pradesh and other provinces where thermal water is used for bathing and therapeutic purposes (Chandrasekharam, 1999).

Consumption of Energy In India

In India the consumption of different energy sources will vary by different energy consuming sectors. The major energy consuming sectors are classified as residential, commercial, industrial, transportation and electric power (Table-1). Among these, the highest energy is consumed in electric power generation.

Table:1 Sector wise Energy Consumption in India.

Sector wise	As on 2008 (in M TOE)	Projections		Annual growth (%)
		2015 (in M TOE)	2020 (in M TOE)	
Residential	40.3	60.5	70.6	3.1
Commercial	10.1	15.1	20.2	4.8
Industrial	257.0	307.4	357.8	2.6
Transportation	50.4	80.6	110.9	5.5
Electric power	226.8	315.0	375.5	3.3

Source: India Energy Book, 2012

The different types of energy consumption in India (as on 2011) are shown in the following table 2

Table: 2. Total Energy Consumption in India.

Energy types	As on 2011(MTs)
Liquids	151.2
Natural gas	40.3
Coal	274.7
Nuclear	5.0
Renewable	60.5
Total	531.7

Source: India Energy Book, 2012

Energy Demand

The growth in global demand for energy has played a key role in causing prices of different energy sources to rise dramatically (World Energy Council, WEC,2008). Demand of energy requirement is directly proportional to the development rate and population growth rate of the country. Presently in 2011-12 the total installed capacity in India is 181.5 GW (including conventional and non conventional sources) with peak load demand of 136 GW. Out of this energy demand only 118 GW is met having a deficit of 13% in energy generation (Arevalo *et al.*, 2007; AEO, 2006).

Table: 3 Indian Primary Energy Demand in the Reference Scenario (mTOE)

	1990	2000	2005	2015	2030	2015-2030
Coal	106	164	208	330	620	4.5%
Oil	63	114	129	188	328	3.8%
Gas	10	21	29	48	93	4.8%
Nuclear	2	4	5	16	33	8.3&
Hydro	6	6	9	13	22	3.9%
Biomass	133	149	158	171	194	0.8%
Other renewable	0	0	1	4	9	11.7%
Total	**320**	**459**	**537**	**770**	**1299**	**3.6%**

Source: International Energy Agency, World Energy Outlook 2007: China and India Insights (Paris, France: OCED/IEA, 2007).

Due to projected increase in population and the country's continued economic growth, primary energy demand in India is expected to increase from 537 M TOE in 2005 to 770 M TOE in 2015 and to 1,299 M TOE by 2030 (Table 3). Over the period 1990–2005, demand grew by 3.5 percent per year. As indicated by the above table, coal is expected to remain the dominant fuel in India's energy mix over the next 25 years. Demand for oil will steadily increase to a projected 328 M TOE by the year 2030, still one-half the projected demand for coal.

Other renewable, mostly wind power, are projected to grow 12 percent per year. Nuclear and hydropower supplies grow in absolute terms, but they make only a minor contribution to primary energy demand in 2030 - 3 % in the case of nuclear and 2 % for hydropower.

Longer-term scenarios from the Integrated Energy Policy Committee indicate that the total coal demand may vary anywhere between 1.5 and 2.5 GT, assuming a coal calorific value of 4000 kcal/kg and 8% GDP growth. The Committee believes that by 2030, annual coal demand will be about 2 GT (Planning Commission, 2006).

Challenges in Using Energy in India

Firstly the usage of coal (Highest share among non-conventional energy sources) is having many problems with respect to its ash content. The high ash content also leads to technical difficulties for utilizing the coal, as well as lower efficiency and higher costs for power plants. Some specific problems with the high ash content include high ash disposal requirements, corrosion of boiler walls and fouling of economizers, and high fly ash emissions (IEA, 2002). The high silica and alumina content in Indian coal ash is another problem, as it increases ash resistivity, which reduces the collection efficiency of electrostatic precipitators and increases emissions.

Though renewable energy has become a necessity in developing country as an alternate method of conventional power sources, it also faces some challenges.

The initial installation cost of renewable energy sources are high compared to conventional energy sources. For example, to set up a solar plant the initial cost of the equipment is high and further it requires maintenance. This is in the case of other renewable energy sources also.

Land issues are multi-dimensional in nature, which involve societal, political and economical aspects associated with a particular developmental project or energy source in this context. The World Commission on Dams estimates that dam construction submerged 4.5 milion hactres of Indian forest land between 1980-2000, and an average dam displaces 31,340 persons and submerges 8748 ha of land (Rangachari *et. al.*, 2000). Because of these issues, land acquiring may be a difficult problem for hydropower generation. Similarly, acquiring land to set up a nuclear power plant may be an issue because of the potential risk involved in the technology. For example, the Chernobyl accident contaminated around 300,000 sq.km of land with radio nucleotides (Barbalace, 1999). This type of accidents causes negative effects on the environment and as well as on human beings.

Another concern area is installing solar cells on the land area. The large amount of land required for solar power plants-approximately one square kilometre for every 20-60 megawatts (MW) generated –poses an additional problem in India.

Proper planning and integration is another aspect to be considered. The renewable energy is source dependent such as water, air, sun etc so the plant has to be planned in places of its availability. Before implementing a technology among people, there have to be social acceptance to implement it. Though many parts of rural and urban India have the awareness of global warming, there is only little/ no awareness about the shortage in non-renewable energy sources and price increase in production of it.

The issue related with usage of biomass is that, incomplete combustion in the traditional stoves contribute not only to the energy inefficiency, but also to substantial emissions of pollutants which can cause severe health damage (Smith and Thorneloe, 1992.

Energy Conservation

Energy Conservation basically means "increasing the productivity of energy and preventing wastage of energy". It is achieved when growth of energy consumption is reduced. Energy conservation can, therefore, be the result of several process or developments, such as productivity increase or technological progress. On the other hand energy efficiency is achieved when energy intensity in a specific product, process or area of production or consumption is reduced without affecting output, consumption or comfort levels. Promotion of energy efficiency will contribute to energy conservation and is therefore an integral part of energy conservation promotional policies.

In a scenario where India tries to accelerate its development process and cope with increasing energy demands, conservation and energy efficiency measures are to play a central role in our energy policy.

The Government of India has formulated an energy policy with the objective of ensuring adequate energy supply at a minimum cost, achieving self-sufficiency in energy supplies and protecting environment from adverse impact of utilising energy resources in an non-judicial manner. The main features of the policy are:

1. Accelerated exploitation of domestic conventional energy resources— oil, coal, hydro   and nuclear power;

2. Intensification of exploration to achieve indigenous production of oil and gas;

3. Management of demand of oil and other forms of energy;

4. Energy conservation and management;

5. Optimisation of utilisation of existing capacity in the country;

6. Development and exploitation of renewable sources of energy to meet energy requirements of rural communities;

7. Intensificationof   resources   and   development   activities   in new and renewable energy resources; and

8. Organisation of training for personnel engaged at various levels in the energy sector; the development and promotion of non- conventional/ alternate/new and renewable sources   of energy such as solar, wind, bio-energy etc. are getting sustained attention from the Department of Non-conventional Energy Sources set-up in 1982.

Steps to Conserve Energy Resources

1. Minimise exploitation of non-renewable energy resources.

2. Emphasis on use of renewable sources of energy.

3. Stop wastage of energy.

4. Creating awareness among people regarding wise and judicious use of energy.

5. More use of bio-mass based energy.

Government Programmes

India has a dedicated Ministry of New and Renewable Energy (MNRE) and a number of ministries have taken on specific renewable technologies to develop and support energy resource management. Many resource assessment programs have been implemented as future projects, including: Wind Resource Assessment Program (WRAP), National level Biomass Resource Assessment Program (NBRAP), and Solar and Wind Energy Resource Assessment (SWERA). RE specific research institutions have emerged, including Centre for Wind Energy Technology (CWET) for wind, Solar Energy Centre (SEC) for solar and National Institute for Renewable Energy (NIRE) for bioenergy. MNRE is also currently in process of creating small enterprises for manufacturing and servicing of RE systems and devices. In September 2007, the GOI launched the Solar Innovation Program in conjunction with Indian Institute of Management (IIM) Ahmadabad (Pursuing Clean Energy Business in India, 2008).

Currently, there are both central and state policies with some application to Re-Development in India. They include the following:

i.   Electricity Act of 2003 which includes Renewable Portfolio Standards (RPS) targets and preferential tariffs for RE are to be set by appropriate commissions;

ii. National Electricity Policy which promotes private participation in RE, reductions in capital cost of RE through competition and benefits of cogeneration;

iii. Tax Act which allows for accelerated depreciation from 80-100%, a ten year tax holiday for infrastructure projects, and income tax exemptions for infrastructure related projects;

iv. 11th 5 year plan (2007-2012) has programme such as the Remote Village Renewable Energy Programme, the Village Energy Security Programme, the Remote Village Solar Lighting Programme and Grid-connected Village Renewable Energy Programmes for Solar thermal and Biogas.

Conclusion

Dissemination of sustainable energy is obstructed by financial, political and technical barriers. Despite that, the utilization of sustainable energy should be highly encouraged in rural areas since it has provided enormous benefits, including ameliorating adverse impacts on environment, boosting economic development and improving peasant's living standards. Therefore, in order to minimize the obstacles and assist the development of sustainable energy in rural areas, it is necessary to first improve energy efficiency.

Biomass, especially woody biomass and energy crops is already an important energy carrier contributing substantially to cover energy demands in many parts of the world. This energy carrier has the potential to contribute even more to provide energy to substitute the use of fossil fuel energy, especially in industrialized countries as well as in developing countries. To cater to the need of biomass demand, more and more land area may be brought enhancing tree and biomass cover. This would faster rural development trough accelerated economic growth.

References

Akella A.K. , M.P. Sharma and R.P. Saini, 2007. Optimum utilization of renewable energy sources in a remote area. *Renewable and Sustainable Energy Reviews,* Vol.11, pp. 894–908.

Arevalo, C.B.M., T.A. Volk, E. Bevilacqua and L.P. Abrahamson, 2007. Development and validation of aboveground biomass estimations for four Salix clones in central New York. *Biomass Bioenerg.,* 31: 1-13.

Annual Energy Outlook 2006.EIA, U.S. Department of Energy.

Annual Energy Outlook 2008.EIA, U.S. Department of Energy.

Barbalace, R.C., Chernobyle nuclear disaster revistited. Environmental Chemistry. Com,1999, accessed on 22 December 2011; http://EnvironmentalChemistry .com/yogi/hazmat/articles/chernobyle.html.

Baruah, D.C. and G.C. Bora, 2008. Energy demand forecast for mechanized agriculture in rural India. *Energ. Policy*, 36: 2628-2636.

Central Electricity Authority, Government of India Ministry of Power,: *www.cea. nic.in* C-WET; http://www.cwet.tn.nic.in/html/information_gi.html.

Demirbas, A., 2003. Energy and environmental issues relating to greenhouse gas emissions in Turkey. *Energ. Convers. Manage.*, Vol. 44, pp. 203-213.

Fthenakis, V. and Kim, H.C., 2009. Land use and electricity generation: A life-cycle analysis. Renew.*Sustain. Energy Rev.*, Vol.13, pp.1465-1474.

GOI, (2001. Economic Survey 2000-2001. Ministry of Finance, Economic Division, New Delhi.

Ganesh Hegde and Ramachandra T.V., 2012. Scope for solar energy in Kerala and Karnataka. National Conference on Conservation and Management of Wetland Ecosystems, 06th- 09th November 2012.

Kumar, A., A .Garg, S. Kriplani and P. Sehrawat, 2008.7th International Conference and Exposition on Petroleum Geophysics, Hyderbad 2008.

"Load Generation Balance Report 2011-12". Central Electricity Authority, Government of India Ministry of Power. May 2011. Retrieved 2011-11-26.

Lund, John W., 2004. 100 years of Geothermal Power Production. *GHC Bull.* September 2004.

McKendry P., 2002a. Energy production from biomass (Part 3): gasification technology, *Bioresource Technology*, 83, 55-63.

[4] MNES. Annual Report: 1999-2000, Ministry of Non-conventional Energy Sources (MNES), Government of India, CGO Complex, Lodhi Road, New Delhi (India), 2000.

McKendry, P., 2001. Energy production from biomass(Part 1): Overview of biomass, *Bioresource Technology*. 83, 37-46.

Manzoor Alam, Jayant Sathaye and Doug Barnes. Urban household energy use in India: efficiency and policy implications.

Mitavachan, H and J. Srinivasan, 2012. Island really a constraint for the utilization of solar energy in India? *Current Science*, Vol.103 (2).

Pokharel, 2006, http://www.overseas- campus. info/ seminar _ program /2006 _ Asian_

Alumni_ Workshop/Asian_Alumni_Workshop_2006_Bali-Indonesia.

Raghavaiah, B. V. , Yugal Agrawal and K.S. Meera, 2011. Renewable Energy Resources and Their Utilization. *Search & Research*, Vol-2(2), pp.164-168.

Ramachanadra, T.V., 2000. Energy alternatives: Renewable energy and energy conservation technologies, Karnataka Environment Research Foundation, Banglore.

Ramachandra T.V., 2003. Ecologically Sound Integrated Regional Energy Planning, Nova Science Publishers, Huntington, NY 11743-6907.

Ramachandra, T.V., Rishabh Jain, and Gautham Krishnadas, 2011. Hotspots of solar potential in India, Renewable and Sustainable Energy Reviews. 15, pp. 3178-3186, doi:10.1016/j.rser.2011.04.007.

Research cell efficiency records. National Center for Photovoltaics (NCPV), National Renewable Energy Laboratory, USA: http://www.nrel.gov/ncpv/, accessed on 22 December 2011.

Rangachari, R., Sengupta, N.,Iyer, R. R., Banerji,P. and Singh, S., Large dams: India's experience. A WCD case study prepared as an input to the World Commission on Dam's, Cape Town, 2000, www.dams.org.

Razdan, P.N., R.K. Agarwal and Rajan Singh, 2008. Geothermal Energy Resources and its Potential in India. *E-Journal Earth Science India*, Vol.1(1), pp.30-42.

Singh, G., 1999. Relationship between mechanization and productivity in various parts of India. A paper presented during the 34th Annual Convention, Indian Society of Agricultural Engineers, CCSHAU, Hisar, December, 16-18.

Shao, H. and L. Chu, 2008. Resource evaluation of typical energy plants and possible functional zone planning in China. *Biomass Bioenerg.*, 32: 283-288.

Singh, J.M., 2002. On farm energy use pattern in different cropping systems in Haryana, India. M.Sc. Thesis, Management University of Flensburg, Germany.

Chapter 8

Microbial Degradation of Lignocelluloses

Avita K. Marihal, S.M. Pradeep and K.S. Jagadeesh

*Department of Agricultural Microbiology, University of Agricultural Sciences,
Dharwad – 580 005, Karnataka*

ABSTRACT

The problem of increasing lignocellulose wastes from plant production (agriculture and forestry) and industrial processes (pulp and paper), has been known for decades. To utilize this waste the intensity of research and the magnitude of capital investment in this field has increased tremendously. The most ambitious of these has been the conversion of lignocelluloses to alternative carriers such as fuel ethanol, acetone, butanol). But the complete and successful utilization of lignocelluloses by man remains a daunting task, whatever may be the intended applications. This review focuses on the current knowledge available on lignocelluloses, its biotechnological applications, genetic engineering biodegradation and composting.

Key words: Polymers, Ligninase, C/N ration, Compost

Introduction

In nature, lignocellulose containing biomass is the major source of renewable organic matter produced by plant photosynthesis. Lignocellulosic wastes are formed in plant production (agriculture and forestry) and industrial processes (pulp and paper). It accounts for about 60% of the total plant biomass produced on earth (Perez *et al.*, 2002). Lignocellulose is physically hard, dense and recalcitrant to degradation. However, it is an extremely rich and abundant source of carbon and chemical energy. Therefore, the recycling of carbon involving lignocelluloses is essential to maintain the global carbon cycle.

The problem of increasing the commercial utility of lignocellulose wastes has been known for decades. In addition to the growing demand for traditional applications (paper manufacture, biomass fuels, composting, animal feed, etc.), novel markets for lignocelluloses have been identified in recent years. The intensity of research and the magnitude of capital investment in this field increased vastly once commercial viability seemed probable for many of these new applications. The most ambitious of these has been the conversion of lignocellulose to alternative energy carriers (*e.g.* fuel ethanol, acetone and butanol) (Kaylen, *et al.*, 2000; Wheals *et al.*, 1999). The pulp and paper industry discovered that lignocellulose biotechnology could improve the process efficiency through savings in money and energy (Breen and Singleton, 1999; Scott *et al.*, 1998). Others aimed at improving digestibility of nutritionally poor forages by exposing these lignocellulosics to white-rot fungi (Agosin and Odier, 1985; Karunanandaa *et al.*, 1992).

The complete and successful utilization of lignocelluloses remains a daunting task, whatever may be the intended applications. Defeating the barriers, which prevent commercial exploitation of lignocelluloses, will be the key to its successful application in biotechnological endeavors. This review focuses on the current knowledge available on lignocellulose biodegradation and composting.

Lignocellulose Composition

Chemical Composition

Cellulose, hemicellulose, and lignin are the main constituents of lignocellulosic materials (Deobald and Crawford, 1997). Apart from these primary polymers, plants comprise other structural polymers (*e.g.* waxes, proteins). Cellulose is a linear polymer of glucose linked through α-1, 4-linkages and is usually arranged in microcrystalline structures, which is very difficult to dissolve or hydrolyse under natural conditions. The degree of polymerization (DP) of cellulose chains or chain length ranges from 500 to 25 000 units (Kuhad *et al.*, 1997; Leschine, 1995). Hemicellulose is a heteropolysaccharide composed of different hexoses, pentoses, and glucoronic acid. Hemicellulose is more soluble than cellulose and is frequently branched with DP of 100 to 200 (Kuhad *et al.*, 1997). Xylan is the most common hemicellulose component of grass and wood. Lignin is a highly irregular and insoluble polymer consisting of phenylpropanoid subunits, namely p-hydroxyphenyl (H-type), guaiacyl (G-type), and syringyl (S-type) units. Unlike cellulose or hemicellulose, no chains containing repeating subunits are present, thereby making the enzymatic hydrolysis of this polymer extremely difficult.

Ecology of Lignocellulose Biodegradation

Lignocellulose degradation is essentially a race between cellulose and lignin degradation (Reid, 1989). This contest is even more extensive and complex in nature (Rayner and Boddy, 1988). Decomposition curves for complex substrates incubated in soil, such as plant residues, usually yield a multislope decomposition curve (Paul and Clark, 1989; Van Veen *et al.*, 1984). Fungi with restricted metabolic capabilities (*e.g.* soft rots like *Mucor* spp.) develop mutualistic relationships with and thrive alongside fungi degrading cellulose and lignin. Microorganisms unable to overcome the lignin or physical barrier can obtain energy from the low molecular weight intermediates released from lignocellulose by the true white-rot fungi. Such complex associations have been observed under natural conditions (Blanchette *et al.*, 1978).

Bacteria and Actinomycetes

Lignocellulose biodegradation by prokaryotes is essentially a slow process characterized by the lack of powerful lignocellulose degrading enzymes, especially lignin peroxidases. Grasses are more susceptible to actinomycete attack than wood (Antai and Crawford, 1981; McCarthy, 1987). Together with bacteria, actinomycetes play a significant role in the humification processes associated with soils and composts (Trigo and Ball, 1994). The enzymatic ability to cleave alkyl-aryl ether bonds enable bacteria to degrade oligomeric and monomeric aromatic compounds released during fungal lignin degradation (Vicuna *et al.*, 1993; Vicuna, 2000; White *et al.*, 1996). Therefore, lignocellulose biodegradation by prokaryotes is of ecological significance, but lignin biodegradation by fungi, especially white-rot fungi, is of commercial importance.

Fungi

Most fungi are capable cellulose degraders. However, their ability to facilitate rapid lignocellulose degradation attracted attention from scientists and entrepreneurs alike. White-rot fungi comprise powerful lignin degrading enzymes that enable them in nature to bridge the lignin barrier and, hence, overcome the rate-limiting step in the carbon cycle (Elder and Kelly, 1994). Of these, *Phanerochaete chrysosporium* is the best studied. New information regarding the identities of the cellulose, hemicellulose or lignin degrading enzymes, their unique catalytic capabilities, the physiological conditions required for optimum secretion or activity etc. is constantly being added to an already impressive volume of work and varies between fungi and bacterial genera, species and even strains. Anaerobic fungi (*Piromyces* spp., *Neocalli-mastix* spp. and *Orpinomyces* spp.) form a part of the rumen microflora. These fungi produce active polymer degrading enzymes, including cellulases and xylanases (Hodrova *et al.*, 1998). Their cellulases are among the most active reported to date and able to solubilise both amorphous and crystalline cellulose (Wubah *et al.*, 1993). These fungi can be used in situations where process principles and design necessitate anaerobic conditions. In such a scenario, ruminant manure will serve as inoculum and this waste product will meet a crucial requirement in biotechnology – cost effectivity versus optimum utility.

Lignocellulose Biodegradation

Lignocellulose is a complex substrate and its biodegradation is not dependent on environmental conditions alone, but also the degradative capacity of the microbial population (Waldrop *et al.*, 2000).

Cellulose and Hemicellulose

The efficient hydrolysis of cellulose requires the concerted action of at least three enzymes: (1) endo-glucanases to randomly cleave intermonomer bonds; (2) exoglucanases to remove mono- and dimers from the end of the glucose chain; and (3) β-glucosidase to hydrolyze glucose dimers (Deobald and Crawford, 1997; Tomme *et al.*, 1995). The concerted actions of these enzymes are required for complete hydrolysis and utilization of cellulose. The rate-limiting step is the ability of endo-glucanases to reach amorphous regions within the crystalline matrix and create new chain ends, which exo-cellobiohydrolases can attack. Although similar types of enzymes are required for hemicellulose hydrolysis, more enzymes are required for its complete degradation because of its greater complexity compared to cellulose. Of these, xylanase is the best studied (reviewed by Kuhad *et al.*, 1997). However, a fundamental difference exists in the mechanism of cellulose hydrolysis between aerobic and anaerobic fungi and bacteria (Leschine, 1995; Tomme *et al.*, 1995).

Aerobic fungi and bacteria characteristically comprise non-complexed cellulase systems, which entail the secretion of the cellulose hydrolysis enzymes into the culture medium. However, anaerobic bacteria (especially Clostridium spp.) and fungi (of the genera Neocallimastix, Piromonas and Sphaeromonas) contain complexed cellulase systems where the cellulose hydrolyzing enzymes are contained in membrane-bound enzyme complexes (called cellulosomes). This fundamental difference has profound implications for biotechnology applications. Biotechnological applications based on anaerobic fungi and bacteria might have an advantage over aerobic systems in terms of hydrolysis efficiency. Complexed cellulase systems allow greater coordination between the different cellulose hydrolyzing enzymes. Their close association will restrict loss of degradation intermediates due to dynamic environmental conditions. In aerobic systems, where active aeration and agitation is required, loss of the secreted enzymes and their degradation intermediates might prove detrimental to overall process efficiency. This apparent contradiction might be offset when the energetics of aerobic and anaerobic microorganisms is compared. In general, aerobic microorganisms gain far more energy from glucose than anaerobic micro-organisms (38 mole ATP vs. 2–4 mole ATP per mole of glucose). Therefore, the apparently "aggressive" cellulose hydrolyzing strategy utilized by aerobes might be beneficial given the potential enormous gain in metabolic energy. However, given the low technical requirements of anaerobic applications as compared to aerobic systems (absence of vigorous agitation to facilitate aeration and flow control technologies), the use of anaerobic fungi and bacteria for use in low cost bioremediation projects might be more attractive given their highly efficient cellulose hydrolysis machinery.

Lignin

Lignin degradation by white-rot fungi is an oxidative process and phenol oxidases are the key enzymes (Kuhad *et al.*, 1997; Leonowicz *et al.*, 1999). Of these, lignin peroxidases (LiP), manganese peroxidases (MnP) and laccases from white rot fungi (*P. chrysosporium, Pleurotus ostreatus* and *Trametes versicolor*) have been best studied. LiP and MnP oxidize the substrate by two consecutive one-electron oxidation steps with intermediate cation radical formation.

Laccase has broad substrate specificity and oxidises phenols and lignin substructures with the formation of oxygen radicals. Other enzymes that participate in the lignin degradation processes are H_2O_2-producing enzymes and oxido-reductases, which can be located either intra- or extracellularly.

Jagadeesh and Geeta (1994) observed maximum (62%) degradation of lignin in pretreated arecanut husk with white rot fungus, *Phanerochaete chrysosporium* under un-amended conditions. Beg *et al.*, (1986) reported that Pleurotus sojor-caju could reduce rice husk lignin to the extent of 40.9 percent. Lara *et al.*, (1989) reported a genetically-improved strain of *Pleurotus ostreatus*, which could degrade 46 percent lignin and 30 per cent hexoses as compared to the parental strains that degraded 44 per cent of both cellulose and lignin after 50 days incubation. Duraw *et al.*, (1987) found 2 to 3 times higher lignolytic activity in Chrysonila sitophila than *Phanerochaete chrysosporium*. Pometto and Crawford, (1985) used Streptomyces viridosporus and *Phanerochaete chrysosporium* for the production of APPL (Acid Precipitable polymeric Lignin), which is an intermediary in the lignin degradation. They also demonstrated that APPL was slowly degraded by Streptomyces viridosporus T-7A and *Phanerochaete chrysosporium*.

Kern, (1989) reported that addition of solid manganese (IV) oxide to cultures of *Phanerochaete chrysosporium* at the beginning of the lignolytic activity was shown to improve production, enzymatic activity and stability of the ligninase production. Anbu *et al.*, (2000) reported that the fungi Geotrichum candidum and P. glomerata were efficient in cellulose degradation. They observed *G. candidum* to be a potential fungus for the degradation of paper, while *P. glomerata* for degradation of cloth. Therefore, hemicellulose degradation is required before efficient lignin removal can commence.

Composting

Composting is the biological conversion of solid organic waste into usable end products such as fertilizers, substrates for mushroom production and biogas. Although composts are highly variable in their bulk composition, composting material is generally based on lignocellulose compounds derived from agricultural, forestry, fruit and vegetable processing, household and municipal wastes. With the course of time, various groups of microorganisms such as bacteria, fungi, actinomycetes become established performing a specific role in the process. Contents of macro and micronutrients are found to increase during the process of composting due to volume reduction and due to loss of organic carbon as CO_2 (Gaur, 1982).

Factors Affecting Composting Process

The process of composting is influenced by many biotic and environmental factors. These factors are essential for efficient decomposition and to obtain quality product (Poincelot, 1974).

Temperature

Temperature is an important factor, which plays an important role in composting of organic wastes. Bhoyar *et al.,* (1979) conducted a study concerning the effect of temperature on mineralization of nitrogen during the aerobic composting of cotton dust and municipal waste. They observed that maximum ammonification occurred in the range of 30°C to 50°C, and at 60°C to 70°C, the rate of nitrification was low.

Temperature monitoring in the composting mass may indicate the amount of biochemical activity taking place. A drop in temperature could mean that the material needs to be aerated or moistened or that decomposition is in last stage. The amount of heat involved varies with the type and quantity of substrate used (Plat *et al.,* 1984). The rate of carbon mineralization of saw dust increased with temperature from 25°C to 40°C with the optimum between 30 to 35°C (Olayinka and Adebaya, 1984). Maximum wood decay by majority of test fungi was recorded in the temperature ranging from 25°C to 40°C (Kirk *et al.,* 1978; Hegarty and Curran, 1985; Kahlonn and Dass, 1987).

Wastewater sludge and municipal solid waste contained huge population of pathogenic microorganisms like bacteria, fungi, viruses and parasites (Bertoldi *et al.,* 1988). In order to kill all the pathogens and weed seeds during the composting process, the temperature must be maintained between 60˚C and 70˚C for 24 hours (Pandey, 1977).

Gaur (1996) observed a great deal of exothermic energy being released during the process, generating substantial amounts of heat, thus raising the temperature in heaps and creating favorable conditions for thermophilic microorganisms. However, if the temperature exceeds 65˚C to 75˚C, the microbial activity is ceased due to thermal killing of organisms; the optimum temperature being 55-60°C for 4 to 5 days.

Gowda (1996) reported that during decomposition of wastes, the microbes consume more O_2 to break down the organic compound and release heat energy through respiration process which caused the temperature to rise in decomposition of organic matter. At high temperatures (40 - 70°C), all thermophilic microbes which are more efficient in degradation, multiply and decompose the materials faster.

Carbon-to-Nitrogen ratio

The C/N ratio of the substrate is an important factor in decomposition. The materials with wide C/N ratio are degraded slowly and narrowed C/N ratio is degraded easily. For decomposition, the C/N ratio of the organic substances should be balanced. The C/N ratio is an important factor to be considered when mixing different kinds of material for composting. Wider C/N ratios of more than 40:1 promote the immobilization of available nitrogen in the compost (Gaur, 1982 and Zibiliske, 1998). A C/N ratio of 30:1 to 40:1 was reported to be desirable. Soil at the rate of five to ten percent can be added to reduce the moisture content and absorb ammonia while composting the materials of low C/N ratio. Soil may also be added to high C/N ratio organic material to buffer acid conditions.

Microorganisms require carbon for growth and nitrogen for protein synthesis. On an average, they utilize 30 parts of carbon per one part of nitrogen. So, C/N ratio of 30 is desirable for composting. If it is above 35, the process becomes inefficient and the compost requires more time for completion. If it is below 26, the excess nitrogen is converted to ammonia and wasted into atmosphere (Poincelot, 1974; Hanna, 1975).

Increasing the C/N ratio by the addition of easily degradable carbon sources to nitrogen rich material is an effective solution for nitrogen losses, as ammonia formed will become rapidly immobilized into new microbial biomass (Fassen and Dijk, 1979; Kirchman, 1985). An associative effect of cellulolytic and lignolytic microorganisms were found to be beneficial in recycling of higher C/N ratio and bulky crop residues (Rasal *et al.*, 1988).

Rajasekaran and Sampath Kumar (1981) and Bishop and Godfrey (1983) opined that with proper adjustment of C/N ratio, agricultural wastes could be profitably recycled with a greater degree of N enrichment. Low C/N ratio resulted in rapid composting accompanied by an increased loss of excess nitrogen from ammonia volatilization.

If there was limitation of nitrogen in organic matter (wider C/N), the degradation was slow because the microbes could not multiply in large numbers for want of nitrogen for synthesis of their cellular protein and needed enzymes for breaking down bigger molecules as in coir pith, lignocellulosic waste, wood, saw dust, dried (fallen) leaves etc. In case of limited nitrogen in organic matter (wider C/N), degradation will be slow as the nitrogen is the building material of cellular proteins and enzymes needed for breaking down the bigger molecules. By narrowing the C/N by mixing suitable nitrogen rich materials from external sources, the rate of decomposition of these materials could be enhanced, on the other hand, if the material was very rich in nitrogen (narrow C/N), the nitrogen content might be lost as gaseous ammonia for want of carbon for immobilization and growth, as in the case with urine. In the above situation, carbon rich materials like cotton waste saw dust, soil, ash etc., can be mixed to conserve nitrogen.

Jagadeesh *et al.*, (1996) found *Phanerochaete chrysosporium*um as an efficient inoculant to degrade red gram straw. It reduced the C/N ratio to favorable extent that could be used either as feed stock for biogas production or as organic manure.

For degradation of wider C/N ratio substrates, it is better to mix nitrogen rich materials like cakes, green leaves etc., to bring down the C/N ratio to at least 50 : 1 (Gowda, 1996). Maleena, (1998) recommended the use of sawdust and other wood wastes in the composting of piggery wastes. Piggery wastes have high nitrogen content and when mixed along with high C/N ratio substrates would avoid loss of nitrogen and also help in composting of high C/N ratio materials.

pH

The compost pH is a good indicator of the development of composting. During the first few days, it descends slowly to values of about 5, and later rises as the material gradually decomposes and stabilizes and finally stays between 7 and 8.

Acidic pH of the compost indicates a lack of maturity due to short composting time or the occurrence of anaerobic process in the heap. The pH should be prevented from rising above 8.5 to minimize the gaseous loss of nitrogen in the form of ammonia (Pandey, 1977). Most of the materials decomposing aerobically will stay with a pH range that is conducive for microbial growth.

Hegarty and Curran, (1985) showed that different fungi used in their study produced greatest weight loss of wood in 5 to 8 pH range. Inbar *et al.*, (1988) found that organic matter with a wide range of pH (3.0 to 11.0) can be composted. The optimum pH levels were between 6 and 8 for composting and between 4 and 7 for end products.

Lignin degradation was optimal at neutral to slightly acidic pH. Whereas lignocellulose degradation with lignin solubilization and acid precipitable polymeric lignin production was promoted at alkaline pH with Streptomyces viridosporus (Pometto and Crawford, 1985), optimum pH for Streptomyces badius and S. viridosporus was found to be 7.5 to 8.5 (Giroux *et al.*,1988).

Moisture and Aeration

For proper and efficient decomposition, these two factors have prominent role to play during composting. Composting is an aerobic process. For multiplication of microbes, sufficient moisture should be provided. However, if the water content is too high, anaerobic conditions set in. Anaerobic process slows down the decomposition of organic matter and produces foul smell.

Poincelot (1974) found that below 40 per cent of moisture, organic matter did not decompose rapidly. Above 60 per cent, it caused anaerobic condition and resulted in foul smell. The optimum moisture content was between 57 to 60 percent. The effect of moisture content and addition of nitrogen on the rate of composting were studied by Inbar *et al.* (1988). They found that moisture content had major effect on oxygen consumption.

Rao *et al.*, (1995) reported that a moisture content of 70 per cent favoured more effective composting as indicated by higher level of mineralization of poplar wood to carbon dioxide.

Gaur (1996) reported that a moisture content of 50-60 % was favorable for composting and this moisture per cent enhanced the bio-degradation of the substrate. The periodic turning was good for enhancing the composting process, since composting was mainly an aerobic process requiring oxygen for oxidation of organic carbon in substrates.

During the process of degradation of carbon rich substrates, about two third of carbon was respired (evolved as carbon dioxide, causing loss of weight of substrate) and remaining one third was combined with nitrogen in living cells (Gaur, 1996). There were no nuisance problems such as bad odour during aerobic composting.

Gowda (1996) reported that in the aerobic decomposition, the organic matter has undergone complete degradation, mineralization and formed non toxic compounds namely carbon dioxide, water, heat (energy), and water soluble mineral forms of nutrients and humus (ligno proteins). Whereas in anaerobic decomposition, due to insufficient oxygen supply, only few anaerobic bacteria decomposed and formed intermediary organic compounds, some of these foul smelled and they were phyto toxic, as well.

Changes in Microbial Population During Composting

Composting occurs due to activities of microorganisms in succession. Bertoldi *et al.,* (1988) reported that bacteria and fungi might reproduce and gain in number with the passage of time and, in turn, could provide conditions during composting process. Pathogenic microorganisms only were reduced and were also unable to grow and multiply in mature compost.

Population of phosphate solubilizer in the enriched compost prepared with the combination of soybean straw and 5 per cent P_2O_5, and other inputs, 1 per cent urea, decomposing culture and 10 per cent pyrite was 158×10^7 cell per g compared to control which showed 71×10^7 cell per g (Bhanawase *et al.,* 1994).

Enrichment of compost by nitrogen rich substrates, mineral additives and microbial inoculation

To improve the quality of compost, it is often inoculated with various microorganisms, enriched with chemical fertilizers and rock phosphate. Supplementing the compost with rock phosphate increased the release of citric acid and water-soluble phosphorus (Mathur *et al.,* 1980; Bangar *et al.,* 1985).

Contents of macro and micronutrients are found to increase during the process of composting due to volume reduction and due to loss of organic carbon as CO_2 (Gaur, 1982). Composting process slows down when the substrates are not easily degradable and also having wider C/N ratio. A raise in temperature of above 40°C could hardly be expected in high C/N ratio materials unless they are mixed with nitrogen rich substrates (Gaur, 1982). Therefore, enrichment with mineral nutrients and microbial cultures can be done to hasten the process (Gaur, 1987).

Kapoor *et al.,* (1983) and Gaur (1979) showed that inoculation with Azotobacter and PSM (*Aspergillus awamori* or *Bacillus polymyxa*) improved the manurial value of compost. Addition of powdered rock phosphate alone resulted in an increase of 28.3 per cent in N content over control. Inoculation of *Aspergillus* awamori amended with rock phosphate showed an increase of 32 percent in N content over the control resulting in gain of 22 kg N per 5 ton of finished product.

Inoculation of four mesophilic cellulolytic fungi, *Aspergillus* sp, *A. niger*, *Trichoderma viridae* and *Penicillium* sp. on composting of jowar stalk: wheat straw (5:3) and jamun leaves, the period of composting was reduced by one month and total nitrogen, available phosphorus and humus content were increased. Inoculation with *A. niger* and *Penicillium* sp. showed the maximum effect (Gaur and Singh, 1982). Inoculation of wheat straw with cellulolytic fungus *Penicillium corylophilum* and N_2 fixing anaerobes *Clostridium butyricum* and moistened with mineral salt solution increased the decomposition rate. N_2 fixation during the utilization of the straw resulted in a gain of 11.5 mg N (g straw lost)$^{-1}$ over a period of 8 weeks at 25°C (Lynch and Harper, 1983).

Gaur and Singh (1982) found that application of rock phosphate at the rate of 5 kg P_2O_5 per ha along with Azotobacter and phosphate dissolving fungi improved the quality of compost. The compost obtained after 3 months of enrichment contained 1.26 percent N and had a C:N ratio of 9 compared to 0.63 and 26.5 respectively in case of initial compost whereas untreated compost maintained under same conditions contained 0.77 per cent N.

Mishra *et al.,* (1982) prepared phosphate enriched compost containing 3.13 percent P by composting cattle dung and farm waste with Mussoorie rock phosphate. The phosphate rich phospho-compost had low water soluble and bicarbonate soluble P but had 50 percent of total P in citrate acid soluble form.

Decomposition of sugarcane trash in open heap method was markedly hastened due to inoculation of culture combinations of Aspergillus sp. Penicillium sp. and Trichurus spiralis along with super phosphate, as compost was ready in 3 months, compared to control which was not ready even after a lap of 5 months. Phosphate solublizing fungi also increased the availability of P, when sugarcane trash (5 to 10 mm size) was amended with rock phosphate (Rasal *et al.,* 1988).

Treatment of Azotobacter alone or with rock phosphate at the rate of 5 kg P_2O_5 per tonne plus Azotobacter increased about 27 percent in nitrogen content of mechanized compost. The inoculation of *A. chroococcum* narrowed the C:N ratio significantly over the uninoculated treatment (Talashilkar, 1987).

Hajra (1988) reported that addition of rock phosphate encouraged cellulolytic activity and in conservation of nitrogen. It was applied at the rate, varying from 0.5 to 5 per cent P_2O_5 basis, depending upon the level of phosphate enrichment. Phosphate solubilizing fungal culture of *Aspergillus* awamori was also applied at the same rate to facilitate dissolution of applied rock phosphate. An increase in citrate soluble P by 64 % was obtained when paddy straw was composted with *Paecilomyces fusisporus* amended with 1 percent rock phosphate.

The effect of multiple inoculation with cellulolytic fungus *Paecelomyces* fusisporus, phosphate dissolving fungus (*Aspergillus awamori*) and *Azotobacter chroococcum* and addition of rock phosphate during composting was studied by Kapoor *et al.,* (1990) and found to increase rate of decomposition, N content and soluble P content.

A study was carried out to prepare nitrogen and phosphorous enriched compost by incorporating Mussoorie rock phosphate, pyrite and urea N at the rate of 10, 10 and 1 per cent of dry weight of compostable material respectively and found that the compost contained 2.0 per cent total N and 1.29 per cent total P after 90 days of decomposition (Singh *et al.,* 1992).

The maturation of compost without any additional nitrogen source is not rapid, proving the usefulness of such addition (Rodriguez *et al.,* 1995). Addition of 0.5 per cent of nitrogen as urea was found to enhance bioconversion of solid wastes (Son, 1995) and addition of bovine blood waste as N source increased the quality of the compost (Ribeiro, 1994). Gaur (1996) reported that the composting could be controlled by activators like cellulolytic fungi, phosphate solublizers and nitrogen fixing bacteria (Azotobacter) which hasten the process of composting.

Sharath and Jagadeesh (2003) showed that at the end of 90 days the fungal consortium converted wood waste into stable biocompost and further enrichment of the biocompost was accomplished by inoculating with Azotobacter and Pseudomomas.

Gowda (1996) opined that if the organic wastes used for composting are poor in nitrogen, they can be enriched by incorporating nitrogen rich organic wastes or 1 % nitrogen as urea. Above 1 %, nitrogen enrichment is not desirable because it will lead to loss of nitrogen through production and volatilization in form of gaseous ammonia due to rapid ammonification and high temperature in the composting material.

Gowda (1996) reported that compost could be profitably enriched with phosphorous sources to utilize low grade and cheaper sources of phosphorous like rock phosphates to improve efficiency of crops to absorb phosphor from soil. The phosphorous enriched compost could be prepared from any organic wastes used for composting. While preparing compost, rock phosphate equivalent to 5 % P_2O_5 (ex: 250 kg of mussorie rock phosphate per ton of organic wastes) could be mixed with 10 percent cow dung slurry, 5 % soil and 5 % well decomposed manure to serve as inoculum for decomposition. Addition of rock phosphate at 5 percent (W/W) hastened the composting process of poultry droppings (Kiruba, 1996).

Thakre and Fulzele (1998) reported the production of enriched compost from agricultural wastes and leafy biomass when composted in heap or pits by employing bio-degradation microbes, mixed with 10 % soil, 10 % farm yard manure. This quick degradation was observed when appropriate turning and mixing of the substrates along with proper moisture control was done. They also observed that the enriched compost was resultant of using additives like 1 % urea, 1 % super phosphate and 1 % murate of potash on three installments.

Biotechnological Applications of Lignocelluloses

The primary objective of lignocellulose pretreatment by the various industries is to access the potential of the cellulose and hemicellulose encrusted by lignin within the lignocellulose matrix. The combination of solid-state fermentation (SSF) technology with the ability of white-rot fungi to selectively degrade lignin has made possible industrial-scale implementation of lignocellulose-based biotechnologies. The advantages and disadvantages of SSF have received attention by Mudgett (1986).

Biopulping

Lignin becomes problematic to cellulose-based wood processing, because it must be separated from cellulose at enormous energy, chemical and environmental expense. Biopulping is, therefore, a solid-state fermentation process during which wood chips are treated with white-rot fungi to improve the delignification process. Biological pulping has the potential to reduce energy costs and environmental impact relative to traditional pulping operations (Breen and Singleton, 1999). The benefits of biopulping was demonstrated by Scott *et al.*, (1998) using 40-ton scale experiments: tensile, tear and burst indexes of the resulting paper were improved (indicative of higher degree of cellulose conservation during pulping process); brightness of the pulp was increased (indicating improved lignin removal); and improved energy savings of 30–38%.

Animal Feed

Cellulose is the most important source of carbon and energy in a ruminant's diet, although the animal itself does not produce cellulose-hydrolyzing enzymes (Czerkowski, 1986). Rumen microorganisms utilize cellulose and other plant carbohydrates as their source of carbon and energy. They convert the carbohydrates in to large amounts of acetic, propionic and butyric acids, which the higher animals can use as their energy and carbon sources (Colberg, 1988). The concept of preferential delignification of lignocellulose materials by white-rot fungi has been applied to increase the nutritional value of forages (Agosin *et al.*, 1985; Akin et al. 1995; Chen *et al.*, 1995; Zadrazil and Isikhuemhen, 1997). This increased digestibility provides organic carbon that can be fermented to organic acids in an anaerobic environment, such as the rumen.

However, upgrading of animal feed by white-rot fungi failed to reach industrial proportions. A possible explanation can be that the animals' instincts prevent them from ingesting mushrooms, for they can contain toxicants or they can be toxic to their rumen micro flora and, hence, toxic to the animal also.

Applications of Genetic Engineering

The scope of lignocellulose-based applications is expanding rapidly towards applications of genetic engineering. Recently, repression of lignin biosynthesis was achieved in *Populus tremuloides* resulting in cellulose accumulation and healthy growth of such transgenic trees (Hu *et al.*, 1999). Cellulose and lignocellulose fibers can be chemically modified to render it useful to miscellaneous applications in the textile industry (Ghosh and Gangopadhyay, 2000). Currently, metabolic engineering is being applied to facilitate simultaneous fermentation of hexoses and pentoses to ethanol (Aristidou and Penttila, 2000). The future might see the application of genetically engineered microorganisms (containing lignocellulases) in biotechnological applications where lignocellulosic wastes serve as the on-site carbon and nutrient source. Commercial byproduct of lignocellulose conversion to fuel ethanol has found application as absorbents of organic pollutants and as enterosorbents (Dizhbite *et al.*, 1999). Therefore, commercializing lignin waste production can offset process costs. Lignocellulose is also of potential medical value. Apart from being essential in the human diet as fiber, lignocellulose can be a source of compounds with biological activity. Such compounds have potential as stimulators of the human immune system and as antiviral agents (Kiyohara *et al.*, 2000; Sakagami *et al.*, 1999).

Potential of Lignocellulose in Space Exploration

Advances in lignocellulose research will enable scientists to contribute to space science/exploration. Space travel will benefit from this research, in the near future as the transport of lignocellulose to space can result in substantial cost savings.

Lignocelluloses can be a feedstock to provide for all basic needs: fuel, energy, feedstock chemicals, food, and water. Recycling of inedible plant material by white-rot fungi (Pleurotus ostreatus) has been investigated in a Closed Ecological Life Support System (CELSS) (Sarikaya and Ladish, 1997). Incineration technology have been proposed as another way of recycling the elemental resources found in spent lignocelluloses to support agriculture in a CELSS (Wignarajah *et al.*, 2000). Lignocellulose can therefore be the "super fuel" of the future – being a compact natural polymer containing enough potential energy to sustain man and machine in space.

References

Agosin E & Odier E 1985. Solid-state fermentation, lignin degradation and resulting digestibility of wheat straw fermented by selected white rot fungi, *Appl. Microbiol. Biotechnol.*, **21**: 397– 403.

Akin, D.E, Rigsby, L,L., Sethuraman, A., Morrison III, W.H., Gamble, G.R & Eriksson, K.E.L., 1995. Alterations in structure, chemistry, and biodegradability of grass lignocellulose treated with the white rot fungi *Ceriporiopsis subvermispora* and *Cyathus stercoreus*. *Appl. Environ. Microbiol.* **61**: 1591–1598.

Anbu, P., Hilda, A. and Kavitha, N.S., 2000. Studies on the biodegradation of paper and cloth by a few fungi. Proceedings of 41st Annual Conference of Association of Microbiologists of India, Jaipur, p.169.

Antai, S.P & Crawford, D.L., 1981. Degradation of softwood, hardwood, and grass lignocelluloses by two *Streptomyces* strains. *Appl. Environ. Microbiol.*, **42**: 378–380.

Aristidou, A & Penttila, M., 2000 Metabolic engineering applications to renewable resource utilization. *Curr. Opin. Biotechnol.* **11**: 187–198.

Bangar, K.C., Yadav, K.S. and Mishra, M.M., 1985. transformation of rock phosphate during composting and the effect of humic acid. *Plant Soil*, **85**: 259-266.

Beg, S., Zafar, S.I. and Shah, F.H., 1986, Rice Husk biodegradation by Pleurotus ostreatus to produce a ruminant feed. *Agric. Wastes*, **17**: 15-21.

Bertoldi, D.M., Zucconi, F. and Cirilini, M., 1988. Temperature, pathogen control and product quality. *Biocycle*, **29**: 43-50.

Bhanawase, D.B., Rasal, P.H., Jadav, B.R. and Patil, P.L., 1994. Mineralisation of nutrients during production of phosphocompost. *J.Indian Soc.Soil Sci.*, **42**:145-147.

Bhoyar, R., Olaniya, M. and Bhide, A., 1979.Effect of temperature on mineralization of Nitrogen during anaerobic composting. Ind. *J. Environ. Health.*, **21**: 23-24.

Bishop, P.L. and Godfrey, C., 1983. Nitrogen composting during sludge composting. *Biocycle*, **24**:34-39.

Blanchette, R.A., Shaw, C.G & Cohen, A.L., 1978. A SEM study of the effects of bacteria and yeasts on wood decay by brown- and white-rot fungi. . *Micros.*, 2: 61–68.

Breen A & Singleton FL 1999. Fungi in lignocellulose breakdown and biopulping. *Curr. Opin. Biotechnol,.* **10**: 252–258.

Breen, A & Singleton, F.L., 1999. Fungi in lignocellulose breakdown and biopulping. *Curr. Opin. Biotechnol.* **10**: 252–258.

Chen, J., Fales, S.L, Varga, G.A & Royse, D.J. 1995. Biodegradation of cell wall components of maize stover colonized by white-rot fungi and resulting impact on in-vitro digestibility. *J. Sci. Food Agric.* **68**: 91–98.

Colberg, P.J., 1988. Anaerobic microbial degradation of cellulose, lignin, oligolignols, and monoaromatic lignin derivatives. In: Zehnder AJB (Ed). Biology of Anaerobic Microorganisms (pp. 333–372). John Wiley & Sons, New York USA.

Czerkowsk,i J.W.,1986 .An Introduction to Rumen Studies. Pergamon Press, Oxford, UK, pp. 9–10.

Deobald, L.A & Crawford, D.L .,1997. Lignocellulose biodegradation. In: Hurst CJ, Knudsen GR, Stetzenbach LD & Walter MV (Eds) Manual of Environmental Microbiology (pp. 730–737). ASM Press, Washington DC, USA.

Deobald, L.A & Crawford, D.L 1997. Lignocellulose biodegradation. In: Hurst CJ, Knudsen GR, Stetzenbach LD & Walter MV (Eds) Manual of Environmental Microbiology (pp. 730–737). ASM Press, Washington DC, USA.

Dizhbite, T., Zakis, G., Kizimia, A., Lazareva, E., Rossinskaya, G., Jurk-jane, V., Telysheva, G & Viesturs, U., 1999 Lignin – a useful bio-resource for the production of sorption-active materials. *Biores. Technol.* **67**: 221–228.

Duraw, N., Rodriguz, J., Ferraz, A. and Campos,V., (1987). Chrysonila sitophila (TFB- 2744) a hyperlignolytic strain. *Biotechnol.* Letters, **9**:357-360.

Elder, D.J & Kelly, D.J 1994. The bacterial degradation of benzoic acid and benzenoid compounds under anaerobic conditions: Unifying trends and new perspectives. *FEMS Microbiol.* Rev. **13**: 441–468.

Fassen, V.H.G. and Dijk, V.H., 1979.Nitrogen conservation during the composting systems. In straw decay and its effect on disposal and utilization. Gross bard E (ed.) John Wiley and Sons, New York, 113-120.

Gaur, A.C., 1979 .Organic recycling prospects in Indian agriculture. *Fertilizer News,* **24**: 49-61.
   Gaur, A.C., 1982. A manual of rural composting, FAO-UNDP Regional project RAS/75/004. Field document No. 15. FAO, UN, Rome, Italy, p.102.

Gaur, A.C., 1996. Biomass potential and composting. Proceedings of National Seminar on Organic Farming and Sustainble Agriculture, University of Agricultural Sciences, Bangalore, pp.27-32.

Gaur, A.C., and Singh, R., 1982. Integrated nutrient supply system. *Fertilizer News,* **27**: 87-98.

Gaur,A.C., 1987. Recycling of organic wastes by improved techniques of composting and other methods. *Res. Conserv.,* **13**: 157-174.

Ghosh, P & Gangopadyay, R., 2000 Photofunctionalization of cellulose and lignocellulose fibers using photoactive organic acids. *Eur. Polymer J.* **36**: 625–634.

Giroux, H., Vialal, P., Bouchard, V. and Lamy, F.,1988. Degradation of kraft indulin lignin by *Streptomyces viridosporus* and *Streptomyces* badis. Appl. Environ. Microbiol., **54**: 3064-3070.

Gowda, T.K.S., 1996. Compost making and enrichment techniques. Proceedings of Natinal Seminar on Organic Farming and Sustainble Agriculture, University of Agricultural Sciences, Bangalore, pp.46-51.

Hajra, J.N., 1988.Potentials and problems. Biofertilizer, pp.249-254.

Hanna, P., 1975.Model for decomposition of organic materials by microorganisms. *Soil Biol. Biochem.*, **7**:161-169.

Hegarty, B.N. and Curran, P.M.T., 1985. The Biodegradation of beech by marine and non-marine fungi in response to temperature, pH, light and dark. *Inter. Biodeg.*, **112**: 11-18.

Hodrova, B., Kopecny, J. & Kas, J, 1998. Cellulolytic enzymes of rumen anaerobic fungi *Orpinomyces joyonii* and *Caecomyces* communis. Res. Microbiol. **149**: 417–427.

Hu ,W.J, Harding, S.A, Lung, J., Popko, J.L., Ralph, J., Stokke ,D.D., Tsai, C.J & Chiang, V.L.,1999. Repression of lignin biosynthesis promotes cellulose accumulation and growth in transgenic trees. *Nature Biotechnol.* **17**: 808–812

Inbar, Y., Chen, Y. and Hardar, Y., 1988. composting of agricultural wastes for their use as a container media stimulation of composting process. *Biol. Wastes,* **26**: 247-259.

Jagadeesh, K.S. and Geeta, G.S., 1994. Extracellular enzyme production by white rot fungi on different agro wastes, Abstracts. Proccedings of National Conference on Fungal Biotechnology, Barkatullah University, Bhopal, p.19.

Kahlonn, S.S and Dass, S.K. 1987. Biological Conversion of paddy straw into fuel. Biological Wastes, **22**:11-21

Kapoor, K.K., Kukreja, K., Bangar, K.C. and Mishra, M.M., 1990. Enrichment of compost by use of microorganisms and the effect on crop yield. Haryana Agrci.Univ. *J. Res.*, **20**: 105-110.

Kapoor, K.K., Yadav, K.S., Singh, D.P., Mishra, M.M. and Tauro, P., 1983. Enrichment of compost by Aztobacter and phosphate solubilizing microorganisms. *Agric. Wastes,* **5**: 125-133.

Karunanandaa, K., Fales, S.L, Varga, G.A & Royse, D.J 1992. Chemical composition and biodegradability of crop residues colonized by white-rot fungi. *J. Sci. Food. Agric.*, **60**: 105–112.

Kaylen M, Van Dyne DL, Choi YS & Blasé M 2000. Economic feasibility of producing ethanol from lignocellulosic feedstocks. *Biores. Technol.*, **72**: 19–32.

Kern, H.W., 1989. Improvement in the production of extracellular lignin peroxidases by Phaneurochaete chrysosporium. Effect of solid manganese (IV) oxide. *App.Microbiol. Biotechnol.*, **32**: 223-234.

Kirchman, H., 1985. Losses, plant uptake and utilization of manure nitrogen during a production cycle. *Acta Agric. Scand. Suplemen.*, p.24.

Kirk, T.K., Schultz, E., Connors, W.J., Lorenz, L.F. and Zeikus, J.G., 1978. Influence of culture parameters on lignin metabolism by *Phanerochaete chrysosporium*. *Archives of Microbiol.*, **117**: 277-285.

Kiruba, C., 1996. Recycling of poultry droppings as manure. M.Sc. (Environmental Sciences) Dissertation, Tamil Nadu Agricultural University, Coimbatore.

Kiyohara, H., Matsumoto, T & Yamada, H .,2000. Lignin-carbohydrate complexes: Intestinal immune system modulating ingredients in kampo (Japanese herbal) medicine, juzen-taiho-to. *Planta Med.* **66**: 20–24

Kuhad, R.C, Singh, A & Eriksson, K.E.L 1997. Microorganisms and enzymes involved in the degradation of plant fiber cell walls. *Adv.Biochem. Eng. Biotechnol.*, **57**: 45–125.

Lara, L.H., Ramirez- Corrillo, R., Eger- Hummeu, G. and Pudillai Hustund, R.J., (1989). Production of *Pleurotus ostreatus* strains for selective degradation of lignin. Abstracts of Fifth International Symposium on Microbial Ecology, held at Kyoto, Japan, pp.5-9.

Leonowicz, A., Matuszewska, A., Luterek, J., Ziegenhagen, D., Wojtas-Wasilewska ,M., Cho, N.S., Hofrichter, M & Rogalski ,J., 1999. Biodegradation of lignin by white rot fungi. *Fungal Genet. Biol.* **27**: 175–185.

Leschine, S.B., 1995. Cellulose degradation in anaerobic environments. *Annu. Rev. Microbiol.*, **49**: 399–426.

Lynch, H.A. and Harper, S.H.t., 1983. Straw as a substrate for cooperative nitrogen fixation. *J. Gen.Microbiol.*, **129**: 251-253.

Maleena, I., 1998. Composting piggery waste: a review. *Biores. Techno.l*, **63**:197-203.

Mathur, B.S.,Sarkar, A.K.and Mishra,B., 1980.Release of nitrogen and phosphorous from compost charged with rock phosphate. *J. Indian Soc. Soil Sci.*, **28**:206-212.

McCarthy, A.J., 1987. Lignocellulose-degrading actinomycetes. *FEMS Microbiol. Rev.*, **46**: 145–163.

Mishra, M.M., Kapoor, K.K. and Yadav, K.S., 1982.Effects of composts enriched with mussoorie rock phosphate on crop yield. *Indian J. Agrci. Sci.*, **52 (10)**: 674-678

Mudgett, R.E., 1986. Solid-state fermentations. In: Demain AL & Solomon NA (Eds) Manual of Industrial Microbiology and Biotechnology (pp 66–83). American Society of Microbiology, Washington D.C., USA

Olayinka, A. and Adebaya, A., 1984. Effect of incubation temperatures and different sources of N and P on decomposition of Saw dust in soil. *Agric. Wastes*, **11**: 293-306.

Pandey, 1997. Environmental Management. Vikas Publishing House Pvt. Ltd., New Delhi, p.390.

Paul, E.A & Clark, F.E., 1989. Soil Microbiology and Biochemistry. Academic Press, Inc. San Diego, USA.

Perez. J., J. Munoz-Dorado, T. de la Rubia., E J. Martinez, 2002. Biodegradation and biological treatments of cellulose, hemicelluloses and lignin: an overview. *Int Microbiol* 5: 53–63.

Plat, J., Sayag, D. and Andre, L., 1984. High rate composting of wool industries wastes. *Biocycle,* 25: 39-42.

Pointcelot, P.R., 1974. A scientific examination of the principles and practice of composting. *Compost Sci.,* 15: 24-31.

Pometto, A.L. and Crawford, D.L., (1985. Catabolic fate of Streptomyces viridosporum T7A produced acid perceptible and Phaneurochaete chrysporium. *Appl. Environ. Microbiol.,* 51: 171-179.

Rajasekaran, P. and Sampath Kumar, M., 1981. Physico-chemical and microbiological properties of plant wastes treated with sewage sludge. *Agric. Wastes.* 3: 262-275.

Rao, N., Hans, G.E. and Reddy, C.A., 1995. Effect of C/N ratio and moisture content on the composting of popular wood. *Biotechnol. Letters,* 17: 889-892.

Rasal, P.H., Kalbhor, H.B., and Patil, P.L., 1988. Effect of cellulolytic and phosphate solubilizing fungi on chick-pea growth. *J. Ind. Soc. Soil Sci.,* 36: 71-74.

Rasal, P.H., Shingte, V.V., Kalbhor, H.B. and Patil, P.L., 1988. A study on production and evaluation of enriched sugarcane trash compost. *J. Maharashtra Agric. Univ.* 13 (1):28-31

Rayner, A.D.M & Boddy,L., 1988. Fungal communities in the decay of wood. *Adv. Microb. Ecol.* ,10: 115–166.

Reid, I.D., 1989. Solid-state fermentations for biological delignifica-tion. *Enzyme Microb. Technol.,* 11: 786–802.

Ribeiro, H., 1994. An evaluation of three municipal solid wastes. Institute Superior de Agronoma, Lisboa, Portugal, p.288.

Rodriguez, A.M., Ferreira, L.J., Fernando, A.L., Virbano, P. and Oliveera, J.S., 1995. Co-composting of sweet sorghum biomass with different nitrogen sources. *Biores. Technol.,* 54: 21-27.

Sakagami, H., Satoh, K., Ida, Y., Koyama ,N., Premanathan, M., Arakaki, R., Nakashima, H., Hatano, T., Okuda, T & Yoshida, T .,1999. Induction of apoptosis and anti-HIV activity by tannin- and lignin-related substances. *Basic Life Sci.* 66: 595–611.

Sarikaya, A. & Ladisch, M.R., 1997. Mechanism and potential applications of bio-ligninolytic systems in a CELSS. *Appl. Biochem. Biotechnol.* 62: 131–149.

Scott GM, Akhtar M, Lentz MJ, Kirk TK & Swaney R 1998. New technology for papermaking: commercializing biopulping. *Tappi J.,* 81: 220–225.

Singh, S., Mishra, M.M., Goyal, S. and Kapoor, K.K., 1992. Preparation of nitrogen and phosphorus enriched compost and its effects on wheat (*Triticum aestvum*). *Indian J. Agric. Sci.,* 62: 810-814.

Son, T.T.N., 1995. Bioconversion of organic waste for sustainable agriculture. M.Sc.(Agri) Dissertation, Tamil Nadu Agricultural University, Coimbatore.

Talashilkar, S.C., 1987. Effect of microbial culture *(Azotobacter chroococcum)* on humification and enrichment of mechanized compost. *Indian J. Agric. Chem.,* **20**: 25-88.

Thakre , R.P.and Fulzele, A.A., 1998. Technology on quicker decomposition of cellulose and lignin in compost. *Biofertilizer Newsletter,* pp. 5- 7

Tomme, P., Warren, R.A & Gilkes, N.R., 1995. Cellulose hydrolysis by bacteria and fungi. *Adv. Microb. Physiol.* **37**: 1–81.

Trigo, C & Ball, A.S., 1994. Is the solubilized product from the degradation of lignocellulose by actinomycetes a precursor of humic substances? *Microbiol.,* **140**: 3145–3152.

Van Veen, J.A, Ladd, J.N & Frissel, M.J., 1984. Modeling C&N turnover through the microbial biomass in soil. *Plant Soil,* **76**: 257–274.

Vicuna R, Gonzalez B, Seelenfreund D, Ruttimann C & Salas L (1993). Ability of natural bacterial isolates to metabolize high and low molecular weight lignin-derived molecules. *J. Biotechnol.,* **30**: 9–13.

Vicuna, R., 2000. Ligninolysis, A very peculiar microbial process. *Mol. Biotechnol.* **14**: 173–176.

Waldrop, M.P, Balser, T.C & Firestone, M.K, 2000. Linking microbial community composition to function in a tropical soil. *Soil Biol. Biochem.* **32**: 1837–1846.

Wheals AE, Basso LC, Alves DMG & Amorim HV (999. Fuel ethanol after 25 years. *TIBTECH ,*17: 482–487.

White, G.F., Russell, N.J & Tidswell, E.C., 1996. Bacterial scission of ether bonds. *Microbiol. Rev.* **60**: 216–232.

Wignarajah, K., Pisharody, S & Fisher, J.W., 2000. Can incineration technology convert CELSS wastes to resources for crop production? A working hypothesis and some preliminary findings. *Adv. Space Res.* **26**: 327–333.

Wubah, D.A, Akin, D.E & Borneman, W.S., 1993. Biology, fiber-degradation, and enzymology of anaerobic zoosporic fungi. *Crit. Rev. Microbiol.*19: 99–115.

Zadrazil, F & Isikhuemhen, O.1997. Solid state fermentation of lignocellulosics into animal feed with white rot fungi. In: Roussos S, Lonsane BK, Raimbault M & Viniegra-Gonzalez G (Eds) Advances in Solid State Fermentation (pp. 23–38). Kluwer Academic Publishers, Dordrecht, The Netherlands

Zibiliske, L.M., 1993. Composting of Organic Wastes. Lewis publishers, Boca Raton, Fla, p.402.

Chapter 9

Phytoremediation: A Quest for Clean and Green Environment

K. P. Kolakar[1], H.C. Lakshman[2], T.C. Taranath[3], S.S. Kamble[4], A.G. Devi Prasad[5]. and A. Channabasava [6]

[1]*Department of Botany, Karnatak Science College, Dharwad – 580 007, Karnataka*
[2,3,6]*DOS in Botany, Karnatak University, Pavate Nagar, Dharwad – 580 003, Karnataka*
[4]*Department of Botany, Shivaji University, Vidyanagar, Kolhapur-416 004, Maharashtra*
[5]*DOS in Environmental Science, University of Mysore, Manasagangotri, Mysore-06, Karnataka*

ABSTRACT

Phytoremediation is an emerging technology that uses various plants to degrade, extract, contain or immobilize contaminants from soil and water. Pistia stratiotes, an aquatic plant has been tested for phytoremediation of domestic waste water. Azolla rubra is a free-floating water fern, which has been used for several decades as a green manure in rice fields owing to the nitrogen fixing ability of Cyanobacterium, Anabaena azollae. Arbuscular mycorrhizal fungi (AMF) belong to the wide spectrum of soil microbiota and are able to improve the growth of the host plant, particularly in soils of low nutritional status. AMF increases tolerance to extreme drought conditions, high soil salinity and heavy metal toxicity. The mycorrhizal plants have particularly advantage over non-mycorrhizal plants because mycorrhizal plants survive soils with deficiency of phosphorous. Research in genetic engineering techniques to implant more efficient accumulator gene in plants is being carried out. For example- seedlings of Brassica juncea introduced with E-coli-gshl-gene showed increased tolerance to cadmium and had higher concentration of phyto-chelatins and glutathione as compared to the wild type seedlings. This chapter thus reviews all the phytoremediation techniques being used in the present day and stresses on creating awareness among the human beings towards the phytoremediation and identification of accumulator plant species.

Key words: Pollution, Pesticide, Mycorrhyzae, Heavy metals, Salt tolerance, Contaminants

Introduction

Phytoremediation is the use of specialized plants to clean up polluted soil. While most plants exposed to high levels of soil toxins will be injured or die, scientists have discovered that certain plants are resistant and an even smaller group actually thrive. Both groups of plants are of interest to researchers, but the thriving plants show a particular potential for remediation because it has been shown that some of them actually transport and accumulate extremely high levels of soil pollutants within their bodies (Conesa *et al.*, 2012). They are, therefore, aptly named as hyper-accumulators. Phytoremediation is an emerging technology that uses various plants to degrade, extract, contain, or immobilize contaminants from soil and water. This technology has been receiving attention lately as an innovative, cost-effective alternative to the more established treatment methods used at hazardous waste sites.

Soil pollution, a very important environmental problem, has been attracting considerable public attention over the last many decades. Now-a-days, more and more people consider that the magnitude of the pollution problem in our soil demands immediate action. One of the factors responsible for so much soil pollution is enormous cost associated with the removal of pollutants from soil by means traditional physicochemical methods. A major environmental concern due to dispersal of industrial and urban wastes generated by human activities is the contamination of soil. Controlled and uncontrolled disposal of waste, accidental and process spillage, mining and smelting of metalliferous ores, sewage sludge application to agricultural soils are responsible for the migration of contaminants into non-contaminated sites as dust or leachate and contribute towards contamination of our ecosystem (Wani *et al.*, 2012). Though several regulatory steps have been implemented to reduce or restrict the release of pollutants in the soil, they are not sufficient for checking the contamination.

The evaluation of plant potential for metal accumulation by several selected plant species is being investigated by growing plants in an inert substrate (quartz sand) in lysimeter pots. The inert substrate is used to eliminate any possible interference with soil components (*e.g.*, clay micelles, organic matter) that might affect the contaminant availability to plants. Replicated treatments in a random scheme are used. Plants are irrigated with nutrient solution, and contaminants are supplied either at low but constant doses of the nutrient, or at higher doses, at specific application times. Leachate is drained from the pots each time nutrient is added. This is done at regular intervals, frequent enough so that measurable leachate volumes are always present. Volume, pH, and conductivity are measured each time leachate is drained, and a sample is taken for metal analysis. Measuring leachate volume permits us to determine evapo-transpiration rates throughout the experiments. Plant leaves and branches are collected at regular intervals for metal analysis. At the end of the experiments (about 4 months each), the pots are disassembled. Samples are collected from the sand substrate, leachate, roots, and above-ground biomass.

The concept of using plants to clean up their environment is not a new one, but most research in this area was strictly in studying those few wild plants that actually grew in waste infested areas. Examples of simpler phytoremediation systems that have been used for years are constructed or engineered wetlands, often using cattails (Typha sps) to treat acid mine drainage or municipal sewage. Our work extends to more complicated remediation cases: the phytoremediation of a site contaminated with heavy metals and/or radionuclides involves "farming" the soil with selected plants to "biomine" the inorganic contaminants (Salt *et al.*, 1995). For soils contaminated with toxic organics, the approach is similar, but the plant may take up or assist in the degradation of the organic (Schnoor *et al.*, 1995). Several sequential crops of hyperaccumulating plants could possibly reduce soil concentrations of toxic inorganics or organics to the extent that residual concentrations would be environmentally acceptable and no longer considered hazardous. The potential also exists for degrading the hazardous organic component of mixed contamination, thus reducing the waste (which may be sequestered in plant biomass) to a more manageable radioactive one.

Role of Rhizosphere in Phytoremediation

Rhizosphere remediation occurs completely without plant uptake of the pollutant in the area around the root. The rhizosphere extends approximately 1 mm around the root and is under the influence of the plant. Plants release exudates in the rhizosphere likely to serve as carbon source for microbes (Bowen and Rovira, 1991). As a result, high microbial build up of 1 - 4 orders of magnitude occur in the rhizosphere compared to bulk soils (Olson *et al.*, 2003). Consequently, rhizosphere microbes can promote plant health by stimulating root growth via production of plant growth regulators, enhance mineral and water uptake. Rhizosphere remediation may be a passive process. Pollutants can be phytostabilized simply by erosion control and hydraulic control. There is also passive adsorption of organic pollutants and inorganic cat ions to plant surface. Pollutant adsorbed to lignin cells is called liginification.

Microbes and plant activities affect pollutant bioavailability. Some bacteria release bio-surfactants (rhamnolipids) that make hydrophobic pollutants more water soluble (Volkering *et al.*, 1998). Plant exudates may also contain lipophilic compounds that increase pollutant water solubility or enhance bio-surfactant-producing bacterial populations (Siciliano and Germida, 1998). Organic pollutants may be degraded in the rhizosphere by root released plant enzymes or through phytostimulation of microbial degradation. Organics such as PAHs and PCBs and other petroleum hydrocarbons have been successfully remediated in the rhizosphere by microbial activity (Hutchinson *et al.*, 2003; Olson *et al.*, 2003). Plants stimulate the entire process by firstly, releasing carbon compounds to facilitate a higher microbial population around root zone. Secondly, secondary plant compounds released from the roots may specifically induce microbial genes involved in the de-gradation or act as co-metabolite to facilitate microbial degradation (Olson *et al.*, 2003; Leigh *et al.*, 2002). Also, roots of leguminous plants that host bacteria species with potential to convert atmospheric N_2 to inorganic N_2 in the soil can improve the C: N ratio of hydrocarbon contaminated soils, which ultimately enhance the process of rhizodegradation. Nwoko *et al.*, (2007) reported sustained plant growth, leaf area and biomass production in *Phaseolus vulgaris* grown on spent engine oil contaminated soil.

Plants as a Phytoremediators

Plants represent an important pathway for the movement of potentially toxic trace elements from soil to human beings. The possible adverse effect of heavy metal pollution and their phytotoxic effects have been reported by several workers (Antonovies *et al.*, 1971; Chiba and Takahashi, 1977; Heale and Ormrod 1983; Leblova *et al.*, 1986). All the heavy metals are potentially toxic at elevated concentrations (Gadd and While 1989). Accumulation of heavy metals in plant parts showed inhibitory and promotory effect on growth. In recent time, the bioremediation became modern thought and safe practice in elimination of heavy metals from the environment, as plants are known to accumulate them (Table 1). Water hyacinth (*Eichhornia crassipes*) is used for pollution treatment and is reported to remove heavy metals (Cirigbo *et al.*, 1982; Selvapathy and Sreedhar, 1991) like Ni, .As, Cd, Pb, Hg, Cu, Mn and Zn. Keeping in view of the role of plants in elimination of heavy metals from soil, the present study aims to evaluate the amount of metals absorbed by the whole plant and accumulation in different plant parts to determine their efficacy of potentiality in bioremediation and also to assess their residues present in edible parts to see their hazardous level for consumption.

Table:1 Plant used in Phytoremediation and their functions.

Name of Plant	Function
Medicago sativa	Symbiosis with hydrocarbon degrading Bacteria
Lycopersicon esculentum	Accumulate Lead; Zinc, and Copper.
Arabidopsis thaliana	carries a 'bacterial gene that transforms Mercury into a gaseous state.
Bladder campion	Accumulate zinc and Cadmium
Brassica juncea	Accumulated Selenium, Sulfur, Lead, Calcium, cadmium, Nickel, Zinc, Copper.
Buxaceae (Boxwood)	Accumulate Nickel
Compositae family plants	Symbiosis with Arthobacteria, accumulate Cesium and Strontium.

Source: Mathew Dempsey, Dec. 1997 Phytoremediation

Major pesticides found as contaminants, such as Atrazine, Alachlor, metolachlor, pendimetalin and trifluralin are most frequently found in pesticide dealership soil and well water (Lone, 2008). Contamination levels are often spatially variable at these sites and high rate of contamination of pesticides no longer in use (DDT, dieldrin, dinoseb etc) still exist. Phytoextraction is a remediation strategy for contaminated soils, which employs plants to remove soil heavy metals through plant uptake and harvest. This is sustainable and comparatively inexpensive process which emerging as a viable alternative to traditional remediation methods. To enhance Phyto-extraction, fast growing plants having high biomass gain and high metal uptake ability are needed.

Role of Aquatic Plants in Phytoremediation

The sustainability of *Pistia stratiotes* for phyto-remediation of domestic waste-water was evident from the results of net primary productivity (NPP), which registered a significant increase in value after seven days of culture. This higher value of NPP - 3.22 $g.m^{-2}.day^{-1}$ was observed in the month of February, while minimum 0.42 $g.m^{-2}.day^{-1}$ in the month of May (Table 2). The observed NPP value proved that rainy and winter seasons are comparatively better than the summer month for phyto-remediation of domestic wastewater through *Pistia stratiotes*.

Table: 2 Monthly variation in net primary productivity f1 day' of *Pistia stratiotes* (oven dry biomass) after 7 days of culture of domestic wastewater during 2004 (initial biomass of *Pistia stratiotes* used for culture = 7,5 g dry weight).

Months	Period	Product $g.m^{-2}$	NPP $g.m^{-2}.day^{-1}$
January	15Th-22nd	10.69	1.52
February	19th to 26th	22.54	3.22
March	17th to 24th	9.53	1.36
April	12th to 19th	4.53	0.67
May	12th to 19th	3.00	0.42
June	16th to 23rd	5.60	0.80
July	06th to 13th	6.93	0.99
Align?!	16th to 23rd	7.80	1.11
September	15th to 22nd	16.01	2.28
October	19th to 26th	7.34	1.04
November	16th to 23rd	17.34	2.47
December	14th to 21st	21.67	3.09
Mean x	-	11.10	1.58
±SD	-	6.65	0.95

Source: Patel D.K, and V.K. Kanungo, 2006.

Azolla rubra is a cosmopolitan free-floating water fern, which has been used for several decades as a green manure in rice fields owing to its nitrogen fixing ability with symbiotic association of *Cyannbacterium, Anabaena azollae*. Today, there is an increasing interest in utilization of the fern for innovative uses such as decontamination in low cost wastewater treatment systems (Ma *et al.*, 2001) and for the accumulation of heavy metals. However, very little work has been done in India regarding to removal abilities of the fern.

Role of Mycorrhizae in Phytoremediation of Polluted Soils

A wide variety of microbial populations live in the natural and agricultural and marginal soils contaminated with Zn, Cu, Mg and Ni etc. Plant roots colonized with mycorrhiza strongly influence the surrounding environment, producing the so called rhizosphere effect. Arbuscular mycorrhizal fungi (AMF) belong to the wide spectrum of soil microbiota and are capable of improving the growth of the host plant, particularly in soils of low nutritional status (Lakshman, 2010). The mycelium of mycorrhizal fungi is more resistant to abiotic agents. AM fungi help in improving the tolerance to extreme drought conditions, high soil salinity and heavy metal toxicity. The mycorrhizal plants particularly have advantage over non-mycorrhizal plants in marginal ecological conditions, as the former plants can survive in phosphorus deficient soils. (Lakshman, 2007).

Alfalfa plants inoculated with *Glomus mosseae* shown to have increased root colonization from 41% to 70% at 3-12 months compared non-inoculated plants. Therefore, the growth of the mycelia of AM fungi could be ecologically significant in helping the plants to develop under stressful situations (Lakshman, 2004). Uranium bioaccumulation increased greatly as a result of inoculation with mycorrhizal fungi. Soil pH could have been the reason for this phyto-availability of uranium. At a pH lower than 5.5, uranium is in the most available form for plants (Ebbs *et al.*, 1998; Shahandeh and Hossner, 2002). The combined inoculation with *Glomus intraradices* and *Streptomyces* sp. strongly stimulated the development of plants. However, in an early study on the interaction of streptomycetes and mycorrhizal fungi, the micro-organisms interacted antagonistically with respect to the growth and phosphorous nutrition of the studied plant species (Krishna *et al.*, 1982). However, Ames *et al.*, (1987) documented that the rhizosphere microbial populations from AM and non-VAM plants differ, and Wyss *et al.*, (1992) reported that mycorrhiza formation with *Glomus mosseae* was significantly depressed in presence of bio-control agents, in particular of *Streptomyces griseoviridis*. In contrast, Ames (1989) tested 12 actinomycetes and reported that seven of them significantly increased the percentage of mycorrhizal root colonization and the density of hyphae, predominantly those of AM. It is today widely accepted that studies on mycorrhizal interactions with rhizosphere bacteria show variable results, depending on the species, on the abiotic environment and on the structure of the soil community (Volkering *et al.*, 1998; Hodge, 2000; Wamberg *et al.*, 2003).

Phytoremediation of Metal Contaminated Soil

Heavy metals constitute major part of the inorganic contaminants released due to various anthropogenic activities. The most common heavy metals at hazardous waste sites are Cadmium (Cd), Chromium (Cr), Copper (Cu). Lead (Pb), Mercurv (Hg), Nickel (Ni) and Zinc (Zn). Migration of contaminants into non-contaminated sites as dust or leachate through the soil, and the spreading of industrial sludge are examples of events that contribute towards contamination of our ecosystems. Remediation of metal compounds presents a different set of problems when compared to organics. Aboulroos *et al.*, (2006) reported that organic chemicals might undergo root sorption, uptake, translocation, metabolic transformation or volatization. Thus organic compounds can be degraded while metals normally need to be physically removed or immobilized (Harms *et al.*, 2003). Heavy metal contaminated soil can be remediated by chemical, physical and biological techniques (Chaney *et al.*, 2000). There are several ways to remediate heavy metal pollution in soil:

i.   by using bacteria;

ii. by adding bonds in soil to solidify and stabilize heavy metals;

iii. by utilization of electricity dynamics;

iv. by thermic absorption;

v. by extraction and washing;

vi. by phytoremediation (Raskin, 2000; Marchiol, 2004; Kramer, 2005; Mukhopadhyay and Maiti, 2010; Mojiri, 2011).

The available techniques can he broadly grouped into two categories: (a) *Ex-situ* techniques, which require removal of the contaminated soil for treatment and/or off site, and (b) *In-situ* techniques, which remediate without excavation of contaminated soil (Jadia and Fulekar, 2008). *In-situ* techniques are favored over the Ex situ techniques due to their lower cost and reduced impact on the ecosystem.

Studies on heavy metal phytoremediation have been very prevalent (Cunningham *et al.*, 1995; Baker, 1999; Henry, 2000; Bio-Wise, 2003; Hooda, 2007; Mathur *et al.*, 2010). There are two main pathways for the export of metals from contaminated areas: through bioaccumulation and further transfer of the organisms outside the area, and by surface or underground hydrological fluxes. These hydrological fluxes can be intercepted by surface water ecosystems, with consequent ecotoxicological and human health risks (Tian *et al.*, 2006). One way of managing the metal fluxes is phytoremediation, that can be coupled with mycoremediation (Neagoe *et al.*, 2004; Channabasava and Lakshman, 2013), and various soil treatments (Khan *et al.*, 2000; Neagoe *et al.*, 2005). Thus, it is an important issue to investigate the effects of the various phytoremediation methods on the leaching of metals to the ground water.

Besides the field monitoring of the sites under remediation, the main instrument for studying the leaching of metals is provided by lysimeters. An area of research is the study of effects of sewage sludge and manure application on soil (Bojakowska and Kochany, 1985; Keller *et al.*, 2002; McLaren *et al.*, 2004; Pirani *et al.*, 2006). It seems that phytoremediation methods involving a strong disturbance of the topsoil, such as mixing with different substrates, enhance the leaching of metals, at least in the first phases of plant development (Neagoe *et al.*, 2006; Iordache *et al.*, 2006). The numerous studies of the mechanisms controlling the transfer of metals to plants and groundwater have led to several attempts to model these processes (Seuntjens *et al.*, 2004; Verma *et al.*, 2006).

Some strains of *Brassica juncea* accumulate heavy metals like chromium when grown in metal contaminated soils. Experiments were conducted on *Brassica juncea* growing in chromium contaminated soil. Small amount of chromium was taken up by the plants and removed from the soil. The use of these plants is being developed by Burt D. Ensley at Phytotech. U.S.A. New Brunswick. Such a process of toxic removal of plants is called as phytoremediation. The use of plants and associated rhizosphere organisms or bioengineered plants to metabolize toxic organic compounds are appears to be more promising in future (Wani *et al.*, 2012). The potential for the use of phtoremediation to clean up contaminated soils have developed by many researchers. Plants such as *Zea mays* and *Brassica juncea* can accumulate 100-500 times higher levels of elements than other crops. Metal hyperaccumulation in shoots or high shoot biomass in phytoremediation of soil metals has been debated (Nwoko, 2010). In conclusion improved hyperaccumulator plants and agronomic technology to improve the annual rate of phytoextraction and to allow recycling of soil toxic metals accumulated in plant biomass is very likely to support commercial environmental remediation in society.

The concept of using plants to remediate soils contaminated with organic pollutants is based on the observation, that disappearance of organic chemicals is accelerated in vegetated soils compared with non-vegetated bulk soils (Schnoor, *et al.*, 1995). Phyto-remediation of organic contaminants has been targeted for three main categories of compounds.

1.   Chlorinated solvents such as dioxin.

2.   Nitro-aromatics such as Trinitrotoluene (TN'T).

3.   Linear halogenated hydrocarbons such as trichloro ethylene (TCE)

Pollutants can be remediated by plants through several natural biophysical and biochemical processes, adsorption, transport and translocation, hyper accumulation and transformation, and mineralization. Although, the correlation between agronomics performance and phyto-remediation potential is yet to be determined, better agronomics performance of the plant may improve Phyto-remediation. The effect of pH, initial metal ion concentration, contact time and adsorbent dose were studied. The adsorption capacity of *Datura stramonium* was dependent on the pH of the nickel solution; maximum nickel removal (78.9%) was obtained at pH of 6.8. The adsorption experimental data fit well with the Longmuir and Freundlich adsorption isotherms. The kinetics of the adsorption process followed the pseudo second-order kinetics model. The results indicate that *Datura stramonium* can be employed as a low cost alternative to commercial adsorbents in the removal of nickel (II) from wastewater.

Industrial effluents are normally considered as toxicants due to the presence of organic and inorganic compounds, acids, alkalis and suspended solids. Such industrial effluents destroy the living organisms and disturb the fragile ecosystem during disposal. Disposal of textile effluent is s major hazard in the ecosystem. There is an urgent need to make use of this effluent for cultivation of certain forest tree species like *Acacia ferruginea* and *Delonix regia*.

Absorption of elements depends upon many phenomena viz; availability of element, requirements of elements, chemical and physical condition of soil and climatic conditions (Paz-Alberto and Sigua, 2013). Plants absorb very little available Fe from the soil due to high pH, as reported by Macfie and Crowder (1987). The higher amount of Cu was accumulated during summer in *A. Senegal*. It may be due to high pH and high calcium cement of the soil as cited Pandit and Pandya (2002). While in *Acacia nilotica* and *A. leucophloea* higher accumulation of Cu was observed during winter. This may be due to the retarded metabolic activity or due to less availability of water. Lowest amount of Cu accumulation was observed during winter in *A. Senegal*. It may be due to high leaf abscission and less availability of Cu in the soil profile. Among *Acacia nilotica* and *A. leucophloea*, lower accumulation of Copper was observed during summer.

Rhizo-filtration is similar to phyto-extraction, but the plant to be used for clean up are raised in greenhouses with their roots in water rather than soil. When the plants have developed a large root system, contaminated water is collected from a waste site and brought to the plants where it is substituted for their water source. The roots take up the water along with the contaminants. As the roots become saturated with contaminants, they are harvested and disposed off. Rhizo-filtration is particularly effective in application where low concentrations of contaminants and large volumes of water are involved.

Complexity of Salt Tolerance

About 7% of the world's total land area is affected by salt, as it is a similar percentage of its arable land (G-hassemi *et al.*, 1995; Szabolcs, 1994). When soils in arid regions of the world are irrigated, solutes from the irrigation water can accumulate and eventually reach levels that have an adverse effect on plant growth. Of the current 230 million ha of irrigated land, 45 million ha are salt affected (19.5) and of the 1,500 million ha under dry land agriculture, 32 million are salt affected to varying degree (2.1%). Salinity is a multi-factorial problem and the use or the breeding of salt resistant crop varieties will require a clear understanding of the complex mechanisms of salt stress resistance, which is still lacking despite intensive research during the last decade (Apse and Blumwald, 2002; Soss *et al.*, 2005).

Most crop plums do not fully express their original genetic potential for growth, development and yield under salt stress and their economic value declines as salinity levels increase (Lauchli and Epstein, 1990; Mass, 1990). Although numerous attempts have been made to improve the salt tolerance of crops by traditional breeding programmes, commercial success has been very limited due to the complexity of the trait: salt resistance is genetically and physiologically complex (Flowers, 2004). At present, major efforts are being directed towards the genetic transformation of plants in order to raise their tolerance (Borsani *et al.*, 2003). Improving salt resistance of crop plants is of major concern in agricultural research. A potent genetic source for the improvement of salt resistance in crop plants resides among wild populations of halophytes (Glenn *et al.*, 1999; Serrauo *et al.*, 1999). These can be either domesticated into new, salt resistant crops, or used as a source of genes to be introduced into crop species by classical breeding or molecular methods.

Role of Genetic Engineering to Improve Phytoremediation

Plant breeding with superior phytoremediation potential and high biomass production can be an alternative method to improve phytoremediation. General plant productivity is controlled by many genes and difficult to promote by single gene insertion. Genetic engineering techniques to implant more efficient accumulator gene into other plants have been suggested by many authors (Brown *et al.*, 1995; Chaney, *et al.*, 2000). Implanting more efficient accumulator genes into other plants that are taller than natural plants increases the final biomass. Zhu *et al.*, (1999), genetically engineered *Brassica juncea* to investigate rate-limiting factors for glutathione and phytochelatin production; they introduced the Escherichia coli–gshl- gene. The γ-ECS transgenic seedlings showed increased tolerance to cadmium and had higher concentrations of Phytochelatins, γ-GluCys, glutathione, and total non protein thiols compared to wild type seedlings. The potential of success of genetic engineering can be limited because of anatomical constraints (Ow, 1996).

Mercury and mercurial compounds are hazardous to all biological organisms. It has been identified that bacteria have evolved with mechanisms for colonizing in mercury contaminated environments and an operon of mercury resistance (mer) genes encoding transporters and enzymes for biochemical detoxification (Summers, 1986). Mer+ bacteria convert organic and ionic mercury compounds to the volatile and less toxic elemental form, Hg (O) which rapidly evaporates through cell surface. Genetically engineered plants with mer A and mer B genes were produced in three plant species A. thaliana (Bizily et al., 2000), N. tobacum and Liriodendron tulipifera L. (Rugh et al., 1998) and have demonstrated that transgenic plants could grow in the presence of toxic levels of mercury (Rugh et al., 2000). To improve the expression of mer genes in plants, the bacterial mer A DNA sequence was modified by reducing the GC content in a 9% block of the protein coding region and adding plant regulatory elements (Rugh et al., 1996). When transferred to A. thaliana and tobacco, the new gene construct (mer A) conferred resistance to 50 mm Hg (II) suggesting that mer A plants enzymatically reduce Hg (II) and evaporate away Hg (0). Three modified mer A constructs (Rugh et al., 1998) were used for transformation of yellow poplar proembryogenic masses, each having different amounts of altered coding sequences. Each of these constructs was shown to confer Hg (II) resistance (Rugh et al., 2000). Transgenic *Populus deltoides* (North American Tree) over expressing mer A9 and mer A18 gene when exposed to Hg (II) evolved 2- to 4-fold Hg (0) relative to wild plant (Che et al., 2003). These transgenic trees when grown in soil with 40 ppm of Hg (II) developed higher biomass. Subcellular targeting of methylmercury lyase was shown to enhance its specific activity for organic mercury detoxification in plants (Bizily et al., 2003).

Arsenic is an extremely toxic metalloid pollutant which is hazardous to human health (Chen et al., 1992; Kaiser, 1998). Dhankher et al., (2002) developed transgenic *Arabidopsis* plants, which could transport oxy-anion arsenate to aboveground, reduce to arsenite and sequester it in thiol peptide complexes. E. coli Ars C gene encoding arsenate reductase (Ars C) which catalyzes the glutathione (GSH) coupled electrochemical reduction of arsenate to the more toxic arsenite. *Arabidopsis* plants transformed with Ars C gene expressed from a light induced soybean rubisco promoter (SRSIp) strongly expressed Ars C protein in leaves, but not in roots and were hypersensitive to arsenate. *Arabidopsis* plants expressing E. coli gene encoding glutamyl cysteine synthetase (g-ECS) with actin promoter was moderately tolerant to arsenic compared to control plants. Plants expressing SRSIp/ArsC and ACT 2p/g-ECS together showed higher tolerance to arsenic. These transgenic plants accumulated 4 to 17-fold greater fresh shoot weight and accumulated 2- to 3-fold more arsenic per gram of tissue than wild plants or transgenic plants expressing g-ECS or ArsC alone.

Future Outlook of Phytoremediation

One of the key aspects to the acceptance of phytoremediation pertains to the measurement of its performance, ultimate utilization of by-products and its overall economic viability. To date, commercial phytoremediation has been constrained by the expectation that site remediation should be achieved in a time comparable to other clean-up technologies. So far, most of the phytoremediation experiments have taken place in the lab scale, where plants grown in hydroponic setting are fed heavy metal diets. While these results are promising, scientists are ready to admit that solution culture is quite different from that of soil. In real soil, many metals are tied up in insoluble forms, and they are less available and that is the biggest problem (Kochian, 1996). The future of phytoremediation is still in research and development phase, and there are many technical barriers which need to be addressed. Both agronomic management practices and plant genetic abilities need to be optimized to develop commercially useful practices. Many hyper-accumulator plants remain to be discovered, and there is a need to know more about their physiology (Raskin, *et al.*, 1994). Optimization of the process, proper understanding of plant heavy metal uptake and proper disposal of biomass produced is still needed. The following aspects need to be achieved:

- use of genetic approaches, introduction of genes responsible for metal accumulation and resistance in the metal accumulators;
- selecting mutants with heavy metal accumulating ability;
- sevelopment of microbial based phytoremediation technologies, such as use of mycorrhizal fungi as inoculants, which will help to absorb more and more heavy metals from the contaminated soils;
- optimize agriculture practices such as irrigation, fertilization, planting and harvest time and amendment application;
- phytoremediation should help to eliminate the need for costly offsite disposal.

Conclusion

Phytoremediation is a way to monitor eco-friendly management of waste generated due to anthropogenic activities. Since last decade, phytoremediation technology has been initiated through filed applications all over the world; it includes phytoremediation of organic, inorganic and radionuclides. This sustainable and inexpensive technology is fast emerging as a viable alternative to conventional remediation methods for the future environmental safety and will be most suitable for a developing country like India. Establishment of commercial applications of phytoremediation technology in India is in its earliest phase.

Approach and outlook to bioremediation by different nations are different. With their vast land and other natural resources, USA can afford larger and expensive projects. Aquaculture and bioremediation are good examples. The Albemarle's Magnolia plant in Arkansas with 54 acres artificial marsh is also a tourist attraction. The rock reed filter marshes at Dugussa's Theodore, Alaska; Biosphere-2 are other examples. Europe is a conglomeration of states- very adjacent to each other, highly populated, many are landlocked and natural resources are limited. Japan is having more acute situation than in Europe. Gremany and Japan have more holistic approach to this aspect of environmental biotechnology and is likely to come out with some newer pathfinders. European ventures are not very much published and details are not available (Nag, 2008). A close watch on their ambient air and effluent quality indicate that they are conscious of their responsibilities and do the needful in proper timeframe to check deterioration of the ecosystem at large. The mandatory installation of automobile exhaust converters is a good example. Bio-treatment of most gaseous and fluid wastes are major remediation targets of the European communities, is likely to take leadership. A few Japanese achievements may be cited here. Since 1990 Japanese Government's official policy demands that the industries will work but pay attention to the preservation of Global environment (not their own alone!) and treat wastes and conserve natural resources.

Fast growing plants with high biomass and good metal uptake ability are needed. In most of the contaminated sites, hardy, tolerant, weed species exist and phytoremediation through these and other non-edible species can restrict the contaminant from being introduced into the food web. The mycorrhizal inoculation technology favors the increased uptake of heavy metals from the highly contaminated plants by releasing certain chemicals such as phytoalexins. However, several methods of plant disposal have been described but data regarding these methods are scarce. For the successful phytoremediation, it is necessary to select the plants which can host preferential mycorrhizal fungi, which develops tolerance capacity in plants against heavy metal stress caused due to mining activities and other anthropogenic activities in the wastelands. However, development of waste lands into agricultural fields by using or adapting various eco-friendly technologies is the need of the hour. Therefore, it is time to develop awareness among the people towards the phytoremediation and more attention need to be paid for the identification of plants species and mycorrhizal fungi to boost the phytoremediation technology and to maintain green and clean environment for future generation.

References

Aboulroos, S.A., Helal, M.I.D. and Kamel, M.M. 2006. Remediation of Pb and Cd polluted soils using in situ immobilization and phytoextration techniques. *Soil Sediment Contam.* 15: 199-215.

Ames, R.N., 1989. Mycorrhiza development in onion in response to inoculation with chitin decomposing actinomycetes. *New Phytol.* 112, 423–427.

Ames, R.N., Mihara, K.L., Bethlenfalvay, G.J., 1987. The establishment of microorganisms in vesicular–arbuscular mycorrhizal and control treatments. *Biol. Fertil. Soils* 3, 217–223.

Antonovies, J., Bradhave, A.D. and Turner, R.G. 1971. Heavy metal tolerance in plants. *Adv.Ecol.Kes.*75:1-85.

Apse, M.P. and Blumwald, E. 2002. Engineering salt tolerance in plants. *Curr. Opin. Biotech.* 13: 146-150.

Baker, A.J. 1999. Metal hyperaccumulatorm plants. A review of the biological resource for for possible exploitation in the Phytoremediation of contaminated soil and water. Ed. Terry, N., Baneulos, G.S., Boca Raton, Florida. CRC press. pp: 85-107.

Bio-Wise. 2003. Contaminated Land Remediation: A Review of Biological Technology, London. DTI

Bizily, S.P., Kim, T., Kandasamy, M.K. and Meagher, R.B. 2003. Sub cellular targeting of methyl mercury lyase enhances its specific activity for organic mercury detoxification in plants. *Plant Physiol.* 131(2):463-471.

Bizily, S.P., Rugh, C.L. and Meagher, R.B. 2000. Phytodetoxification of hazardous organomercurials by genetically engineered plants. *Nat. Biotechnol.* 18:213-217.

Bojakowska, I., Kochany, J., 1985. Studies on the removal of heavy metals from sludges by leaching and uptake by plants. *Plant Soil* 86, 299–302.

Borsani, O., Valpuesta, V. and Bcrtella, M.A. 2003. Developing salt tolerant plants in a new century: a molecular biology approach. *Plant cell tissue Organ Cult.* 73: .101-115.

Bowen, G.C. and Rovira, A.D. 1991. The rhizosphere- the hidden half of the hidden half. In the Roots-the hidden half, eds. Y Waisel, A Eshel, U Kaffkafi, NYMarcel Dekker. pp. 641-649.

Brown, S.L., Chaney, R.L., Angle J.S. and Baker A.J.M. 1995. Zinc and cadmium uptake by hyperaccumulator *Thlaspi caerulescens* grown in nutrient solution. *Soil Sci. Soc. Am.* J. 59:125–133.

Channabasava A. and Lakshman, H.C. 2013. AM fungus and mine spoil consortium: A microbial approach for enhancing proso millet biomass and yield. *Int J Pharm Bio Sci* 2012 Oct; 3(4): (B) 676 – 684.

Chaney, R.L., Li, Y.M., Angle, J.S., Baker, A.J.M., Reeves, R.D., Brown, S.L., Homer, F.A., Malik, M. and Chin, M. 2000. In Phytoremediation of contaminated soil and water. (ed Terry N. and G. Banelos) – Lewis Publishers, Boca Raton, FL. pp. 129–158.

Che, D., Meagher, R.B., Heaton, A.C.P., Lima, A., Rugh, C.L. and Merkle, S.A. 2003. Expression of mercuric ion reductase in Eastern cottonwood (*Populus deltoids*) confers mercuric ion reduction and resistance. *Plant Biotechnol. J.* 1:311.

Chen, C.J., Chen, C.W., Wu, M. and Kuo, T.L. 1992. Cancer potential in liver, lung, bladder and kidney due to ingested inorganic arsenic in drinking water. *Br. J. Cancer.* 66:888-892.

Chiba, S. and Takahashi, K. 1977. Accumulation of heavy metal pollution in agriculture land (2). Absorption of cadmium and growth retardation in forage crops. *Bull. Shikoku Agri. Exp. Station.* 30:49-73.

Ciribgo, F.E. ., Smith, R.W. and Shore, F.L. 1982. Uptake of arsenic, cadmium, lead and mercury from polluted waters by the water hyacinth (*Eichhornia crassipes*). *Environ. Pollut.*,(Ser;B),27: 31-36.

Conesa, H.M., Evangelou, M.W.H., Robinson, B.H., and Schulin, R. 2012. Critical review of current state of phytoremediation to remediate soils: still a promising tool.

Cunningham SD, Berti WR, Huang JW (1995). Phytoremediation of contaminated soils. *TIBTECH*. 13: 393-397.

Dhankher, O.P., Li, Y., Rosen, B.P., Shi, J., Salt, D. and Senecoff, J.F. 2002. Engineering tolerance and hyperaccumulation of arsenic in plants by combining arsenate reductase and g-glutamylcysteine synthetase expression. *Nat. Biotechnol.* 20:1140-1145.

Ebbs, S., Brady, D., Kochian, L., 1998. Role of uranium speciation in the uptake and translocation of uranium by plants. *J. Exp. Bot.* 49, 1183–1190.

Flowers, TJ. 2004. Improving crop salt tolerance. *J. Exp.* Bot. 55: 307-319.

Gadd, G.M. and White, C. 1989. Heavy metal and radionuclide accumulation and toxicity in fungi and yeasts. In: Metal Microbe Interaction (eds) Polle, R.K. and Gadd, G.M, IRL Press, 19-38.

G-hassemi, F., Jakeman, A.J. and Nis, H.A. 1995. Salinization of land and water resources". Human causes, extent management and case studies. USNW press, Sydney, Australia.

Gleen, E., Brown, J.J. and Blumwald, E. 1999. Salt tolerance and crop potential of halophytes. *Crti. Rev. Plant Sci.* 18: 277-255.

Harms, H., Bokevn, M., Kolb, M., and Bock, c. 2003. Transformation of organic contaminant by different plant systems. In: Phytoremediation; Transformation and control of contaminant. (Ed) McCutcheon, S.C., Schnoor, J.L. NY Wiley. pp: 285-316.

Heale, E.L. and Onnord, DJ'. 1983. Effect of nickel and copper on seed germination, growth and development of seedlings of *Acer pinnata, Betula papyrifera, Picea abies* and *Pinns banksiena* Reclamation and revegetation research.

Henry, J.R. 2000. An Overview of the Phytoremediation of Lead and Mercury. U.S. Environmental Protection Agency Office of Solid Waste and Emergency Response Technology Innovation office Washington, D.C.

Hodge, A., 2000. Mycrobial ecology of arbuscular mycorrhiza. *FEMS Microbiol. Ecol.* 32, 91–96.

Hooda, V. 2007. Phytoremediation of toxic metals from soil and wastewater. *J. Environ. Biol.* 28:367-371.

Hutchinson, S.L., Schwab, A.P. and Banks, M.K. 2003. Biodegradation of petroleum hydrocarbon in the rhizosphere. In: phytoremediation: transformation and control of contaminants, ed. McCutcheon SC, Schnoor JL, NY Wiley. pp. 355-386.

Iordache, V., Neagoe, A., Bergmann, H., Kothe, E. and Bu¨chel, G. 2006. Factors influencing the export of metals by leaching in bioremediation experiments. 23. Arbeitstagung in Jena, Lebensnotwendigkeit und Toxizita¨t der Mengen-, Spuren- und Ultraspurenelemente, pp. 288–295.

Jadia, C.D. and Fulekar, M.H. 2008. Phytotoxicity and remediation of Heavy metals by fibrous root grass (Sorghum). *J. Appl. Biosci.* 10: 491-499.

Kaiser, J. 1998.Toxicologists shed new light on old poisons. Science. 279:1850-1851.

Kramer, U. 2005. Phytoremediation: novel approach to cleaning up polluted soils. Curr. *Opin.Biotechnol.* 16: 133-141.

Keller, C., McGrath, S.P. and Dunham, S.J. 2002. Trace metal leaching through a soil–grassland system after sewage sludge application. *J. Environ. Qual.* 31, 1550–1560.

Khan, A.G., Kuek, C., Chaudhry, T.M., Khoo, C.S. and Hayes, W.J. 2000. Role of plants, mycorrhizae and phytochelators in heavy metal contaminated land remediation. *Chemo-sphere* 41, 197–207.

Kochian, L. 1996. In International Phytoremediation Conference, Southborough, MA. May 8-10

Krishna, K.R., Balakrishna, A.N. and Bagyaraj, D.J. 1982. Interaction between a vesicular–arbuscular mycorrhizal fungus and *Streptomyces cinnamomeous* and their effects on finger millet. *New Phytol.* 92, 401–405.

Lakshman, H.C. 2004. Prevalence of AM fungi in stock mined spoils. *Nature Environ. And Ecology.* 79:31-36.

Lakshman, H.C. 2007. Mycorrhizal fungi: A boon for sustainable agriculture. In: organic farming and mycorrhiza in agriculture (Ed) Trivedi, P.C. I.K. International publishing house, New Delhi. 290pp.

Lakshman, H.C. 2010. The importance of mycorrhizal plants in adverse conditions. In: Bioinoculants for integrated plant growth. MD publishers, New Delhi. 548pp.

Lauchli, A. and Epstein, E. 1990. Plant response to saline and sodic conditions, Iin: Tanji, K.K. (Ed). Agricultural salinity assessment and management. NY, ASCE manual No.71 pp: 113-137.

Leblova, S., Mucha, A. and Sprihauzlova. E. 1986. Compartmentation of cadmium, lead and Zinc in seedlings of maize (*Zea mays*) and induction of metallothion. *Biol. Czechos.* 4:77-78.

Leigh, M.B., Fletcher, J.S., Fu, X. and Schmitz, F.J. 2002. Root turnover: an important source of microbial substances in rhizosphere remediation of recalcitrant contaminants. *Enviro Sci. Techn.* 36: 579-1583.

Lone, M.I., He, Z.l., Stoffella, and Yang, X.E. 2008. Phytoremediation of heavy metal polluted soil and water: progress and perspectives. *J.Zhengjiang Uni. Sci.* B 9(3): 210-220.

Ma, L.Q., Komar, K.M., and Tu, C. 2001. A fern that accumulates arsenic. *Nature.* 409:579.

Maas, E.V. 1990. Crop salt tolerance. In: Tanji, K.K.. (Ed). Agricultural salinity assessment and management. NY, *ASCE manual* No-71. Pp:262-304.

Macfie. S.M. and Crowder, A.A. 1987. Soil factor influencing ferric hydroxide plague formation on mot of *Typha latifolia* L. 102 (2): 177-184.

Marchiol L, Assolari S, Sacco P, Zerbi G. 2004. Phytoextraction of Heavy Metals by canola (*Brassica napus*) and radish (*Raphanus sativus*) grown on multicontaminated soil. *Environ. Pollut.* 132:21-27.

Mathur, N., Singh, J., Bohra, S., Bohra, A., Mehboob., Vyas, M., and Vyas, A. 2010. Phytoremediation Potential of Some Multipurpose Tree Species of Indian Thar Desert in Oil Contaminated Soil. *Adv. Environ. Biol.* 4(2):131-137.

McLaren, R.G., Clucas, L.M., Taylor, M.D. and Hendry, T. 2004. Leaching of macronutrients and metals from undisturbed soils treated with metal-spiked sewage sludge. 2. Leaching of metals. Aust. *J. Soil Res.* 42, 459–471.

Mojiri, A. 2011. The Potential of Corn (*Zea mays*) for Phytoremediation of Soil Contaminated with Cadmium and Lead. *J. Biol. Environ. Sci.* 5:17-22.

Mukhopadhyay, S. and Maiti, S.K. 2010. Phytoremediation of metal mine waste. *Appl. Eco. Environ*. Res. 8:207-222.

Nag, A. 2008. Text book of agricultural biotechnology. PHI Learning Pvt. Ltd. New Delhi, India. pp: 150-151.

Neagoe, A., Ebena, G. and Carlsson, E., 2005. The effects of soil amendments on plant performance in an area affected by acid mine drainage. *Chem. Erde* 65 (S1), 115–130.

Neagoe, A., Iordache, V., Mascher, R., Knoch, B., Kothe, E. and Bergmann, H. 2006. Lysimeters experiment using soil from a heavy metals contaminated area. 23. Arbeitstagung in Jena, Lebensnotwendigkeit und Toxizita"t der Mengen-,Spuren- und Ultraspurenelemente, Friedrich-Schiller Uni-versita"t, pp. 568–575.

Neagoe, A., Mascher, R., Iordache, V., Voigt, K., Knoch, B. and Bergmann, H., 2004. The influence of vesicular arbuscular mycorrhizaGlomus intraradiceaeon mustard (Sinapis alba L.) grown on a soil contaminated with heavy metals. Mengen und Spuren Elemente 22, 597–606.

Nwoko, C.O. 2010. Trends in phytoremediation of toxic elemental and organic pollutants. *African Journal of Biotechnology.* 9 (37): 6010-6016.

Nwoko, C.O., Okeke, P.N., Agwu, O.O. and Akpan, I.E. 2007. Performance of *Phaseolus vulgaris* L. in a soil contaminated with spent-engine oil. *Afr. J. Biotechnol.* 6(16): 1922-1925.

Olson P.E., Reardon K.F., Pilon-Smits E.A.H. 2003. Ecology of rhizosphere bioremediation. In Phytoremediation: transformation and control of contaminants, ed. McCutcheon SC, Schnoor JL, Wiley NY. pp. 317-354.

Ow, D.W. 1996. Heavy metal tolerance genes-prospective tools for bioremediation. – *Res. Conserv. Recycling.* 18; 135–149.

Pandit B.R. and Ushma Pandya, 2002. Statistical study of relation between pH and minor elements in wasteland soils of Bhavnagar district. *Ad. Plant Sci. India* (accept for June, 2002).

Paz-Alberto, A and Sigua, G.C. 2013. Phytoremediation: A green Technology to remove environmental pollutants. *American Journal of Climate change.* 3: 71-86.

Pirani, A.L., Brye, K.R., Daniel, T.C., Haggard, B.E., Gbur, E.E. and Mattice, J.D. 2006. Soluble metal leaching from a poultry litter–amended Udult under pasture vegetation. *Vadose Zone* J. 5, 1017–1034.

Raskin, I, Kumar, P.B.A.N., Dushenkov, S. and Salt, D. 1994. Bioconcentration of heavy metals by plants. – *Current Opinion Biotechnology* 5; 285-290.

Raskin, I. and Ensley, B.D. 2000. Phytoremediation of Toxic Metals: Using Plants to Clean Up the Environment. John Wiley & Sons, Inc., New York.

Rugh, C.L., Bizily, S.P. and Meagher, R.B. 2000. Phytoremediation of environmental mercury pollution. In: Raskin I, Ensley BD, editors. Phytoremediation of toxic metals using plants to clean up the environment. New York7 Wiley. p. 151-571.

Rugh, C.L., Senecoff, J.F., Meagher, R.B. and Merkle, S.A. 1998. Development of transgenic yellow poplar for mercury Phytoremediation. *Nat. Biotechnol.*16:925– 8.

Rugh, C.L., Wilde, D., Stack, N.M., Thompson, D.M., Summers, A.O. and Meagher, R.B. 1996. Mercuric ion reduction and resistance in transgenic Arabidopsis thaliana plants expressing a modified bacterial merA gene. *Proc. Natl. Acad. Sci.* USA. 93:3182-3187.

Salt, D.E., Blaylock, M., Chet, I., Dushenkov, S., Ensley, B., Nanda, P., and Raskin, I. 1995. Phytoremediation: A novel strategy for the removal of toxic metals from the environment using plants. *Biotechnol.* 13(5): 468-474.

Schnoor, J.L., Light, L.A., McCutcheon, S.C., Wolfe, N.L., and Carriera, L.H. 1995. Phytoremediation of organic and nutrient contaminants. *Environ. Sci. Technol.* 29: 318-328.

Selvapathy, P. and Sreedhar, P. 1991. Heavy metal removal by water hyacinth. *J. IPHE*, India, 3: 11-17.

Serrauo, R., Mulet, J.M., Rios, G, Marquez, J.A, De Larrijloa, I,F, Leubs, M.P., Meadizabal, I., Pascula-Ahuir, A., Proft, M., Ros, R., and Montesmos, C. 1999. Glimpse of the mechanisms of ion homeostasis during salt stress. J.Exp. Boi. 50: 1023- 1036.

Seuntjens, P., Nowack, B. and Schulin, R. 2004. Root-zone modeling of heavy metal uptake and leaching in the presence of organic ligands. *Plant Soil* 265, 61–73.

Shahandeh, H. and Hossner, L.R. 2002. Role of soil properties in phytoaccumulation of uranium. *Water Air Soil Pollut.* 141, 165–180.

ShelrorL S.G. 1983. Alfalfa *Madicago Sativa* L. Establishment of mine- mill tailings, I-Plant analysis of Alfalfa growing on Iron and Copper trailin. *Plant and Soil* 73(2): 227-237.

Siciliano, S.D. and Germids, J.J. 1998. Bacterial inoculants of forage grasses enhance degradation of 2-chlorobenzoic acid in soil. *Environ. Toxicol. Chem.* 16: 1098-1104.

Soss, L., Llanese, A., Reinoso, H., Reginato, M., and Luna, V. 2005. Osmotic and specific ions effects on the germination of *Prosopis strombulifera*. *Ann. Bot.* 96(2): 261-267.

Summers, A.O. 1986. Organization, expression and evolution of genes for mercury resistance. *Annu. Rev. Microbiol.* 40:607-634.

Szabolcs, I. 1994. Soils and Salinisation. In: Pessarakli, M. (Ed). Handbook of plant and crop stress. New York, 'Marcel Dekker. Pp: 3 -11.

Tian, G., Granato, T.C., Pietz, R.I., Carlson, C.R. and Abedin, Z. 2006. Effect of long-term application of biosolids for land reclamation on surface water chemistry. *J. Environ. Qual.* 35, 101–113.

Verma, P., George, K.V., Singh, H.V., Singh, S.K., Juwarkar, A. and Singh, R.N., 2006. Modeling rhizofiltration: heavy-metal uptake by plant roots. *Environ. Mod. Assess.* 11,387–394.

Volkering F., Breure A.M. and Rulkens, W.H. 1998. Microbiological aspects of sulfactant use for biological soil remediation. *Biodegradation*, 8: 401-417.

Wamberg, C., Christensen, S., Jakobsen, I., Muller, A.K. and Sørensen, S.J. 2003. The mycorrhizal fungus (Glomus intraradices) affects microbial activity in the rhizosphere of pea plants (*Pisum sativum*). *Soil Biol. Biochem.* 35, 1349–1357.

Wani, S.H., Sanghera, G.S., Athokpam, H., Nongmaithem, J., Nongthongbam, R., Naorem, B.S. and Athokpam, H.S. 2012. Phytoremediation: Curing soil problems with crops. *African journal of Agricultural Research.* 7(28): 3991-4002.

Wyss, P., Boller, T., Wiemken, A., 1992. Testing the effect of biological control agents on the formation of vesicular arbuscular mycorrhiza. *Plant Soil* 147, 159–162.

Zhu, Y. L., Pilon-Smits, E.A.H., Tarun, A.S., Weber, S.U., Jouanin, L. and Terry, N. (1999): Cadmium tolerance and accumulation in Indian mustard is enhanced by over expressing glutamylcysteine synthetase. *Plant Physiology.* 121:1169-177.

Chapter 10

Phytoremediation of Heavy Metals with Special Reference to *Vetiveria zizanioides*

K. J. Thara Saraswathi and M.N. Shivakameshwari

Department of Microbiology and Biotechnology, Bangalore University,
Bangalore – 560 056
Department of Botany, Bangalore University, Bangalore – 560 056

ABSTRACT

Phytoremediation involves the use of plants to remove, transfer, stabilize and/or degrade contaminants in soil, sediment and water. Vetiveria zizaniodes belongs to the family Poaceae. The vetiver grass due to its unique morphological and physiological characteristics is widely known for their effectiveness in erosion and sediment control. The plants are highly tolerant to extreme soil conditions including heavy metal toxicity. The deep rooting system of vetiver makes it tolerant to extreme drought conditions and thus is difficult to dislodge by strong current. This paper reviews the heavy metal accumulation capacity and growth performance ability of the vetiver grass. The studies have proven to be highly tolerant to adverse conditions hence these plants can serve as an appropriate candidate for phytoremediation and can be used for rehabilitation of mine tailings, garbage landfills and industrial waste dumps which are acidic or alkaline, high in heavy metal contamination.

Key words: Soil pollutant, Toxic element, Contamination, Chelate

Introduction

The global problem concerning contamination of environment as a consequence of human activity is increasing. Most of the environmental contaminants are chemical byproducts and heavy metals. The human activities that contaminate soils with large quantities of heavy metals are industrial and mining industries, fuel burning and fuel production, intensive agriculture and sludge dumping. Heavy metals accumulated in soil can affect flora, fauna and human living in the vicinity or downstream of the contaminated sites. Both metal and non-metal mining activities generate huge quantity of waste rocks, which damages the aesthetics of the area. Particularly, in case of metal mining, activities such as crushing, grinding, washing, smelting and all the other process used to extract and concentrate metals, generate a large amount of waste rocks and tailings which scars the landscape, disrupt ecosystems and destroys microbial communities.

Waste materials or spoils that remain after the extraction of usable ores are dumped on the surrounding land is the source of toxic metals, leave the land devoid of topsoil, nutrients and supportive microflora and vegetation, thus remains barren. The rapid increase in population coupled with fast industrialization and intensive agricultural practices causes serious environmental problems, including the production and release of considerable amounts of toxic wastes into the soil environment. Soil pollutants are classified into two groups: inorganic and organic. The major components of inorganic pollutants are heavy metals such as lead, arsenic, cadmium, copper, zinc, nickel and mercury which are added into the environment via disposal of urban and industrial sewage sludge and via agrochemical usage. Vast areas of soil around the world are contaminated with organic pollutants mainly pesticides, petroleum hydrocarbons, polycyclic hydrocarbons and explosives.

Heavy metals are defined as metals with density higher than 5 g mL^{-1}. Approximately fifty three of the ninety naturally occurring elements are heavy metals (Weast, 1984), but not all of them are of biological importance. Based on the solubility under physiological conditions, 17 heavy metals are available for living cells and are of importance to organism and ecosystem (Weast, 1984). Among these metals, Fe (Iron), Mo (Molybdenum) and Mn (Manganese) are important as micro nutrients. Zn (Zinc), Ni (Nickel), Cu (Copper), V (Vanadium), Co (Cobalt), W (Tungsten) and Cr (Chromium) are toxic elements with high or low significance as trace elements. As (Arsenic), Hg (Mercury), Ag (Silver), Sb (Antimony), Cd (Cadmium), Pb (Lead) and U (Uranium) have no known function as nutrients and seem to be more or less toxic to plant and micro organism (Godbold and Huttermann, 1985; Breckle, 1991; Nies, 1999). Heavy metals are ubiquitous environmental contaminants in the industrialized society.

Soil pollution by heavy metals persists in soil much longer than in other compartments of the biosphere. Sources of heavy metal contaminants in soils include metalliferous mining and smelting, metallurgical industries, sewage sludge treatment, warfare and military training, waste disposal sites, agricultural fertilizers and electronic industries. Since most of the current technologies cannot selectively remove heavy metals, many contaminated sites can be remediated only by using labor- intensive and costly excavation and land filling technology. Many sites around the world remain contaminated with no remediation in sight, simply because it is too expensive to clean them up with the available technologies.

Toxic heavy metals cause DNA damage, and their carcinogenic effects in animals and humans are probably caused by their mutagenic ability. Exposure to high levels of these metals has been linked to adverse effects on human health and wildlife. Lead poisoning in children causes neurological damage leading to reduced intelligence, loss of short term memory, learning disabilities and coordination problems. The effects of arsenic include cardiovascular problems, skin cancer and other skin effects, peripheral neuropathy (WHO, 1997) and kidney damage. Cadmium accumulates in the kidneys and is implicated in a range of kidney diseases (WHO, 1997). The principle health risks associated with mercury are damage to the nervous system, with such symptoms as uncontrollable shaking, muscle wasting, partial blindness, and deformities in children exposed in the womb (WHO, 1997).

Heavy metal contaminated soil can be remediated by chemical, physical or biological techniques (McEldowney *et al.*, 1993). Chemical and physical treatments irreversibly affect soil properties, destroy biodiversity and may render the soil useless as a medium for plant growth. Methods used for decontamination have often been done by chemically treating the contaminants, burying and removing them from the site. In recent years, public concerns relating to human health and ecological threats caused by soil pollution have led intensive research of new economical remediation technologies. Most of the conventional remedial technologies like leaching of pollutant, vitrification, electro-kinetic treatment, excavation and off-site treatment are expensive and technically limited to relatively small areas. Moreover, they deteriorate the soil fertility, which subsequently causes negative impacts on the ecosystem. It is highly desirable to apply suitable remedial approaches to polluted soil, which can reduce the risk of metal contamination. The excavation and disposal of soil is no longer considered to be a permanent solution. The demand for soil treatment techniques is consequently growing and the development of new low-cost, efficient and environmentally friendly remediation technologies has generally become one of the key research activities in environmental science and technology.

In selecting the most appropriate soil remediation method for a particular polluted site, it is of paramount importance to consider the characteristics of the soil and the contaminants. At present, various approaches have been suggested for remediation of metal-contaminated sites. Some of these technologies like soil washing using particle size separation and chemical extraction with aqueous solutions of surfactants and mineral acids are in full-scale use (Kuhlman and Greenfield, 1999; Mann, 1999) and technologies addressed under "Phytoremediation" are still largely in the development phase. Toxic metals and other contaminants can be isolated and contained to prevent their further movement, i.e. by leaching through soil or by soil erosion. This can be achieved by capping the site with asphalt or other impermeable material to prevent infiltration of water, by planting permanent plant cover (phytostabilization) or by covering the site with unpolluted soil.

Phytoremediation

Phytoremediation involves the use of plants to remove, transfer, stabilize and/or degrade contaminants in soil, sediment and water (Hughes *et al.*, 1997). The idea that plants can be used for environmental remediation is very old and cannot be traced to any particular source. A series of fascinating scientific discoveries, combined with interdisciplinary research, has allowed phytoremediation to develop into a promising, cost-effective, and environmentally friendly technology. The term phytoremediation ("phyto" meaning plant, and the Latin suffix "remedium" meaning to clean or restore) refers to a diverse collection of plant-based technologies that use either naturally occurring, or genetically engineered, plants to clean contaminated environments (Cunningham *et al.*, 1997; Flathman and Lanza, 1998).

Some plants which grow on metalliferous soils have developed the ability to accumulate massive amounts of indigenous metals in their tissues without any symptoms of toxicity (Reeves and Brooks, 1983). Phytoremediation is a publicly appealing (green) remediation technology. However, phytoextraction can be effectively applied only for soil contaminated with specific and potentially toxic metals and metalloids *viz.*, Ni, Zn and As which are readily bioavailable for plants. Establishment of vegetation cover thus can fulfil the objectives of stabilization, pollution control, visual improvement and removal of threats to human being. The use of plants for purifying contaminated soils and the methods adopted are expensive.

Phytoremediation is an alternative method that utilizes plants to clean up a contaminated area. The method is relatively easy to implement and can reduce remedial cost restoring the habitat. It was not until the 1990s, that the concept of phytoremediation emerged as a new technology that uses plants to reduce, remove, degrade or immobilize environmental toxins primarily those of anthropogenic origin with the aim of restoring contaminated sites to a condition useable for private or public applications.

Plants have shown the capacity to withstand relatively high concentrations of organic chemicals without having any toxic effects; they uptake and convert chemicals rapidly to a lesser toxic metabolite in many cases. In addition, they stimulate the degradation of organic chemicals in the rhizosphere by the release of root exudates, enzymes and the build-up of organic carbon in the soil. For metal contaminants, the plants show a potential for phytoextraction (uptake and recovery of contaminates into above-ground biomass), rhizofiltration (filtering metals from water into root systems), or stabilizing waste sites by erosion control and evapotranspiration of large quantities of water (phytostabilization) and so on. All phytoremediation processes are not exclusive and may be used simultaneously. Phytoremediation method is a cost effective, environmental friendly and an aesthetically pleasing approach having long term applicability. This method is well suited for use at very large field sites where other methods of remediation are not cost effective or practicable. However, the plant species used in this means must grow well under toxic levels of heavy metal contamination and need to produce high biomass.

The selection of trace element tolerant plant species is a key factor to the success of remediation in the degraded mine soil. For long term remediation, metal tolerant species are used for re-vegetation of mine tailings (Lan *et al.*, 1997) and herbaceous legumes can be used as pioneer species to solve the problem of nitrogen deficiencies in mining waste lands because of their nitrogen fixing ability. Metal uptake capacity by Caryophyllaceae species (Dianthus, Minuartia, Scleranthus and Silene) were tested in metalliferous soil (containing Cu, Pb, Zn, Cd, Ni, Cr, Fe, Mn, Ca, Mg) in northern Greece (Konstantinou and Babalonas, 1996). The study stated that *Scleranthus perennis* showed highest Cu concentration (205 mg/kg), *Minuartia* cf. *Bulgarica* hyper-accumulated Pb (1175 mg/kg). The Ca concentration in plants was in most cases much higher than those in soil, whereas the contrary was true for Mg. As a result the Ca/Mg ratio which was in almost all cases lower than 1 mg in soil was much increased in the plants.

The plant species used for heavy metal accumulation are categorised as "metal excluder", "metal indicator" and "metal accumulator". The metal excluder species prevent the metal from entering the aerial parts of plant, maintain low and constant metal concentration over a broad range of metal concentration in soil, and restrict metal accumulation in roots. The plant may alter its membrane permeability, change metal binding capacity of cell walls, or exclude more chelating substances. The metal indicator species is one that actively accumulate metal in their aerial tissue and generally reflect metal level in the soil. They tolerate the existing concentration level of metals by producing intracellular metal binding compounds (chelators) or alter metal compartmentalization pattern by storing metals in non-sensitive parts (Ghosh and Singh, 2005). Metal indicator species are used for mine prospecting to find new ore bodies. The metal accumulator species concentrates the metal in their aerial parts to the level far exceeding than soil and known as hyper-accumulators. The metal accumulators are herbaceous or woody plants which accumulate and tolerate without visible symptoms a hundred times or greater metal concentrations in shoots than those usually found in non-accumulators.

The process of metal accumulation involves several steps; one or more of which are responsible for the hyper accumulation in plants:

Solubilisation of Metal From Soil Matrix

Many metals are found in soil-insoluble form. The plants use two methods to desorb metals from the soil matrix like acidification of rhizosphere through the action of plasma membrane proton pumps and secretion of ligands capable of chelating the metal. The plants have evolved these processes to liberate essential metals from the soil, but soils with high concentrations of toxic metals will release both essential and toxic metals to solution (Lasat, 2000).

Uptake of Metal into the Root

Soluble metals can enter into the root system by crossing the plasma membrane of the root endodermal cells or can enter root apoplast through the space between cells. While it is possible for solutes to travel up through the plant by apoplastic flow, the more efficient method of moving up is through the vasculature of the plant called xylem. To enter into the xylem, the solutes must cross the Casparian strip, a waxy coating which is impermeable to the solutes unless they pass through the cells of endodermis. Therefore, to enter the xylem, metals must cross a membrane probably through the action of membrane pump or channel. Most of the toxic metals are thought to cross these membranes through pumps and channels intended to transport essential elements. Excluder plants survive by enhancing specificity for the essential element or pumping the toxic metal back out of the plant (Hall, 2002).

Transport of Metal to the Leaves

Once loaded into the xylem through the flow of xylem sap, the metal is transported to the leaves, where it is loaded into the cells of the leaf, again crossing a membrane. The cell types where the metals are deposited vary between hyper accumulator species.

Detoxification and/or Chelation of Metal

At any point along the pathway, the metal could be converted to a less toxic form through chemical conversion or by complexation. Various oxidation states of toxic elements have very different uptake, transport, and sequestration or toxicity characteristics in plants. Chelation of toxins by endogenous plant compounds can have similar effects on all of these properties as well. As many chelators use thiol groups as ligands, the sulfur (S) biosynthetic pathways have been shown to be critical for hyper accumulator function (Van Huysen *et al.*, 2004) and for possible phytoremediation strategies.

Sequestration and Volatilization of Metal

The final step for accumulation of most of the metal is by the sequestration of these metals away from any cellular processes which it might disrupt. Sequestration usually occurs in the plant vacuole where the metal/metal-ligand must be transported across the vacuolar membrane. Metals may also remain in the cell wall instead of crossing the plasma membrane into the cell, as the negative charge sites on the cell walls may interact with polyvalent cations (Wang and Evangelou, 1994).

An alternative remediation technique for metal-contaminated site is phytostablization also called "inplace inactivation" or "phytorestoration". It is a type of phytoremediation technique that involves stabilizing heavy metals with plants in the contaminated soil. To be a potentially cost-effective remediation technique, the plants selected need to tolerate high concentrations of heavy metals and stabilize heavy metals in the soil by plant roots with some organic or inorganic amendment such as domestic refuse, fertilizer, and others. The success of reclamation scheme is dependent upon the choice of plant species and their methods of establishment. There are some important considerations when selecting plants for phytostabilization.

- The plants should be tolerant of soil metal level as well as other inherent site conditions like soil pH, salinity, soil structure, water content, lack of major nutrients and organic materials.

- The plants chosen for phytostablization should be poor translocators of metal contaminants to above ground plant tissues that could be consumed by humans or animals.

- The plants must grow quickly to establish ground cover having dense rooting system and canopy and to have relatively high transpiration rate to effectively dewater the soil (Raskin & Ensley, 2000).

Phytoextraction

This process reduces soil metal concentrations by cultivating plants with a high capacity for metal accumulation in shoots (Barcelo and Poschenrieder, 2003). The plants must extract large concentration of heavy metals into their roots, translocate the heavy metals to above ground shoots or leaves and produce large quantity of plant biomass that can be easily harvested. When plants are harvested, contaminants are removed from the soil. Recovery of high price metals from the harvested plant material may be cost effective (eg. phytomining of Ni, Tl or Au). Alternatively, the dry matter can be burnt and the ash disposed of under controlled conditions. Phytoextraction is also known as phytoaccumulation, phytoabsorption and phytosequestration. Phytoextraction can be divided into two categories: continuous and induced (Salt *et al.*, 1998). Continuous phytoextraction require the use of plants that accumulate particularly high levels of toxic contaminants throughout their lifetime (hyper accumulators), while induced phytoextraction approaches enhanced toxin accumulation at a single time point by addition of accelerants or chelators to the soil.

Rhizofiltration

This technique is used for cleaning the contaminated surface water or waste water such as industrial discharge, agricultural runoff or acid mine drainage by absorption or precipitation of metals onto roots or absorption by roots or other submerged organs of metal tolerant aquatic plants. For this purpose, plants must not only be metal resistant but also have a high absorption surface and must tolerate hypoxia (Dushenkov *et al.*, 1995). Contaminant should be those that sorb strongly to roots such as hydrophobic organics, lead, chromium (III), uranium and arsenic (V). Plants like sunflower, Indian mustard, tobacco, rye, spinach and corn are studied for their ability to remove lead from effluent with sunflower having the greatest ability (Raskin and Ensley, 2000).

Phytostabilization

It refers to the holding of contaminated soil and sediment in place by vegetation and to immobilize toxic contaminants in the soil. It is also known as in-place inactivation or phytoimmobilization. Phytostabilization occurs through sorption, precipitation, complexation or metal valence reduction (Ghosh and Singh, 2005). Metals do not ultimately degrade thus capturing them in situ become the best

Table 1: Plant species used for accumulation of heavy metals in soil

Process	Mechanism	Media	Contaminants	Typical plants
1.Phytoextraction	Hyper accumulation	Soil, Brownfields, Sediments	Metals (Pb, Cd, Zn, Ni, Cu) with EDTA addition for Pb, Se.	Sunflowers, Indian mustard, Rape seed plants, Barley, Hops, Crucifers, Serpentine plants
2.Rhyzofiltration	Rhizosphere accumulation	Groundwater, Water and Wastewater in Lagoons or Created Wetlands	Metals (Pb, Cd, Zn, Ni, Cu) Radionuclides (^{137}Cs, ^{90}Sr, ^{238}U) Hydrophobic organics	Aquatic plants: Emergents (bulrush, cattail, pondweed, arrowroot, duckweed), Submergents: (algae, stonewort, parrot feather, *Hydrilla*)
3.Phytostabilization	Complexation	Soil, Sediments	Metals (Pb, Cd, Zn, As,Cu, Cr, Se, U) Hydrophobic organics (PAHs, PCBs, dioxins, furans, pentachlorophenols, DDT, Dieldrin)	Phreatophyte trees to transpire large amounts of water for hydraulic control; Grasses with fibrous roots to stabilize soil erosion; Dense root systems are needed to sorb/bind contaminants
4.Phytovolatalization	Volatization by leaves	Soil, Groundwater, Sediments	Mercury, Selenium, Tritium	Poplar, Indian mustard, Canola, Tobacco plants
5.Phytodegradation	Degradation in plant	Soil, Groundwater, Landfill leachate, Land application of wastewater	Herbicides (atrazine, alachlor) Aromatics (BTEX) Chlorinated aliphatics (TCE) Nutrients (NO_3'', NH_4^+, PO_4^{3-}) Ammunition wastes (TNT, RDX)	Phreatophyte trees (Poplar, willow, cottonwood). Grasses (Rye, Bermuda, Sorghum, Fescue). Legumes (Cloevr, Alfalfa, Cowpeas)
6.Rhyzodegradation	Degradation by plant rhizosphere microorganisms	Soil, Sediments, land application of Wastewater	Organic contaminants (pesticides, aromatics and polynuclear aromatic hydrocarbons [PAHsl]	Phenolics releasers (Mulberry, Apple, Orange), Grasses with fibrous roots (Rye, Fescue, Bermuda) for contaminants 0-3 ft deep; Phreatophyte trees for 010 ft; Aquatic plants for sediments

alternative at sites with low contamination levels or at vast contaminated areas where a large scale removal action or other *in-situ* remediation is not possible. Plants with high transpiration rate, such as grasses, sedges, forage plants and reeds are useful for phytostabilization by decreasing the amount of ground water migrating away from the site carrying contaminants. Combining these plants with hardy, perennial, dense rooted or deep rooting trees (popular, cottonwoods) can be an effective combination (Berti and Cunningham, 2000).

Phytovolatization

It involves the use of plants to take up contaminants from the soil transforming them into volatile form and transpiring them into the atmosphere. Selenium (Se) is taken up by some of the plants and volatilized. The axenic cultured isolate of single celled freshwater microalgae (Chlorella sp.) have metabolized the toxic selenate to volatile dimethylselenide at exceptionally high rates when transferred from mineral solution to water for 24h than those similarly measured for wetland macroalgae and higher plants (Neumann *et al.*, 2003). Hyper-volatilization of selenate by microalgae cells may provide a novel detoxification response. Uptake and evaporation of Hg is achieved by some bacteria. The bacterial genes responsible have already been transferred to Nicotiana or Brassica species and these transgenic plants may act as tool for cleaning Hg contaminated soil (Meager *et al.*, 2000).

Phytodegradation

It involves the uptake, metabolisation and degradation of contaminants within the plant or degradation of contaminants in the soil sediment, sludge, groundwater or surface water by the enzymes produced and released by plant. Phytodegradation does not depend on microorganisms associated with the rhizosphere. The process is also known as phytotransformation and is a contaminant destruction process. The major water and soil contaminant trichloroethylene (TCE) was found to be taken up by hybrid poplar trees (*Populas deltoids nigra*), which breaks down the contaminant into its metabolic components (Newman *et al.*, 1997).

Rhizodegradation

Rhizodegradation is the breakdown of organics in the soil through microbial activity of the root zone (rhizosphere). Enhanced rhizosphere degradation uses plants to stimulate the rhizosphere microbial community to degrade organic contaminants (Kirk *et al.*, 2005). Grasses with high root density, legumes and alfalfa that fix nitrogen and having high evapotranspiration rate are associated with different microbial populations. Significantly higher populations of total heterotrophs, denitrifers were found in rhizosphere soil around hybrid poplar trees in a field plot than in non-rhizosphere soil (Jordahl *et al.*, 1997).

Phytorestoration

It involves complete remediation of contaminated soils to fully functioning soil (Bradshaw, 1997). In particular, this division of phytoremediation uses plants that are native to a particular area in an attempt to return the land to its natural state.

Vetiveria zizanioides (L.) Nash

Vetiveria zizanioides (L.) Nash belongs to the family Poaceae. The vetiver grass is originated from Indian sub-continent and commonly cultivated in flood plains and stream banks. The plants are also found throughout the tropical and subtropical regions of Africa, Asia, America, Australia, and Mediterranean Europe. The vetiver grass due to its unique morphological and physiological characteristics is widely known for their effectiveness in erosion and sediment control. The plants are highly tolerant to extreme soil condition including heavy metal toxicity. The plant is extensively grown for phytoremediation of heavy metals from soil such as of aluminium (68%-87%), arsenic (100-250 ppm), cadmium (20 ppm), copper (50-100 ppm), chromium (200-600 ppm), nickel (50-100 ppm) and manganese (> 578 ppm).

Vetiver grass has stiff and erect stem and can stand up to relatively deep water flow. A dense hedge of these plants are formed acting as sediment filter and water spreader. The new shoots emerging from the base of these plants help to withstand heavy traffic and heavy grazing pressure. The grass has no stolon and consists of massive, finely structured root system; growing very fast and bear a high value for its essential oil. The deep rooting system of vetiver makes it tolerant to extreme drought conditions and is difficult to dislodge by strong current. Vetiver is sterile and non-invasive and highly resistant to pest attack, disease and fire. The new shoots developing from underground crown protects the plants from fire and frost.

The grass is fast-growing and tolerates various environmental conditions including soil pH value 3.0 to 10.5 and temperature 14 to 55 °C (Truong and Baker, 1997). High concentrations of heavy metal in the soil contaminated with multiple elements do not affect the plant's growth. The ability of these plants to accumulate high concentration of heavy metals in the roots especially Pb (1094 mg/kg of dry root weight), Fe (24737 mg/kg of dry root weight) (Wilde *et al.,* 2005) and Zn (1162 mg/kg of dry root weight) (Yang *et al.,* 2003) has been studied. Due to the unique morphological and physiological characteristics, these plants are known for their tolerance to extreme soil condition including heavy metal toxicity. These plants also show high efficiency in absorbing dissolved nutrients and heavy metals from polluted water. The plants can survive in a soil environment containing high concentrations of wide range of heavy metals and accumulate the metals into root and shoot (Truong, 1999b). Studies have been carried out to evaluate the ability of the plant to remove heavy metals such as arsenic, lead, copper, zinc, cadmium, mercury from contaminated soils.

The metal accumulation capacity and growth performance ability of vetiver grass has been studied. Most of the heavy metals were accumulated by roots and translocation of metal from roots to shoots was restricted (Yang *et al.,* 2003). Studies have shown that vetiver grass act as typical heavy metal excluder (Shu *et al.,* 2004). Therefore, this plant is more suitable for phytostabilization of toxic mined land showing relatively high level of metal accumulation in roots. Thus, vetiver grass makes the best choice of species used for phytoremediation including phytostabilization and phytoextraction of metal contaminated soils.

Heavy Metal: Lead

Vetiver grass has been studied to evaluate phytoextraction of different heavy metals. The plant is tolerant to extremely high levels of lead in soil with 100% survival rate and showed good growth performance even at high Pb concentrations. The tolerance of this plant for Pb was upto 10750 mg/kg in artificially contaminated soil and 9020 mg /kg in naturally contaminated mine soil (Rotkittikhun *et al.*, 2007). These values were the highest soil Pb. Vetiver grass survived and showed noticeable growth in soil with lead concentration upto 5000 mg/ kg. This grass cultivated in firing range soil contaminated with 3281.6 mg / kg (Wilde, Brigmon, and Dunn, 2005) and mining tailing soil with 4164 mg / kg (Yang *et al.*, 2003) were studied.

A significant accumulation of lead concentration in the shoots and roots of vetiveria increased significantly with increasing Pb concentration in soils (Rotkittikhun *et al.*, 2007; Wong *et al.*, 2007; Chen, Shena, and Lib, 2004). Accordingly, the concentration of Pb accumulation in the root increased from 19.4 to 449 mg/ kg DW as the level of Pb in the soil was increased from 0 to 1000 mg (Wong *et al.*, 2007). Similarly, the accumulation of Pb in the shoots varied from 6.98 to 17.4 mg kg-1 DW with increased soil contamination (Wong *et al.*, 2007). It was studied that the grass accumulated Pb level of upto 0.22% of its total dry matter. The studies made by Antioncha *et al.*, (2007) showed that 0.4% and 1% of Pb accumulated in shoots and roots respectively within 30 days of treatment. Lead concentrations in vetiver plant tissue samples of upto 1390-1450 mg /kg DW were detected after nine months of cultivation in firing range soils.

Most of the absorbed lead in vetiver plants tends to accumulate in roots and generally, only a small portion of lead is moved to the shoots. When Pb enters the plant roots, it immediately gets exposed to the solution in the intercellular spaces that has high phosphate concentrations, relatively high pH and high carbonate-bicarbonate concentrations. Under these conditions, Pb precipitates out of solution in the form of phosphates or carbonates that can be seen in electron micrographs of roots from plants grown hydroponically in Pb solutions. The formation of insoluble Pb compounds reduces Pb translocation in plants (Cunningham and Berti, 2000). The translocation ratio of Pb from roots to shoots that was defined as the percentage of shoot Pb concentration versus root Pb concentration reflected the efficiency of Pb phytoextraction. The translocation value decreased as the concentration of Pb in the soil increased. The studies showed that the amount of Pb fixed in roots is much higher than that of Pb translocated into shoots. The majority of accumulated Pb in roots makes vetiver a useful plant for phytostabilization. The process of using vetiver plants for immobilsation of soil contaminant under in situ condition ranged from 12.5 to 375 mg/ kg DW in shoots and from 18.7 to 4940 mg /kg DW in roots in the soil spiked with 0, 100, 1000 and 10000 mg / kg of Pb (Rotkittikhun *et al.*, 2007).

The vetiver plant also acts as lead hyper accumulator. According to Shah and Nongkynrih (2007), the plant is a Pb hyperaccumulator if it accumulate Pb of atleast 0.1% of DW equivalent to 1000 mg / kg DW. The vetiver plants grown in soil amended with 10000 mg / kg of Pb accumulated upto 4940 and 359 mg / kg

DW in roots and shoots respectively. The accumulation of Pb and its translocation from roots to shoots in vetiver is improved by adding chelating agents. Chelating agents increase the mobility and bioavailability of metal in the soils and also increase its accumulation (Wu, Hsu, and Cunningham, 1999; Ebbs and Kochian, 1998; Blaylock *et al.*, 1997; Huang *et al.*, 1997; Huang and Cunningham, 1996). Among chelating agents, EDTA has been shown to be the most efficient in mobilizing Pb from various soil (Shen *et al.*, 2002; Wenzel *et al.*, 2002; Huang *et al.*, 1997). The application of EDTA significantly increased the biomass of vetiver plants (Wilde, Brigmon, and Dunn, 2005). The presence of EDTA in Pb contaminated soil resulted in increased Pb concentration in shoots and roots of Vetiveria. The increase in Pb translocation from shoots to roots by application of EDTA is potentially useful for the phytoextraction of lead from contaminated soils. However, EDTA and EDTA-heavy metal complexes may leach into groundwater, causing further environmental pollution in surrounding areas (Romkens *et al.*, 2002; Barona, Aranguiz, and Elias, 2001; Kedziorek *et al.*, 1998). To prevent migration of mobilized heavy metal down the soil profile, a phytoremediation program need to be designed with optimum chelate concentration, time and location of the chelate application including root system of plants (Chen, Shena and Lib, 2004). Vetiver grass has very long (3-4 m), massive and complex root system and can penetrate to deeper layers of soil (Dalton, Smith, and Truong, 1996; Truong, 2000; Pichai, Samjiamjiaras and Thammanoon, 2001), thereby preventing heavy metal leaching. The soil matrix with planted vetiver showed re-adsorption of 98.12% of initially applied Pb compared to 61.55% without vetiver plants. The high level of Pb retained in soil matrix in presence of vetiver was due to the significant decrease in water content of the soil because of high transpiration from the plant. Such process displayed by this grass play an important role to immobilize heavy metals from soil matrix (Chen, Shena and Lib, 2004). The efficiency of lead phytoremediation using vetiver together with EDTA can be further enhanced by regular removal of above-ground part of plant which facilitates Pb translocation from roots to shoots.

The effect of mycorrhizae on growth and uptake of N, P, Zn and Pb by vetiver grass as host were studied in green house trial. The host plants with arbuscular mycorrhizal fungi (AMF), *Glomus mosseae* and *G. Itraradices* spores significantly increased the growth and uptake of P. Mycorrhizal colonization increased the Pb and Zn uptake by vetiver plants under low soil metal concentrations (0 and 10 mg/kg) whereas under higher concentration (100 and 1000 mg/kg) it decreased Pb and Zn uptake. The P concentration in soil was negatively correlated with mycorrhizal colonization as well as Zn or Pb concentrations. The studies showed that inoculation of host plants with AMF protect them from the potential toxicity caused by increased uptake of Pb and Zn, and the degree of protection varied according to the fungus and host plant combination. The remediation of Pb contaminated soil using vetiver grass showed that these plants are not only retains the contaminant but reduces the chance of spreading.

Heavy Metal: Zinc

The tolerance of vetiver plants to zinc in the soil has been studied. The uptake of Zn was between 1583 and 4377 mg/ kg (Hoang, Tu, and Dao, 2007; Chiu, Ye, and Wong, 2006; Zhuang *et al.*, 2005; Yang *et al.*, 2003). The plant survived at different contamination levels in the soil containing 2472 mg /kg of Zn or less (Chiu, Ye, and Wong, 2006; Zhuang *et al.*, 2005). The vetiver plants accumulated high concentrations of Zn in both roots and shoots. The Zn uptake was higher than 10000 mg/ kg DW in the shoots and roots after daily irrigation with 250 ml of 653 ppm Zn (ZnCl2) solution for 30 days (Antiochia *et al.*, 2007). Hence, vetiver serve as Zn hyper accumulator (Barker and Brook, 1989). However, the ability of vetiveria to accumulate Zn depends on the fraction containing diethylene triamine pentaacetic acid (DTPA) which is the extractable Zn or plant available Zn. It was observed that high level of soluble Zn (Antiochia *et al.*, 2007; Chiu, Ye, and Wong, 2006) produced higher plant uptake than that of non-readily available fraction of Zn (Wong *et al.*, 2007; Zhuang *et al.*, 2005 and Yang *et al.*, 2003). Hence, the portion of bio-available Zn plays a major role in bioremediation when compared to the level of soil contamination.

The heavy metal translocation from roots to shoots was dependent on the fraction of water soluble Zn available to the plant and usually get retained in the roots. However, when soil contains high level of soluble Zn, the translocation ratios from root to shoot was as high as 100% (Antiochia *et al.*, 2007). A possible approach for enhancing the Zn bioavailability is by applying chelating agents. Nitrilotriacetic acid (NTA) serve as the most effective chelating agent to improve water solubility of Zn in the soil (Chiu, Ye, and Wong, 2006). At application levels of 10 and 20 m mol/ kg of NTA the extractable Zn in the soil was 2 to 15 times higher than with the application of N-(2-hydroxyethyl) iminodiacetic acid (HEIDA), N-hydroxy-ethylene-diamine-triacetic acid (HEDTA), trans-1,2-cyclo-hexylene-ditrilo-tetra-acetic acid (CDTA), diethylene- triamine-penta acetic acid (DTPA), ethylene-diamine-triacetic acid (EDTA), citric acid, ethylenebis (oxy-ethylene-trini-trilo) tetra-acetic acid (EGTA), and malic acid (Chiu, Ye, and Wong, 2006).

Multi Heavy Metal

Glasshouse and field studies have demonstrated that vetiver plants are tolerant not only to high concentrations of individual heavy metal in soils but also in combinations of several heavy metals. In glasshouse studies, these plants could survive and grow well on contaminated soil containing total Pb of 1155-3281.6 mg/kg, Zn of 118.3-1583 mg/kg, Cu of 68-1761.8 mg/kg (Chiu, Ye, and Wong, 2006; Wilde, Brigmon, and Dunn, 2005). In field studies, the vetiver plants survived cultivation on mine tailing soil containing total Pb of 2078-4164 mg /kg, Zn of 2472-4377 mg/kg, Cu of 35-174 mg/kg, and Cd of7-32 mg/kg (Zhuang *et al.*, 2005; Shu *et al.*, 2004; Yang *et al.*, 2003; Shu *et al.*, 2002). Organic matter (domestic refuse and sewage sludge), inorganic fertilizer and a combination of organic matter and inorganic fertilizer greatly enhanced the survival rate, growth and biomass of vetiver plants cultivated in soil contaminated with combination of heavy metals (Chiu, Ye, and Wong, 2006; Wilde, Brigmon, and Dunn, 2005; Yang *et al.*, 2003; Shu *et al.*, 2002).

The vetiver plants are extensively studied and proved to be highly tolerant to extremely adverse conditions. Hence these plants serve as an appropriate candidate for phytoremediation and can be used for rehabilitation of mine tailings, garbage landfills, and industrial waste dumps which are acidic or alkaline, high in heavy metal contamination.

References

Bradshaw A.D. (1997). What do we mean by restoration? In Restoration Ecology & Sustainable Devlopment. pp.8-16.Cambridge University Press.

Godbold DL, Hüttermann A.(1985), Effect of zinc, cadmium and mercury on root elongation of Picea abies (Karst.) seedlings, and the significance of these metals to forest.*Environmental Pollution* 38, 375–381.

Jordahl, J., Foster, L., Alvarez, P.J., and Schnoor, J. (1997). Effect of Hybrid Poplar Trees on Microbial Populations Important to Hazardous Waste Bioremediation. *Environ. Toxicol. Chem.*16:1318-1381.

Konstantinou, M., Babalonas, D. (1996), Metal uptake by Caryophyllaceae species from metalliferous soils in northern Greece. Plant Systematics and Evolution 203: 1-10.

Shah, K. and J.M. Nongkynrih, (2007). Metal hyperaccumulation and bioremediation. Biol. *Plantarum*, 51: 618-634.

Shu, W., Zhao, Y., Yang, B., Xia, H., Lan, C. (2004), Accumulation of heavy metals in four grasses grown on lead and zinc mine tailings. *J. of Env. Sciences* 16: 730-734.

Truong, P.N. and Baker, D. (1997). The role of vetiver grass in the rehabilitation of toxic and contaminated lands in Australia. International Vetiver Workshop, Fuzhou, China, Oct. 1997.

Wilde, Brigmon, and Dunn, (2005). Phytoextraction of lead from firing range soil by Vetiver grass. Epub ,61(10):1451-7.

Wong, J.L., and Wessel, G.M. (2007), FRAP analysis of secretory granule lipids and proteins in the sea urchin egg. In: Exocytosis and Endocytosis, A. Ivanov, ed. (Totawa, NJ: Humana Press).

Yang, B., Shu, W., Ye, Z., Lan, C., Wong, M. (2003). Growth and Metal Accumulation in Vetiver and two Sesbania Species on Lead/Zinc Mine Tailings. *Chemosphere* 52: 1593-1600.

Chen, S.-H., and J. Dudhia, (2000). Annual report: WRF physics. Air Force Weather Agency, 38 pp. [Available online at http:// wrf-model.org.]

CUNNINGHAM, S.D.; BERTI, W.R. and HUANG, J.W.(1995). Phytoremediation of contaminated soils.*Trends in Biotechnology*, vol. 13, no. 9, p. 393-397.

Lan JY, Skeberdis VA, Jove T, Zheng X, Bennett MV, Zukin RS (2001), Activation of metabotropic glutamate receptor 1 accelerates NMDA receptor trafficking. *J Neurosci* 21:6058–6068

Roongtanakiat, N.; Tangruangkiat, S.; and Meesat, R. (2007). Utilization of vetiver grass (Vetiveria zizanioides) for removal of heavy metals from industrial wastewater. *Science Asia* 33: 397-403.

Truong, P.N. and Hengchaovanich, D. (1999). Vetiver Grass Technology: Potential applications and benefits in the protection of farm and forestry lands, infrastructure and the environment in Viet nam. Report to TVN.

Zhuang Q, Melillo J M, Kicklighter D W, Prinn R G, McGuire A D, Steudler P A, Felzer B S and Hu S . (2004). Methane fluxes between terrestrial ecosystems and the atmosphere at northern high latitudes during the past century: A retrospective analysis with a process-based biogeochemistry model Glob. *Biogeochem.* Cycles 18 GB3010

Chapter 11

Applications of Nanotechnology in Environmental Science

T.C. Taranath[1], Veena Rokhade[2], Nandini Banad[3] and A.G. Devi Prasad[4]

[1,2,3]*DOS in Botany, Karnatak University, Pavate Nagar,*
Dharwad – 580 003, Karnataka
[4]*DOS in Environmental Science, University of Mysore, Manasagangotri,*
Mysore-06, Karnataka

ABSTRACT

The principle objective of this paper is to shed light on the capability of nanotechnology to deliver environmental benefits both in production processes and in product. New nanotechnologies have been developed to enhance environmental protection, improve pollution detection and remediation.

Nano-scale catalysts significantly reduce the use of raw materials and minimize energy consumption in chemical industries. Manufactured nano-materials significantly improve efficient and low cost methods for energy transformation and storage (i.e., fuel cells and lithium ion batteries), enabling low emission and fuel consumption cars. Water purification can be made more effective by using nano-sized/nano-structured materials. In addition nano materials can be substitute to toxicological hazardous substances (flame retardants and toxic inhibitors). Nano-coating with implantation materials having bio compatible surfaces or easy to clean surfaces with bioacidic or anti-adhesion properties is also being carried out.

Key words: Nano-remediation, Trichloroethene, Silver nanocrystals, Nanoparticle

Introduction

Nanotechnology is a field of applied science dealing with the control of matter at dimensions roughly 1 to 100nm, which is one-billionth of a meter. The appealing feature of nanoparticles is that they can be engineered to function in ways that naturally occurring materials do not. Their larger surface area per unit volume and enhanced chemical reactivity can be exploited in novel applications.

Researchers anticipate profound effects of nanotechnology on industry and technology, human health, social and economic development and the environment. In public and private investments, application of nanotechnology is significantly increasing because of its potential to transform sectors as diverse as medicine, energy, water supply and transportation.

Nanotechnology is poised to become a major element in the global economy. In 2004, nanotech products accounted for less than 0.1 per cent of revenue from manufacturing. By 2014 they are projected to account for 14 per cent, totaling US $2.6 trillion - a figure that will match the information technology and telecommunication industries combined (Lux Research 2004).
Nanotechnology has the potential to contribute to the targets set for achieving the UN millennium development goals, particularly in the areas of affordable energy, clean water, human health and efficient ways to harness renewable energy. This can help to reduce dependency on conventional energy sources and support greater energy self-sufficiency, an important goal for developing nations. Nanofiltration may improve access to safe and affordable drinking water and basic sanitation.

Nanomaterial based sensors and membranes will have a positive impact on the detection of contaminants and the cleaning of water pollutants. Enhanced fuel cells, solar panels and high capacity batteries will save primary energy consumption; allowing a more effective use of negative regenerative energy sources, which will also help in constructing cars with lower carbon dioxide emission.

Importance of Nanoparticles

The application of nanotechnologies and the use of manufactured nanomaterials have potential to significantly reduce the environmental impact of technical processes and products. In addition, the use of nanomaterials may lead to innovation in other sectors; for example, automotive, aeronautic and energy sectors. Some examples of positive environmental effects using nanomaterials are given below. There are many other promising areas and applications for manufactured nanoparticles.

Nanoparticles have the potential to deliver environmental benefits both in production processes and in product. New nanotechnologies poised to enhance environmental protection and improve pollution detection and remediation. A growing number of nanoparticles are 'functionalized', meaning that their surfaces are designed to trigger specific chemical or biological reaction (table1). This offers novel mechanisms for targeted delivery of drugs in humans and animals or of pesticides and fertilizers for crop. Targeted delivery facilitates more effective use of substance in far lower concentration, so it has potential to reduce use of chemicals and materials, particularly those with negative environmental impacts such as pesticides.

- Catalytic steps are widely used in the chemical industry. Nanoscale catalysts significantly reduce the use of raw materials and minimize side streams and energy consumption. One striking example is the automotive catalysts, reducing hydrocarbon, nitrogen oxide and carbon monoxide emissions by 90%. In mobile as in stationary applications, nanomaterial based catalysts and filters will lead to cleaner combustion processes and hence a reduction of emissions. Also, by using nanomaterials, the efficiency of regenerative energy sources like solar cells can be improved. Manufactured nanomaterials significantly improve efficient and low cost methods for energy transformation and storage (i.e., fuel cells and lithium ion batteries), enabling low emission and fuel consumption cars.

- Water purification can be made more effective by using nanosized/ nanostructured materials. Membranes and highly sensitive nanostructured sensors enhance the early recognition of pollutants before damage can occur. In addition, nanomaterials can substitute (eco-) toxicological hazardous substances (i.e., flame retardants and toxic corrosion inhibitors).

- Useful for the health sector is the coating of implantation materials with biocompatible surfaces or easy-to-clean surfaces due to nanocoatings with biocidic or anti-adhesion properties. New drug delivery systems based on manufactured nanomaterials to cure, e. g., neurodegenerative diseases are of strong interest for the pharmaceutical sector.

Table:1 Properties of Nanomaterials

SI. No	Properties	Examples
1.	Chemical	Higher surface-to-volume ratio makes particles highly reactive, increasing their efficiency as catalysts for desired chemical reactions.
2.	Electrical	Increased electrical conductivity in ceramics and magnetic nanocomposites, increased electrical resistance in metals.
3.	Mechanical	Improved hardness and toughness of metals and alloys, ductility and super plasticity of ceramics.
4.	Optical	Increased conversion efficiency of light to electrical charge in Photoelectronic devices such as solar panels.
5.	Sterical	The spatial arrangement of atoms in a substance affects chemical reactions and facilitates increased selectivity. For example, hollow spheres can be used to transport and control the release of specific drugs.
6.	Biological	Increased permeability through biological barriers (membranes, blood-brain barrier, etc.), improved biocompatibility (i.e., the quality of NOT having toxic or injurious effects on biological systems).

Application of Nanoparticles in Environmental Problems

1. Nanoparticles in Waste Water Treatment

There are four classes of nanoscale materials that are being evaluated as functional materials for water purification:

(a) Dendrimer

(b) Metal-containing nanoparticles,

(c) Zeolites

(d) Carbonaceous nanomaterials.

These have a broad range of physicochemical properties that make them particularly attractive as separation and reactive media for water purification. Nanomaterials reveal good result than other techniques used in water treatment because of its large surface area. It is suggested that these may be used in future at large scale water purification.

a) Dendrimer in Waste Water Treatment

Reverse Osmosis (RO) membranes have pore size of 0.1-1.0 nm and thus are very effective in retaining dissolved inorganic and organic solutes with molar mass below 1000 Da (Zeman and Zydney, 1996). Nanofilter (NF) membranes are used in removing hardness (e.g., multivalent cations) and organic solutes with molar mass between 1000-3000 Da for e.g., natural organic material (Zeman and Zydney, 1996). However, high pressure is required to operate both RO and NF membranes. Conversely, Ultrafine (UF) membranes require lower pressure (200-700 kPa). Unfortunately, they are not very effective at removing dissolved organic and inorganic solute with molar mass below 3000 Da. Advances in macromolecular chemistry such as the invention of dendritic polymers are providing unprecedented opportunities to develop effective UF processes for purification of water contamination by toxic metal ions, radionuclide, organic and inorganic solutes, bacteria and viruses.

b) Metal Nanoparticles

Nanoparticles have two key properties that make them particularly attractive as sorbents. Nanoparticles can also be functionalized with various chemical groups to increase their affinity towards target compounds.

Stemenov *et al.* showed that MgO nanoparticles and magnesium (Mg) nanoparticles are very effective biocides against Gram-positive and gram-negative bacteria (*Escherichia coli* and *Bacillus megaterium*) and bacterial spores (*Bacillus subtilis*).

Silver nanocrystals are used as an anti-microbial, anti-biotic and anti-fungal agent when incorporated in coatings, nanofibers, first aid bandages, plastics, soap and textiles, treatment of certain viruses, self cleaning fabrics, as conductive filler and in nanowire and certain catalyst applications. Ag (I) and silver compounds have been used as antimicrobial compounds against coliforms found in waste water (Jain, and Pradeep,2005). Silver nanoparticles, nanodots or nanopowder are spherical or flake shaped high surface area metal particles having high antimicrobial activity (Furno, F. *et. al*, 2004 and Moran, J. R. *et al.*, 2005) which are used on wounds. Ag nanoparticles were reported as active biocides against Gram-positive and Gram-negative bacteria including *Escherichia coli, Stephylococcus aureus, Klebsiella pneumoniae* and *Pseudomonas aeruginosa* (Jain, P. Pradeep, T. 2005; Sons, W.K. *et al.*, 2004).The nanoparticles of gold coated with palladium are very effective catalysts for removing tri-chloroethene (TCE) from groundwater 2,200 times better than palladium alone.

However, Zinc oxide nanoparticles have been used to remove arsenic from water, even though bulk zinc oxide cannot absorb arsenic. Some adsorption process for wastewater treatment have utilized ferrite and a variety of iron containing minerals, such as akaganeite, feroxyhyte, ferrihydrite, goethite, hematite, lepidocrocite, maghemite and magnetite. Adsorption of organic to the nanoparticles media was extremely rapid. More than 90% of the organics is adsorbed within 30 minutes. However, the smaller size of magnetic nanoparticles, which are 2-3 orders of magnitude smaller than a bacterium, provide extra benefits compared to magnetic beads. When their surface is appropriately elaborated, magnetic nanoparticles can also provide efficient binding to the bacteria because their high surface/ volume ratio simply offers more contact areas.

Ferrite is a genetic term for a class of magnetic iron oxide compounds. Ferrites display a property of spontaneous magnetization. They are crystalline materials, soluble only in strong acid. Iron atoms in iron ferrite ($FeO.Fe_2O_3$) can be replaced by many other metal ions without seriously altering its spinal structure. Various ferrites and natural magnetite were used in batch modes for actinide and heavy metal removal from wastewater. Nanoscale iron particles are typically 20-40nm with specific surface area (SSA) in the 30-50 m^2g^{-1} range. Other recent studies have demonstrated the magnetic enhanced removal of cobalt and iron from simulated ground water. The magnetic field-enhanced filtration/sorption process differs significantly from magnetic separation processes used in the processing of minerals and more recently, for water treatment and environmental applications. The use of iron ferrite and magnetite in wastewater treatment has a number of advantages over conventional flocculent precipitation techniques for removal of metal ion removal.

c) Zeolite nanoparticle

Zeolites are effective sorbents and ion exchange media for metal ions. NaP1 zeolites ($Na_6Al_6Si_{10}O_{32}12H_2O$) have a high density of Na^+ ion exchange site. They can be inexpensively synthesized by hydrothermal activation of fly ash with low Si/Al ratio at 150°C in 1.0-2.0 M NaOH solution (Brittany, L., *et al.*, 2006). NaP1 zeolites have been evaluated as ion exchange media for removal of heavy metals from acid mine waste waters. Alvarez-Ayuso *et al.* reported the successful use of synthetic Nap1 zeolite to remove Cr (III), Ni (II), Zn (II), Cu (II) and Cd (II) from metal electroplating wastewater (Alvarez, *et al.*, 2003).

d) Carbonaceous nanoparticle

Carbonaceous nanomaterials can serve as high capacity and selective sorbents for organic solutes in aqueous solutions. A number of polymers that exhibit anti-bacterial properties were developed for this purpose including soluble and insoluble pyridinium-type polymers which are involved in surface coating (Li, 2000), azidated poly vinyl chloride (Lakshmi, *et al.*, 2002) which can be used to prevent bacterial adhesion of medical devices, PEG polymers that can be modified on polyurethane surfaces and also prevent initial adhesion bacteria to the biomaterial surfaces (Lin, 2002) and polyethyleni- mine (PEI) (Park, *et al.*, 1998) that exhibit high antibacterial and antifungal activity. High activity of polycationic agent is related to absorption of positive changed nanostustures on to negative by charged cell surfaces of the bacteria. This process is brought through to be responsible for the increase of cell permeability and many disrupt the cell membranes.

Cross link polycations are prepared as nanoparticles. These are formed from PEI crosslinking and alylation followed by methylation in order to increase degree of amino groups substitution (Graveland, and Kruif, 2006). Because of its positive change and hydrophobicity, PEI nanoparticles have attracted attention as possible antimicrobial agents. Studies on PEI nanostructured compounds are made to evaluate its antimicrobial properties as a function of hydrophobicity, molecular weight, particles size and change that can play a significant role in antibacterial effect of the tested compound. The antibacterial activity is evaluated against *Steptoccoucus mutans*, cariogenic bacteria. Various PEI nanoparticles from 100nm to 1 micron in diameter are prepared having different degree of cross linking, particle size and zeta potential that are achieved by alkylation with a bromoalkane followed by methylation. Their antibacterial effect is examined against *Steptoccoucus mutans* in direct contact with bacteria. One important feature of the antibacterial agent is to maintain antibacterial activity over a long time. However, only the PEI nanoparticles samples including long chain alkyl demonstrated high antibacterial effect against *Steptoccoucus mutants* for more than four weeks (Park, K. D. *et al.*, 1998).

2. Remediation

The nanoscale particles may hold the potential to cost-effectively address some of the challenges of site remediation (Masciangioli and Zhang, 2003: EPA, 2003d, e). Two factors contribute to the nanoparticles' capabilities as an extremely versatile remediation tool. The small particle size (1-100nm) of the nanoparticles contributes their capabilities as an extremely versatile remediation tool. In comparison, a typical bacterial cell has a diameter on the order of 1μm (1000nm). Nanoparticles can be transported effectively by the flow of groundwater. Due to this attribute, the nanoparticles-water slurry can be injected under pressure and/or by gravity to the contaminated plume where treatment is needed. The nanoparticles can also remain in suspension for extended periods of time to establish an *in situ* treatment zone.

Nanoremediation methods entail the application of reactive nanomaterials for transformation and detoxification of pollutants. These nanomaterials have properties that enable both chemical reduction and catalysis to mitigate the pollutants of concern. For nanoremediation *in situ*, no groundwater is pumped out for above-ground treatment, and no soil is transported to other places for treatment and disposal (Otto *et al.* 2008). Nanomaterials have highly desired properties for in situ applications. Because of their minute size and innovative surface coatings, nanoparticles may be able to pervade very small spaces in the subsurface and remain suspended in groundwater, allowing the particles to travel farther than larger, macro-sized particles and achieve wider distribution. However, in practice, current nanomaterials used for remediation do not move very far from their injection point (Tratnyek and Johnson 2006). Many different nano-scale materials have been explored for remediation, such as nano-scale zeolites, metal oxides, carbon nanotubes and fibers, enzymes, various noble metals and titanium dioxide. Of these, nanoscale zero-valent iron (nZVI) is currently the most widely used.

nZVI: nZVI particles ranges from 10 to 100nm in diameter, although some vendors sell micrometer-scale iron powders as "nano-particles". Typically, noble metal (e. g., Palladium, Silver, Copper) can be added as a catalyst. The second metal creates a catalytic synergy between itself and Fe and also aids in the nanoparticles' distribution and mobility once injected into the ground (Saleh *et al.* 2007; Tratnyek and Johnson 2006; U. S. EPA 2008).

The environmental chemistry of metallic or zero-valent iron has been extensively documented (EPA, 2003c). Contaminants such as tetrachloroethene (C_2Cl_4), a common solvent, can readily accept the electrons from iron oxidation and be reduced to ethane in accordance with the following stoichiometry:

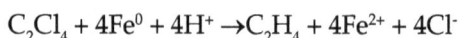

$$C_2Cl_4 + 4Fe^0 + 4H^+ \rightarrow C_2H_4 + 4Fe^{2+} + 4Cl^-$$

The underlying chemistry of the reaction of Fe with environmental pollutants (particularly chlorinated solvents) has been extensively studied and applied in micrometer-scale ZVI PRBs (Matheson and Tratnyek 1994). There are two main degradation pathways for chlorinated solvents: beta elimination and reactive chlorination. Beta elimination occurs most frequently when the contaminant comes into direct contact with the Fe particles. The following example shows the path way of trichloroethene (TCE).

$$TCE + Fe^0 \rightarrow \text{Hydrocarbon products} + Cl^- + Fe^{2+}/Fe^{3+}$$

Under reducing conditions fostered by nZVI in groundwater, the following reaction takes place:

$$PCE \rightarrow TCE \rightarrow DCE \rightarrow VC \rightarrow \text{Ethene}$$

Where PCE is perchloroethylene, DCE is dichloroethylene, and VC is vinyl chloride (Tratnyek 2003, U. S. EPA 2008b).

Recent laboratory research has largely established nanoscale iron particles as effective reductants and catalyst for a wide variety of common environmental contaminants including chlorinated organic compounds and metal ions (Lien, 2000; Lien and Zhang 1999, 2001; Zhang *et al.* 1998). Examples are given in table 2. For halogenated hydrocarbons, almost all can be reduced to benign hydrocarbon by the nano-Fe particles. Ample evidence indicates that iron based materials have been successful in the transformation of many other contaminants, including anion (Example NO_3^-, $Cr_2O_7^{2-}$), heavy metals, (*e. g.*, Ni^{2+}, Hg^{2+}), and radionuclides (*e. g.*, UO_2^{2+}).

The laboratory study was conducted as a part of a project to evaluate the potential of using nanoscale iron particles for in situ remediation of chlorinated organic solvents [e.g., 1,1,1-trichloroethane(TCA), trichloroethene (TCE)] found in the soil and ground water an US Naval site. Ground water and soil samples were collected at the site and were shipped to Lehigh University for various laboratory tests. The sites were conducted during the period of August 2002 to March 2003. Experiments were designed to determine the concentrations of organic contaminants, to investigate changes of groundwater chemistry as a result of the addition of the nano iron particles, and to examine the efficiency of the nanoparticles for dechlorination of major chlorinated organic compounds found in the groundwater and soil.

Table: 2 Common environmental contaminants that can be transformed by nanoscale iron particles

Chlorinated methanes
Carbon tetrachloride(CCl_4)
Chloroform ($CHCl_3$)
Dichloromethane (CH_2Cl_2)
Chloromethane (CH_3Cl)
Chlorinated benzenes
Hexachlorobenzen (C_6Cl_6)
Pentachlorobenzen (C_6HCl_5)
Tetrachlorobenzen ($C_6H_2Cl_4$)
Trichlorobenzen ($C_6H_3Cl_3$)
Dichlorobenzen ($C_6H_4Cl_2$)
Chlorobenzen (C_6H_5Cl)
Pesticides
DDT ($C1_4H_9Cl_5$)
Lindane ($C_6H_6Cl_6$)
Organic dyes
Orange II ($C1_6H_{11}N_2NaO_4S$)
Chrysoidine ($C_{12}H_{13}ClN_4$)
Tropaeolin O ($C1_2H_9N_2NaO_5S$)
Acid orange
Acid red
Heavy metal ions
Mercury (Hg^{2+})
Nickel (Ni^{2+})
Silver (Ag^+)
Cadmium (Cd^{2+})

In the 1990s, Fe at the nanoscale was synthesized from Fe (II) and Fe (III) to produce particles ranging from 10 to 100nm, initially using borohydride as the reductant. Zhang (2003) tested nZVI for the transformation of a large number of pollutants, most notably halogenated organic compounds commonly detected in contaminated soil and groundwater. The author reported that nanoscale Fe particles are very effective for the transformation and detoxification of a variety of common environmental pollutants, including chlorinated organic solvents, organochlorine pesticides, and polychlorinated biphenyls (PVBs). According to Zhang (2003), Fe-mediated reactions should produce an increase in pH and a decrease in the solution redox potential created by the rapid consumption of oxygen, other potential oxidants, and the production of hydrogen. Although batch reactors produce pH increases of 2-3 and an oxidation reduction potential (ORP) range of -500 to -900mV, it is expected that the pH and ORP would be less dramatic in the field applications where other mechanisms reduce the chemical changes (Zhang 2003). Previous work showing an increase of pH by 1 and an ORP in the range of -300 to -500 mV supports this assessment (Elliott and Zhang 2001; Glazier *et al.* 2003). Zhang (2003) also showed that modifying Fe nanoparticles could enhance the speed and efficiency of the remediation process.

Initially, Fe nanoparticles have a core of ZVI and an outer shell of Fe oxides, which suggest the following redox reactions:

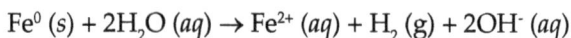

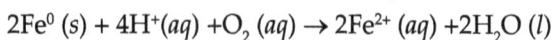

Fe^0 (s) + $2H_2O$ (aq) $\rightarrow$ Fe^{2+} (aq) + H_2 (g) + $2OH^-$ (aq)

$2Fe^0$ (s) + $4H^+$(aq) +O_2 (aq) $\rightarrow$ $2Fe^{2+}$ (aq) +$2H_2O$ (l)

Where s is solid, aq is aqueous, g is gas, and 1 is liquid (Matheson and Tratnyek 1994).

Although Fe nanoparticles have been shown to have a strong tendency to form micro scale aggregates, possible because of their weak surface charges, coating can be applied to change the surface properties. These different forms of Fe could be useful for the separation and transformation of a variety of contaminates, such as chlorinated organic solvents organochlorine pesticide, PCBs, organic dye, various inorganic , and the metals As (III) (trivalent arsenic), Pb (II) (bivalent lead), copper [Cu (II)(bivalent copper)], Ni (II) (bivalent nickel),and Cr (VI)(Hexavalent chromium)(Sun *et al.*, 2006).

Nanoremediation, particularly use of n ZVI, has site-specific requirements that must be met in order for it to be effective. Adequate site characterization is essential, including information about site location, geological condition, and the concentration and types of contaminants. Geological, hydrogeological, and subsurface conditions include composition of the soil matrix, porosity, hydraulic conductivity, groundwater gradient and flow velocity, depth to water table, and geochemical properties (pH, ionic strength, dissolved oxygen, ORP, and concentration of nitrate, nitrite, and sulfate). All of these variables need to be evaluated before nanoparticles are injected to determine whether the particles can infiltrate the remediation source zone, and whether the conditions are favorable for reductive transformation of contaminants. The sorption or attachment of nanoparticles to soil and aquifer materials depends on the surface chemistry (i.e., electrical charge) of soil and nanoparticles, groundwater chemistry (*e.g.*, ionic strength, pH, and presence of natural organic matter), and hydrodynamic conditions (pore size, porosity, flow velocity, and degree of mixing or turbulence). The reaction between the contaminants and the nZVI depend on contact or probability of contact between the pollutant and nanoparticles (U. S. EPA 2007, 2008b).

In field tests, Henn and Waddill (2006) found that, with the use of nZVI, decrease in parent pollutant compound concentrations (TCE and Trichloroethane) were accompanied by increases and subsequent decreases in daughter product concentrations (cis-1,2-DCE, 1, 1-dichloroacetic acid, 1, 1-DCE, and vinyl chloride). The nanoscale Fe created conditions for abiotic degradation for about 6 to 9 months, followed by biological degradation as the primary degradation process. Both processes had significant impacts on the degradation on contaminants (Henn and Waddill 2006). Cao *et al* (2006) found that nZVI particles in an aqueous solution, reduced perchlorate to chloride almost completely without producing intermediate degradation product.

Fe oxide nanoparticles have been shown to bind As irreversible up to 10 times more effectively than micrometer-sized particles. Based on their super-paramagnetic properties, the Fe particles and bound As can be separated from the water with a magnetic field. Laboratory has shown 99% removal of As using 12nm-diameter Fe oxide nanoparticles (Rickery and Morrison, 2007). Kanel *et al*. (2006) concluded that nZVI can reduce As (V) to As (III) in a short period of time at neutral pH. They also found that a high amount of nZVI was needed to completely remove As (V), possibly because of the presence of dissolved organic carbon, sulfate, and phosphate.

In addition to groundwater remediation, nanotechnology holds promise in reducing the presence of NAPLs. Recently, a material using nanosized oxides (mostly calcium) was used *in situ* to clean up heating oil spills from under ground oil tanks. Primarily results from this redox-based technology suggest faster, cheaper methods and ultimately, lower overall contaminant levels compared with previous remediation methods. Most of these sites have been in New Jersey, with clean up conducted in consultation with the New Jersey Department of Environmental Protection (Continental Remediation LLS, 2009).

Remediating Pollution

Nanotechnology-based solutions may also help reduce or prevent pollution and toxic emission at source. Nanostructured catalysts based on metal oxides or metal nanoparticles show promise in reducing industrial and vehicle emissions (Rickerby and Morrison 2006). For example, a variety of precious metal nanoparticles have the ability to oxidize poisonous carbon monoxide (CO) in vehicle exhausts, transforming them into less harmful carbon dioxide (CO_2). At the nanoscale, various particles demonstrate impressive capabilities to remediate pollutants. Nanoparticles of titanium dioxide (TiO_2) absorb energy from light and then oxidize nearby organic molecules; this property of photocatalysis is exploited to make coatings that attract and oxidize pollutants, such as vehicle and industrial emissions (Strini *et al.* 2005). These properties can be exploited to create self-cleaning surface (e.g., self-cleaning glass or walls that can trap particles of air pollution).

Several nanostructured materials show promise for cleaning water and groundwater. Nanoporous membranes that filter pathogens and other undesirable material are now commercially available. Some scientists propose to remediate ground water pollution by using nanoparticles of iron as a chemical reductant; in the process the iron oxidizes and becomes rust, a naturally occurring substance. Taking advantage of the high surface area of nanoparticls, magnetic iron nanocrystals are used to remove arsenic from drinking water. This method reportedly reduces, by more than 100-fold, the amount of waste produced by standard techniques. Another innovative approach involves coating the surface of iron oxide particles with molecules that selectively bind to pollutant molecules or ions. Introduced into water, the coated particles attract the pollutant and then a magnetic field is used to concentrate and recapture the bound pairs.

3. Gold nanoparticles used as a catalyst for the reduction of hazardous and toxic pollutant

The catalytic function of biomatrix-gold nanomaterial was substantiated by carrying out the reduction of a hazardous and toxic pollutant - aqueous 4-nitrophenol (4-NP), which has a peak at 317 nm in the UV-visible spectral range. In the absence of any catalyst, the peak at 400nm remained unaltered even for 2 days. Plant samples carrying gold nanoparticles were cryoground in liquid nitrogen and added to the 4-NP solution. Following sonication, the crushed biomatrix-nanomaterial induced fading of the characteristic yellow colour of the 4-NP solution on incubation. The decolourization was quantitatively monitored using a spectrophotometer, and a successive decrease in the peak intensity was observed over time. The decrease in the peak arises as a result of the adsorption of 4-nitrophenolate ions by the crushed nanomaterials, which produces a colourless solution at the end. When sodium borohydride was added to the colourless solution, it showed a new peak at ~290nm because of the reduction of 4-nitrophenolate species to 4-aminophenol (4-AP) (Praharaj S *et al.*, 2004). A control experiment with roots devoid of gold nanoparticles did not give any signature of 4-AP under the same experimental condition. In that case, only the decrease in the absorbance of 4-NP peak was observed as a consequence of the adsorption of nitrophenolate ions by the solid mass. The generation of 4-AP conforms the catalytic activity of the gold nanoparticles for the reduction of 4-NP in aqueous solution (Praharaj S *et al.*, 2004). This observation authenticates the presence and surface activity of gold nanoparticles in the biomatrix. The exploitation of the in situ generated gold nanoparticles for a catalytic reaction is a definite departure from the conventional methods, where separation of nanoparticles from a biological system has been a difficult task.

4. Antibacterial activity of silver nanoparticles

Microorganisms (bacteria, yeast and fungi) play an important role in toxic metals remediation through reduction of metal ions; this was considered interesting as nanofactories. Recently, it was found that aqueous chloroaurate ions may be reduced extracellularly using *Fusarium oxysporum*, to regenerate extremely stable gold or silver nanoparticles in water. These particles can be incorporated in materials and cloth making them sterile. The sterile materials are important in hospitals, where wounds are often contaminated with microorganisms, in particular, *Staphylococcus aureus*. A new generation of dressing incorporating antimicrobial agents like silver was developed to reduce or prevent infections. Extracellular production of silver nanoparticles by *Fusarium oxysporum* strain and its bactericidal effect in cotton and silk cloth against *Staphylococcus aureus* were studied. In the silver reduction, approximately 10 g of *Fusarium oxysporum* biomass was taken in a conical flask containing 100ml of distilled water, kept for 72hrs at 28°C and then the aqueous solution components were separated by filtration. To this solution, $AgNO_3$ (10^{-3}M) was added and kept for several hours at 28°C. In order to incorporate silver nanoparticles in the cloth, these were immersed in the filtrate fungal, centrifuged and dried. The reductions of metal ions occur by a nitrate-dependent reductase and quinine extracellur process. Antibacterial activity was observed when silver nanoparticles were incorporated in cotton cloth. However, in silk cloth this activity was not observed demonstrating a low incorporation of the silver nanoparticles due to possibly to the size of the silk pores. This work demonstrates the possible use of biological synthesized silver nanoparticl;es and its incorporation in cloths leading them to sterilization. Moreover, these particles could have innumerable applications, in different areas as receptors, catalysis, biolabelling and others.

5. Silver nanoparticles used for the vegetables and fruit preservation

Edible coatings or films could serve as moisture, lipid and gas barriers. Alternatively, they could improve the textural properties of food or serve as carrier of functional agents such as colors, flavors, antioxidants, nutrients, and antimicrobials. Different microorganisms are responsible for the spoilage of fruits and vegetables, thus decreasing their quality and self life. In recent years, the antimicrobial coatings process without the use of preservative has been gaining more importance in controlling and preventing food-borne microbial outbreaks. Many naturally occurring enzymes such as lysozymes, lactoperoxidases, and glucose oxidase have also been used as antimicrobial agents for food preservation (Davidson, 2001). The main draw back of these enzymes is their lower stability and activity unless maintained at their optimum conditions (pH and temperature).

The use of protective nanocoating and stable packing has become a topic of great interest in the field of food nanotechnology because of their potential for increased shelf life of many food products (Ahvenainen, 2003). Over the past decade, there has been a strong push towards the development of silver-containing materials for commercial use that exhibit antimicrobial or antibacterial properties. Silver nanoparticles have a broad spectrum of antibacterial activity against Gram-negative and Gram-positive bacteria, and there is also minimal development of bacterial resistance (IP, *et al.*, 2006). Because chemical synthesis method produces toxic substances as byproduct, there was a big challenge to develop a new synthesis of nanoparticles that does not use toxic chemicals in the synthesis protocols. Recently, scientists have looked to microorganisms as possible eco-friendly nanofactories for the synthesis of nanoparticles such as CdS, gold, and silver.

They synthesized biogenic silver nanoparticles using *Trichoderma viride* and demonstrated that silver nanoparticles incorporated with polysaccharide films along with glycerol gives both surface protection and antimicrobial edible coatings for vegetables and fruits. This prevents weight loss and protein loss and also precludes microbial spoilage, thus increasing the shelf life of vegetables and fruits.

6. Nanoparticles in environmental monitoring

One way in which advances in nanotechnology may benefit the environment (both indoor and outdoor) is through detection devices that are less expensive and more sensitive - in some cases thousands or millions of times more sensitive than existing devices. For example, new protein-based nanotech sensors can detect mercury at concentration of approximately one part in 10^{-15} or one-quadrillionth, a task previously impossible (Bontidean 1998). Using nanoparticles of europium oxide, a highly sensitive method has been developed to measure the pesticide Atrazine, a frequent ground water contaminant (Feng *et al.* 2003).

Many new nanotechnology-based monitoring devices operate on site and in real-time, simultaneously measuring broad range of pollutant and toxic agent. Rapid detection allows for swift response, thereby minimizing damage and reducing remediation costs.

Monitoring Air Pollution Hot Spots

Thin layers of nanocrystalline metal oxide are the key component of solid-state gas sensors for air quality monitoring.

Nanotechnology can be used to improve monitoring of air and water quality. For example, miniaturized quality monitoring devices selectively detect carbon monoxide (CO) and nitrogen dioxide (NO_2) by measuring changes in electrical conductivity that occur when these gas molecules are present on the surface. Other gases such as methane, ozone and benzene can also be detected. In some applications, nano-based sensors outperform conventional air pollution monitoring devices. They provide faster response with real-time analytical capability, greatly improved geographical resolution, simplified operation and lower running costs. They are ideal for monitoring localized pollution peaks in urban areas.

To verify the safety of drinking water, it is necessary to monitor pollutants (pesticides, antibiotics, natural toxin, carcinogens, industrial waste, etc.) down to the level of one nanogram (i.e., one-billionth of gram) per litre. A new biochemical sensor uses an integrated optical chip to analyze water from various sources by means of a miniaturized immunoassay system. In approximately 20 minutes, the sensor can detect and provide data on more than 30 different substances. The devices can be re-used up to 500 times before the surface chemistry needs to be regenerated (Rickerby *et al.* 2000).

Saving Energy and Resources

Some new nanocatalysts can be used at room temperature. This is a huge advantage over traditional catalysts, which typically operate at high temperatures and require greater energy input. Nanotechnology may transform energy production, storage and consumption by providing environmentally sound alternatives to current practices. Several technologies can enhance the efficiency of current energy sources and reduce carbon dioxide (CO_2) emission-including nanostructured catalysts for fuel cells, improved electrode materials in lithium ion batteries (Tarascon and Armand 2001), and nanoporous silicon and TiO_2 in advanced photovoltaic cells (Stalmans *et al.* 1998, Pizzini *et al.* 2005). Nanoscale optically selective coatings for windows can reduce energy consumption while also improving indoor air quality. The use of high strength and light weight nanomaterials may extend the lifespan of conventional materials such as plastics and save energy in transportation and other areas. For example, carbon nanotubes are molecular scale cylinders of carbon that exhibit novel properties such as extraordinary strength, unique electrical properties and highly efficient heat conductivity. This makes them potentially useful in electronics, optics and other applications of material science. They will likely become widely used in common consumer products.

Windows that Save Energy

Windows are inefficient from an energy standpoint. During hot months, sun shining through glass increases indoor temperatures and the need for cooling. In cool season, windows leak a significant portion of the indoor heat, wasting energy. Depending on the country, a significant amount of energy may be used to heat or cool buildings.

Nanoscale window coatings are promising in reducing energy consumption and CO_2 emission. Coatings, tailored to warm climates, allow visible light to pass through glass but block infrared wavelength. In cool climates, coating makes more efficient use of light and heat by hindering their radiation back to the outside world. Other coatings still in development can respond to change in the weather or the angle of the light. At present, reflective coatings are expensive to produce. Although they are less effective, so-called 'Absorptive coatings' provide a more affordable alternative. A coating that contains nanoparticles of the compound lanthanum hexaboride (LaB_6) is already on the market and is used to make more cost-effective solar glazing.

Lithium Batteries Increase Safety and Last Longer

The power output of rechargeable lithium batteries can be increased by 50 per cent using nanostructured electrodes containing lithium cobalt oxide. These batteries are intrinsically safer; they have faster charge/discharge rates and better accommodate the expansion caused by migrating lithium ions during charging. Lithium batteries are already being used to power a wide range of devices, many of which operate in remote locations and extreme environments-from oceans to outer space.

Nanomaterials Trap and Transform Solar Energy

Various nanomaterials such as nanostructured cadmium and copper indium diselenide are proving effective in solar energy technologies, including photovoltaic cells. Thin layers of semiconductor materials can be applied to inexpensive bases, such as glass, plastic, or metal, to create photovoltaic cells. Compared to conventional silicon solar cells, less semiconductor material is required and manufacturing costs are significantly reduced.

Better Storage for Emission-Free Fuels

New vehicles in development operate by converting hydrogen fuel into electrical energy, producing water as a product. Thus, they offer the promise of eliminating greenhouse gas emissions in the transportation sector. However, hydrogen gas is highly flammable and presents considerable storage and transport problems. Nanomaterials that facilitate storage include metal hydrides (chemical compounds formed when hydrogen gas interacts with metals). Some metal hydrides react at near room temperature and at pressures only a few times greater than that of the Earth's atmosphere, making them suitable candidates for hydrogen storage. However, they have relatively slow absorption and desorption rates. Nanostructured materials can reduce this problem by providing fast diffusion paths for hydrogen.

Current Research Trends in Nanoparticles to Environmental Problems

Current research and development seek to rapidly exploit the novel applications of nanomaterials. It is widely accepted that the traditional methods of measuring dose exposure are of little use for predicating toxicological effects of nanoparticles. Standard monitoring instruments cannot always detect nanoparticles in environmental samples. Thus, characterizing their behavior and novel properties, and tracing their impacts, presents a real scientific and technical challenge.

There is a need for studies that examine the long term effects of nanoparticles on different environments and their resident organism. Nanoparticles are used in remediating toxic compounds into simple compounds. These particles are used in the waste water treatment, vegetables and fruit preserving. A balanced approach is required to maximize benefits minimizing risks.

References

Ahvenainen, R. *Novel Food Packing Techniques*; CPR Press: Boca Raton, FL, 2003.

Alvarez A E, Sanchez A G and Querol X (2003). Purification of metal electroplating waste waters using zeolites. *Water Res.* 37: 4855-4862.

Baughman R H, Zakhidov A A and De Heer W A (2002). Carbon nanotubes-the route   toward applications. *Science* 29: 787-792.

Bontidean I (1998). Detection heavy metal ions at femtomolar levels using protein-based biosensors *Anal. Chem.* Vol. 70: 4162-4169.

Brittany L, Carino V, Kuo J, Leong L and Ganesh R (2006). Adsorption of organic compounds to metal oxide nanoparticles.

Cao J, Elliot D, Zhang W-X (2005). Perchlorate reduction by nanoscale iron particles. *J Nanopat Res* 7: 499-506.

Comini E, Faglia G and Sberveglieria G (2001). CO and NO$_2$ response of Tin oxide silicon doped thin films. *Sensors Actuators, B* 76: 270-274.

Davidson, P. M. chemical preservatives and natural antimicrobial compounds. *In Food Microbiology Fundamentals and Frontiers*; Doyle, M. P., Beuchat, L. R., Montville, T. J., Eds.; ASM Press: Washington, DC, 2001; pp 593-627.

Elliott D W and Zhang W X (2001). Field assessment of nanoscale bimetallic particles for groundwater treatment. *Environ Sci. Technol.* 35: 4922-4926

EPA (US Environmental protection agency), (2003a). Technology innovation office. Permeable reactive barrier.

EPA (US Environmental protection agency), (2003b). Workshop on nanotechnology and the environment. August 28-29, 2002. Arlington, Virginia. 50-51 EPA/600/R-02/080.

Feng J, Shan G, Maquiera A, Koivunen M E, Bing Guo, Hammock B D and Kennedy J M (2003). Functionalized Europium oxide nanoparticles used as a fluorescent label in an immunoassay for atrazine. *Analytical Chemistry* 75: 5285-5286.

Frechet J M J and Tomalia D A (2001). Dendrimers and other dendritic polymers. New York: Wiley and Sons.

Furno F, Morley K S, Bong B, Sharp B L, Arnold P L, Howdle S M, Bayston R, Brown P D, Winship P D and Reid HL (2004). Silver nanoparticles and polymeric medical device: A new approach to prevention of infection. *J Anti. Chemo.* 54: 1019-1024.

Glazier R, Venkatakrishnan R, Gheorghiu F, Walata L, Nash R, Zhang W (2003). Nanotechnology takes root. *Civil Eng* 73(5): 64-69.

Global Dalogue on nanotechnology and the poor opportunities and risks (2006). Overview and comparison of conventional and nano-based water treatment technologies. The Meridian Institute.

Graveland B J F and Kruif C G (2006). Unique milk protein based nanotubes: Food and nanotechnology meet. *Trends Food Sci. Technol.* 17: 196-203.

Henn K W and Waddill D W (2006). Utilization of nanoscale zero-valent iron for source remediation-a case study. *Remediation* 16(2): 52-77.

Hillie T M, Manusinghe Hope M and Deranyagia Y (2006). Nanotechnology water and development. The Meridian Institute.

IP, M.; Lui, S. L.; Poon, V. K. M. ; Lung, I, ; Burd, A. *J. Med. Microbiol.* 2006, 55, 59-63.

Jain P and Pradeep T (2005). Potential of silver nanoparticle- coated polyurethane foam as an antibacterial water filter. *Biotech.Bioeng.*,90:59-63.

Lakshmi S, Kumar S S P and Jayakrishnan (2002). Bacterial adhesion onto azidated poly (vinyl chloride) surfaces. *J Biome. Mat. Res.* 61: 26-32.

Li G (2000). A study of pyridinium-type functional polymers. Behavioral features of the antibacterial activity of insoluble pyridinium-type polymers. *J App. Pol. Sci.* 78: 676-684.

Lien H (2000). Nanoscale bimetallic particles for dehalogenation of halogenated aliphatic compound. Unpublished dissertation, Lehigh University, Bethlehem, Pennsylvania.

Lien H and Zhang W (1999). Reaction of chlorinated methanes with nanoscale metal particles in aqueous solution. *J Environ. Eng.* 125(11): 1042-1047.

Lien H and Zhang W (2001). Complete dechlorination of chlorinated ethenes with nanoparticles. *Colloids Surfaces A* 191: 97-105.

Lin J (2002). Bactericidal properties of flat surfaces and nanoparticles derivatized with alkylated polyethylene imines. *Biotec. Prog.* 18: 1082-1086.

Lux Research (2004). Revenue from nanotechnology enabled products to equal IT and Telecom by 2014, exceed biotech B 10 times. Press release 25 Oct 2004

Masciangioli T and Zhang W X (2003). Environmental technologies at the nanoscale. *Environmental science and technology* 37: 103A-108A.

Matheson L J and Tratnyek P G (1994). Reductive dehalogenation of chlorinated methanes by iron metal. *Environ Sci. Technol.* 28(12): 2045-2053.

Moran J R, Elechiguerra J L, Camacho A, Holt K, Kouri J B, Ramirez J T and Yacaman M J (2005). The bactericidal effect of silver nanoparticles. *Nanotech* 16: 2346-2353.

Muir J, Smith G, Masens C, Tomkins D and Cortie M (2004). The nanohouse TM-An Australian initative for the future of energy efficient housing in nanotechnology in construction. The Royal society of chemistry, Cambridge.

Nachhaltigkeitseffekte durch Herstellung und Anwendung nanotechnologischer Produkte, Schriftenreihe des IÖW 177/04, Berlin.

NanoRoad SME: http://www.nanoroad.net.

Nanotechnologien in der Schweiz: Herausforderungen erkannt Bericht eines Dialogverfahrens Zentrum für Technikfolgenabschätzung TA-P 8/2006 d, Bern, 2006, ISBN-Nr. 3-908174-25-2.

Oelerich W, Klassen T and Bormann R (2001). Metal oxides as catalysts for hydrogen sorption of nanocrystaline Mg based alloys and compounds. 315: 237-242.

Otto M, Floyd M and Bajpai S (2008). Nanotechnology for site remediation. *Remediation* 19(1): 99-108.

Panigrahi S, Kundu S, Ghosh S K, Nath S and Pal T (2004). General method of synthesis for metal nanoparticles. *J Nano. Res.* 6: 411-414.

Park K D (1998). Bacterial adhesion on PEG modified polyurethane surface. *Biomaterials* 19: 851-859.

Pizzini S, Accam M and Bretti S (2006). From electronic grade to solar grade silicon: chance and challenges in *Phys. Stal. Sci* 202: 2928-2942.

Praharaj S, Nat S, Gosh SK, Kundus S, Pal T. immobilization and recovery of Au nanoparticles from an ion exchange resin: resin-bound nanoparticles matrix as a catalyst for the reduction of 4-nitrophenol. *Lanhmuir* 2004; 20: 9889-9892. [PubMed: 15518467].

Publifocus "Nanotechnologien und ihre Bedeutung für Gesundheit und Umwelt"

Rickerby D G and Morrison M (2007). Nanotechnology and the environment: a European perspective. *Sci. Technol. Adv Mater* 8:19-24.

Saleh N, Sirk K, Liu Y Q, Phenrat T, Dufour B and Matyjaszewski K (2007). Surface modification enhance nanoiron transport and NAPL targeting in saturated porous media. *Environ Eng Sci.* 24(1): 45-57.

Schelm S and Smith G B (2003). Dilute LaB6 nanoparticles in polymer as optimized clear solar control glazing. *Applied Physics Letters* 82(24): 4346-4348.

Son W K, Youk J H, Lee T S and Park W H (2004). Preparation of antimicrobial ultrafine cellulose acetate fibres with silver nanoparticles. *Macromol. Rapid Commun* 25: 1632-1637.

Sondi I and Sondi B S (2004). Silver nanoparticles as antimicrobial agent: A case study on E-coli as a model for Gram-negative bacteria. *J Coll. Interf. Sci.* 275: 177-182.

Stalmans L, Poortmans J, Bender H, Caymax M, Sad K, Vazsuny E, Nis J and Mertens R (1998). Porous silicon in crystalline silicon solar cells: A review and the effect on the internal quantum efficiency. *Prog. Photovolt Res. Appl.* 6: 233-246.

Stoimenov P K, Klinger G L, Marchin and Klabunde K J (2002). Metal oxide nanoparticles as bactericidal agents. *Langmuir* 18: 6679-6686.

Strini A, Cassese S and Schiavi L (2006). Measurements of benzene, toluene, etnebenzene and o-xylene gas phase photodegradtion by titanium dioxide dispersed in cementitius materials using a mixed flow reactor. *Applied catalysis B* 61: 90-97.

Sun Y P, Li X, Cao J, Zhang W and Wang H P (2006). Characterization of zero-valent iron particles. *Adv Colloid Interface Sci* 120: 47-56.

Tarascon JM and Armand M (2001). Issues and challenges facing rechargeable Li batteries. *Nature* 414: 359-367.

Tratnyek P G and Johson R L (2006). Nanotechnologies for environmental cleanup. *Nanotoday* 1(2): 44-48.

Tratnyek P G, Scherer M M, Johnson T L and Matheson L J (2003). Permeable reactive barrier of iron and other zero-valent metal. In: Chemical degradation methods for wastes and pollutants, Environmental and Industrial application.

U S EPA (2007). Nanotechnology white paper. EPA 100/B-07/001. Washington, D C: U S. Environmental protection agency.

U S EPA (2008).Nanotechnology for site remediation: Fact sheet. EPA 542-F-08-009. Washington, D C: U S Environmental protection agency.

Zeman L J and Zydney A L (1996). Microfiltration and ultra-filtration. New York: Marcel dekker principles and applications.

Zhang W, Wang C and Lien H (1998). Catalytic reduction of chlorinated hydrocarbons by bimetallic particles. *Catal. Today* 40(4): 387-395.

Zhang WX (2003). Nanoscale iron particles for environmental remediation: An overview nanoparticles research 5(3/4): 323-332.

Zhang W-X, Elliott D W (2006). Application of iron nanoparticles for groundwater remediation. *Remediation* 16(2): 7-21.

to recognize levels of existing nanoparticles in the major environmental recipients and some other properties of these recipients, as to know how ENPs may interact with the environment. The nanoparticles are classified into following three types:

i.   Natural nanoparticles

ii. Incidental nanoparticles

iii. Engineered nanoparticles

i) Natural Nanoparticles

Natural nanoparticles have existed before beginning of life on earth. All life forms that have developed have been exposed to at least some types of nanoparticles during their evolution, and they have thus developed mechanisms to tolerate their presence (Buffle 2006). The most common natural nanoparticles are soil colloids, which are constituted of silicate clay minerals, iron or aluminum oxides, hydroxide or humic organic matter, including black carbon. Airborne nanocrystals of sea salts formed from evaporation of sea water sprays are the most common and wide spread natural nanoparticles.

ii) Incidental Nanoparticles

These are largely either of anthropogenic or pyrogenic origin. One of the most abundant ENPs, carbon black, is also formed incidentally e.g. during fires and are thus not new to the environment.

iii) Engineered Nanoparticles

Engineered nanoparticles are particles that are produced by man have specific nanotechnological properties. They can be made of single elements like Carbon (C) or Silver (Ag) or a mixture of elemental molecule. Few ENPs are described below.

a) Fullerenes: These are made up of pure carbon and represent a newly discovered (1985) carbon allotrope. The simplest fullerene, C_{60} is a ball made up of 60 carbon atoms and resembles orientation of patches on surfaces of a football. They are also known Buckminster fullerenes or "Bucky balls", named after the architect Richard Buckminster Fuller who constructed a range of geodesic spheres and domes with structures resembling the fullerenes.

The next stable fullerene is C_{70} and is having an oblong form. Fullerenes are hydrophobic and soluble in organic solvents, like toluene (2.8 mg/l). Fullerenes containing atoms, ions or small clusters inside their spherical structure are called endo-fullerenes. A common functionalization is hydroxylation which renders the molecules more hydrophyilic (Brant *et al.* 2007b). Fullerenes may form highly stable hydrophilic nano-size aggregates (n-C_{60}; 100-200 nm diameters) that can exist as stable aqueous suspensions at low ionic strength (Brant *et al.* 2006; Deguchi *et al.* 2001).

b) Carbon nanotubes (CNTs): Carbon nanotubes are fibrous fullerenes consisting of rolled–up graphene sheets that may or may not be capped at the ends by a half fullerene sphere. They can be made by laser ablation using graphite as a starting material or made from CH_4 or other carbon containing gases through chemical vapor deposition. Resulting CNTs may consist of a single wall (SWCNTs), double wall (DWCNTs) or multiple wall of carbon (MWCNTs).

SWCNTs are believed to have superior mechanical strength, thermal and electric conductivity compared to MWCNTs. The hydrophobicity and fibrous structure of CNTs make them prone to aggregation in bundles termed nanofibres, nano-ropes and nanowires, depending on their dimensions.

c) **Carbon Black (CB):** They may be considered as an ENP, though CB has been produced and used industrially since long before the era of nanotechnology. CB is a by–product of combustion, and is also formed naturally during fires. CB has been used as reference substances in studies of toxicity and ecotoxicity of CNT and fullerenes (Cheng *et al.* 2007) and displays no or low toxic effects.

d) **Quantum-dots (QD):** QD are fluorescent semi-conductor nanocrystals, commonly made up of a 3-6 nm diameter core of CdS, CdSe, PbSe, CdHg or a range of other metals and coated by an organic polymer (to protect against oxidation and to permit linking biologically active moieties, like antibodies).

Other mineral nanoparticles can be made up of single elements (Ag, Au, Cu, Fe, etc.), compounds (SiC, Si_3N_4, WC etc.), single–metal oxides (Al_2O_3, TiO_2, ZnO, etc) and multi-element oxides ($MgAl_2O_4$, $SrTiO_3$ etc.). Long list of these particles includes ENPs from different preparation methods, with different particle sizes, different degrees of purity, and different coatings.

ENPs based on organic carbon which escape, has lead to the concern for potential environmental impacts due to the fact that they are composed of degradable polymers, and unless prepared to resist biodegradation will have a short life span once entering the environment (Avella *et al.* 2005).

Characterization of Nanoparticles

Characterization of nanometer sized materials is particularly challenging due to the fact that materials in this size range fall into a gray area where they may in many cases behave either as small particles or as large solutes. (Brant *et al.* 2007a)

The ENPs are usually not homogeneous in size. Thus, it is often best characterized by an average particles size range. ENPs have defined dimensions, like fullerenes, where the diameters are determined by the number of atoms that they contain (e.g. 0.7 nm for C_{60}). Also single walled carbon nanotubes have rather narrow ranges of well defined diameters (1-3nm), while their length can vary a lot. For ENPs to remain suspended as particles in a liquid, most ENP suspensions must be stabilized with surfactants.

1. **Aspect ratio:** It is the ratio between the longest diameters of a particle to the shortest perpendicular diameter. It describes one of the most important parameter regarding the shape of a particle and is probably affecting both mobility and uptake of ENPs in organisms.

2. **Surface area:** The surface area of ENPs is the total area of both external and internal surfaces available from a particle. This is an important

parameter because all interaction between ENPs and other surfaces or solutes will relate to the particle's surface. Consequently the reactivity of ENPs increases strongly with decreasing particle size.

3. **Crystalinity:** It refers to more or less stable three dimensional atom arrangements, and is a property that defines many oxide ENPs as crystal structure. It may have several direct and indirect effects on chemical and morphological parameters like surface area, charge, aspect ratios etc.

4. **Degradability or Persistence of ENPs:** It relates to their chemical composition, with organic ENPs being biodegradable, other allotropes being extremely persistent and mineral ENPs being more or less prone to weathering by oxidation or dissolution.

5. **Elemental Composition:** Elemental composition is a description of elements which an ENP is capable of contributing major contaminants, which is often given by the manufacture.

6. **Surface charge:** It is a measure of a particle's propensity to interact with changed surfaces and ions. Surface charge of ENPs may be either pH dependant, as in oxide minerals or may be fixed as in clays (where this change results from crystal lattice defects and atomic substantiation, Rose *et al.* 2007).

7. **Point of zero change (PZC):** It is the pH at which the positive and negative charge are balanced so that there is no net change making particles mobile in an electric field. The PZC determines if a given ENP is changed or unchanged, meaning that suspension pH may determine its mobility.

I) Analytical techniques for detection of nanoparticles

a) Qualitative methods for detection of nanoparticles

Elemental composition of bulk ENPs is measured with standard chemical analysis, like Inductively Coupled Plasma Mass Spectrometry (ICP-MS). These may be present in very low amounts and are thus difficult to detect. Determination of such surface constituents may require more specialized analytical methods, depending on their nature and concentration.

The morphology of pure nanomaterials is frequently described using High Resolution-Scanning Electron Microscopy (HR-SEM) or High Resolution-Transmission Electron Microscopy (HR-TEM). These methods permit a direct visualization of nanoparticles and can thus provide data on size, shape and structure. HR-SEM can be used for particles as small as 10-20 nm, while HR-TEM can have resolution below 1nm. TEM coupled with EDX as well as Selected Area Electron Diffraction (SAED) provides the sufficient sensitive data on elemental distribution at a very low spatial scale. Atomic Force Microscopy (AFM) and Scanning Tunneling Microscopy (STM) have resolution down to the atomic level, and permit both three dimensional imaging of nanometer scale surface and the measurement of forces between surface at the piconewton scale, both in liquid and gaseous environment.

Static light and X-ray scattering experiments may be used to characterize particles in suspension with respect to size and shape. Agglomeration state particle size in dilute suspensions may also be determined indirectly by Dynamic light scattering (DLS), based on Brownian motions of nanoparticles.

Surface change is often measured as zeta potential for particles dispersed in a liquid. It is essential that the modification of the surface change is determined as a function of pH and ionic strength.

b) Quantitative methods for detection of nanoparticles

Very small changes in concentration can be detected for rare elements like Ag, Au, Ce etc by Chemical analyses in samples where their background levels are very low. But such analyses do not distinguish between particles and dissolved ions.

Q-dots or aromatic structure like fullerenes and CNTs which have intrinsic fluorescent properties, can be detected based on autofluorescence (Cherukuri *et al.* 2006; Mureau *et al.* 2007). Fullerenes may be extracted from environmental matrices using organic solvents, and can be quantified by chromatography and UV-vis spectroscopy or HPLC.

CNTs may be detected by UV-VIS spectroscopy at high concentrations, and their extreme persistence can permit acid digestion of a surrounding matrix to avoid interference in subsequent analyses. SWCNTs can also be detected by Raman Spectroscopy due to specific resonance properties which are not present in MWCNTs. The FTIR measurement can be utilized to study the presence of protein molecule in the solution. Isotopic labeling of ENPs in laboratory experiments is a powerful tool for tracing and quantification. Fullerenes for example, may be produced with either stable (Fortneret *et al.* 2007) or radioactive isotopes (Masumoto *et al* 1999; Yamago *et al.* 1995) and can then be followed by GC-MS and Scintillation counting respectively.

ENPs with magnetic properties (Containing *e.g.*: Fe, Ni, Co, Mn, etc.) may be extracted, detected or even quantified using a strong magnetometer like the superconducting quantum interference device (SQUID).

Behavior of Engineered Nanoparticles

Human toxicity studies, with a main focus on workers and their environments are primarily considering dispersion and uptake from air but also direct uptake through ingestion, dermal exposure and in some special cases related to medical use injections. Both in this and other scenarios for spreading ENPs, water may serve as a transport medium and a temporary reservoir for ENPs.

Natural organic matter can be divided into two major classes; non-humic substances (polysaccharides, amino acids, etc.) and humic substances. The latter comprises organic macromolecules formed by partial degradation and transformation of recalcitrant plant and microbial polymers. Soluble humic substances may be subdivided into fulvic acids which are water soluble at all pH, and humic acids which are defined as humic substances that are insoluble at low pH (<2) but soluble at alkaline pH.

Dispersed colloids are particles in the ENP range (1-200nm) that are kept in a stable suspension in water. They do not precipitate by gravitation due to their small size, a certain surface charge and the fact that electrostatic interactions, vander-waals forces and steric forces maintain a stable suspension. Any changes in such a system with respect to pH, ion concentrations etc., may destabilize the (O' Melia 1972) particle stability, affecting the concentration of cations (coagulants), means that increasing salt concentration free nanoparticles will start to aggregate.

The production, use and imperfect waste treatment of nanomaterials will inevitably lead to losses where nanomaterials end up in the environment i.e. in water sediments and soil (Nowack and Bluceli 2007). Three aspects seem to be important when assessing the impact of ENPs as pollutants ending up in the environment:

i.   **Mobility (transport and transfer)**: their ability to move from one place to another (*e.g.*: from a spill site to an uncontaminated site) or from one recipient to another (*e.g.*: from soil to drinking water or food plants or crops).

ii. **Ecotoxicity**: the possible harm that ENPs can cause to organisms living in water, sediments and soils that they enter.

iii. **Modification**: mode and extent of ENPs modified by contact with the environmental compartments.

The studies on attachment and transport of ENPs through porous media was started by taking an applied approach using simplified model systems with glass beads or quartz sand as a solid phase and pure water as a transport medium. Lecoanet and Wiesner (2004) were the first to study the mobility of ENPs in porous media. They used columns packed with glass beads (APS 355 µm) and assessed mobility of three types of fullerenes (fullerol, nC_{60} and SWCNTs) TiO_2 and SiO_2 at two different flow rates. A high mobility was observed in case of fullerol and surfactant modified SWCNTs. The lowest mobility was observed in case of nC_{60}.

A major determinant for particles mobility is the stability of its suspension. If destabilized, particles suspension will aggregate, which in turn may lead to massive deposition (Gimber *et al.* 2007). Several factors that affect ENP surface potential may destabilize a suspension, including pH changes, increased ion concentration, dilution or degradation of stabilizing agents etc. (Guzman *et al.*, 2006).

Degradability or persistence of ENPs is highly relevant in the context of environmental impact of ENPs, as persistent particles with negative effects may be able to continue having negative effects for a very long time. C-based ENPs like fullerenes and CNTs can theoretically be degraded and re-enter the biogeochemical cycle as CO_2. The presence of Al_2O_3 nanoparticls facilitated disruption of inter-particle attractive forces and similar mechanical degradation of aggregates was also observed in soil samples containing aggregated nano-Fe_2O_3.

Organic coatings and surfactants, which are used to disperse ENPs for various applications, get easily degraded in the environment. It is thus likely that organically coated ENPs will be prone to degradation of the coating and release their core particles upon such degradation. This has been observed e.g. for lipid–coated SWCNTs that were ingested by the crustacean *Daphnia magna* and stripped of its lipid coating during passage through the intestines (Roberts *et al.* 2007).

Toxicity and Ecotoxicity

i) Toxicity

Toxicity of fine and ultra fine particles in ambient air (atmospheric pollution, coal dust, asbestos, etc.) has a long history which has been exploited to give "nanotoxicity" studies a flying start (Oberdorster *et al.*, 2005). Indeed, large amounts of nanoparticles are naturally present in the atmosphere including carbon-nanotubes and other fullerene related nanocrystals (Murr *et al.*, 2004) before the nanotechnology era. Great deal of understanding has been made about what happens when incidental nanoparticles deposit in mammalian airways and lungs inducing oxidative stress, inflammation and cardiopulmonary disease (Handy and Shaw 2007; Oberdorster *et al.*, 1995).

Toxicity Mechanisms of Metal Oxide Nanoparticles

a) Reactive Oxygen Species (ROS)

Currently, the best developed paradigm for nanoparticles toxicity for eukaryotes is generation of ROS. ROS production is especially relevant in the case of nanoparticles such as TiO_2 with photo-catalytic properties. ROS have been shown to damage cellular lipid, carbohydrates, protein and DNA, leading to inflammation and oxidative stress response. Lipid peroxidation is considered the most dangerous since it leads to alterations in cell membrane properties which in turn disrupts vital cellular functions.

The oxidative stress mechanisms are also linked to a number of human pathologies and aging. Experimentally nanoparticles have been shown to induce oxidative stress responses *in vitro* in keratinocytes, macrophages and blood monocytes. Recent *in vitro* data revealed ROS mediated potential neurotoxicity of nano TiO_2. In human lung epithelial cells, the chemical composition of nanoparticles was the most decisive factor determining the formation of ROS in exposed cells to iron, cobalt, manganese and titanium containing silica nanoparticles and respective pure metal oxide nanoparticles. During their life cycle, engineered nanoparticles might also produce ROS upon interaction with abiotic and biotic environmental factors. Indeed damaging effects of TiO_2 nanoparticles on bacteria have been shown to be enhanced by sunlight or UV illumination. TiO_2 nanoparticles in combination with UV-light have been shown to inactivate algae *Anabaena microcystis* and *Melosira,* also have been shown to destroy the cell surface architecture of blue-green algae *chroococcus* sp. However, TiO_2 nanoparticles have shown toxicity to *Bacillus subtilis* and *E. coli* in the dark. Analogously in fish cells *in-vitro,* hydroxyl radicals were generated by TiO_2 nanoparticles in the absence of UV light. ROS could be also involved in toxicity of ZnO nanoparticles to bacteria as shown for *E. coli.*

b) Release of Metal Ions

In the case of metal containing nanoparticles, the release of metal ions and their species may be a key factor in their ecotoxicity as shown for oxide nanoparticles using *in-vitro* cell cultures; solubility of nanoparticles strongly influenced their cytotoxicity. In human lung epithelial cell *in vitro* it shows Trojan-horse type mechanisms i.e. metal oxide nanoparticles enter the cells but not the respective ionic forms. The nanoparticles are not only act as transfer vectors in the environment, but they also facilitate the entry of nanoparticle sorbed pollutants into cells of organisms potentiating toxic effects.

Release of (toxic) heavy metal ion species from metal containing nanoparticles under environmental conditions should be taken into account in their ecotoxicity evaluation. Metal solubility may be influenced by organisms; initially insoluble forms of heavy metals may become bioavailable due to the direct contact between bacteria and soil particles.

ii) Ecotoxicity

Ecotoxicity measurements are conducted on different tropic levels including microorganism, plants, invertebrates and vertebrates. Test systems have been standardized for different organism for different exposure conditions. Thus, protocol approved by the OECD or ISO on how to test the adverse effects of heavy metals or pesticides on earthworms, daphnia or zebra fish can be obtained.

Many of these tests employ microorganisms (mainly bacteria, but also fungi, protozoa and algae) have the advantage that they are ubiquitous and highly diverse (filling a range of habitats and functions), small (permitting miniaturized tests) and with short generation times.

In case of microorganism, it is easy to have culture and easy to extract DNA form them. The latter permits identification based on sequencing of DNA and more importantly to describe whether or not certain genes related to toxicity protection or stress have been activated. Identification of multiple microorganisms in a single sample through molecular methods permits us to describe the composition of complex microbial communities and changes in composition due to a suspected harmful agent. Measurements of such changes are often far more sensitive than tests based on isolation, pure culture and testing of individual microorganism.

a) Bioassays taking bioavailability into account

Ecotoxicity strictly means toxicity to environmentally relevant organisms, while the term 'bioassay' implies that the toxicity or stress caused by a compound has been measured in an environmental matrix pertinent to the habitat where the organism live in nature. Exposing a fish to ENPs in pure water may thus be an ecotoxicity test, while exposing it to ENPs in water containing salts, dissolved organic carbon (DOC) and other colloidal materials would constitute a bioassay.

The modification of ENP constituents, like ions, natural colloidal and other changed surfaces, after entering environmental matrix, are likely to affect not only mobility, aggregation etc., but also modify toxicity characteristics (Lyon *et al.* 2007). Further an ENP that interacts with a charged surface of a large particle may not be available for absorption in the same way or to the same extent as a freely suspended ENP, rendering its bioavailability, consequently a far lower exposure to ENPs may be observed in an environmental matrix compared to what is experienced *in vitro*.

Toxicity data are essential to describe the potential negative effects of ENPs. In the context of environment, it is important to know the steps involved in ENPs interaction while entering the environment in order to predict as to where the ENPs end up and to assess their toxicity.

b) Ecotoxicity of C_{60} fullerenes

C_{60} fullerenes have been among the first ENPs to be investigated with respect to ecotoxicity, starting with the pioneer work of Eva Obserdoster and colleagues. In Oberdoster's first report of fullerene toxicity (Oberdorster 2004) it has been shown that low concentrations of C_{60} (0.5mg/l) could cause oxidative damage (lipid peroxidation in the brain) and enzyme changes (glutathione reduction in gills) in fish (large mouth bass; *Micropterus salmoides*). In these experiments, C_{60} had been dissolved using the organic solvent tetrahydrofuran (THF) and mixed with water before evaporating the solvent to obtain an aqueous suspension of colloidal C_{60} aggregates called nC_{60}. Later it was demonstrated that nC_{60} dissolved with THF trap the solvent within its aggregate structure and may release this for prolonged periods of time giving rise to toxicity response that should not be attributed to the C_{60} but rather to THF (Brant *et al.* 2005).

These authors also exposed fathead minnow (*Pimephales promelas*) to the same fullerene preparation and at 0.5mg/l this led to 100% mortality in the case of THC-C_{60} within 8-16h, whereas water stirred with nC_{60} at the same concentration induced no observable effects even after 48h. Exposure of another fish species zebra fish (*Danio rerio*) to nC_{60} at 1.5mg/l delayed embryo and larval development, and resulted in decreased survival and hatching rates and caused pericardial oedema (Zhu *et al.* 2007).

Toxicity to daphnia of nC_{60} prepared without THF showed delayed molting and reduced offspring production at 2.5mg/l, on the embryo chorion was nanoscaled while the size of SWCNTs agglomerates was micro scaled longer, indicating that the chorion of zebrafish embryos was an effective protective barrier to SWCNTs agglomerates.

Lam *et al.*, (2006) examined histopathological alteration in mice at 7 and 90 days following exposure to SWCNTs manufactured by different methods and containing varying amounts of residual catalytic metals. Carbon black and quartz particles were employed as low and high pulmonary toxicity particles control, respectively. They demonstrated that all three types of SWCNT products induce dose-dependent lung lesions that were characterized by granulomas.

The most comprehensive comparative toxicity assessments of SWCNTs in rats are by examining the ability of these nanoparticles to induce pulmonary inflammation as well as alter lung cellular proliferation, and pathology following intratracheal installation of SWCNTs.

According to Oberdorster *et al.*, (2006), SWCNTs are less toxic to aquatic organisms than fullerenes.

In a study on MWCNT toxicity of unicellular protozoa *Stylonychia mytilus*, it was observed that CNTs were ingested, redistributed during the dividing process of the cells, and excreted from the cells (Zhu *et al.*, 2006). Also, exposure to MWCNTs at concentration higher than 1mg/l induced a dose-dependent growth inhibition of the cells.

A recent study showed that when nanotubes are coated with organic lipids they become more accessible to the H_2O flea *Daphina magna* (Roberts *et al.* 2007). The fleas ingest the materials and strip the lipid layer for food, eventually causing the uncoated nanotubes to block their digestive tracts and kill them.

c) Ecotoxicity of metal nanoparticles

Silver nanoparticles: These are mainly produced for antiseptic applications and they have been well documented for bactericidal (Fu *et al.* 2006; Jeong *et al.* 2005) and cytotoxic effects (Broydich-Stolle *et al.*, 2005), including specific effects on mitochondria and generation of ROS (Hussain *et al.* 2005). A recent study on uptake of Ag nanoparticles into zebrafish embryos *in vivo* showed an NOEC as low as 0.19nM (Lee *et al.*, 2007). Ecotoxicity data on nano silver are scarce, especially with regard to aquatic forms. Toxicity of silver nanoparticles to bacteria has been examined, but the mechanisms of toxicity have not been fully elucidated. In particular it is unclear whether toxicity is specifically related to nanoparticles properties or is due to the effect of Ag^+ ions. Toxic effects of Ag nanoparticles may be related to damages at cell membrane, to oxidative stress, or to interactions of Ag^+ ions with proteins and enzymes.

Copper nanoparticles: (particles with a biphasic size distribution broadly peaking at 80 and 450nm) have been investigated for toxicity towards zebra fish (*D. rerio*) and compared to toxicity response towards soluble Cu ions ($CuSO_4$) in dechlorinated tap H_2O with a hardness of 142 $CaCO_3$/l and a pH of (Griffitt *et al.* 2007) 8.2. In this study Cu ENPs were less toxic than Cu ions with static LD_{50} after 48h of 0.25mg/l for Cu ions and a corresponding value of 1.56 mg/l for Cu ENPs. Exposure to both Cu formulations caused similar gill injuries and elevated gill Cu concentration, but Cu ENPs resulted in higher expression levels of several genes related to oxidative stress in gill compared to Cu ions and Cu ENPs are used as an antimicrobial agent in similar way as Ag.

Gold nanoparticles: Data on eco-toxicity of gold nanoparticles are not available for aquatic microorganisms. A recent cyto-toxicity study on Au nanoparticles ranging in size from 0.8 to 15nm showed that several human cell types were sensitive to the smallest gold particles (≥ 1.4nm) with EC_{50} values for cell death within 12h ranging from 30 to 56 μm.

d) Ecotoxicity of Oxide nanoparticles

Titanium Oxide: *Daphnia magna* was treated with THF-titanium oxide after the 48 hr exposure; it was found that LC_{50} value is 5.5 mg/l where as treatment with TiO_2 prepared by sonication showed a maximum mortality of 9% for any TiO_2 concentrations up to 500 mg/l (Lovern and Klaper, 2006).

The ecotoxicity of TiO_2 SiO_2 and ZnO ENPs to gram positive (*Bacillus subtilis*) and gram negative (*Escherichia coli*) bacteria in water suspensions containing citrate and low PO_4 concentration was investigated by Adams *et al*, (2006). Exposure to ENPs in sunlight for 6 hours was found to be harmful to various degrees, with antibacterial activity increasing with particle concentration (tested range 10-500 mg/l). Antibacterial activity generally increased from SiO_2 to TiO_2 to ZnO and *B. subtilis* was most sensitive to such effects.

In an ecotoxicity experiment with algae (*Desmodesmus subspicatus*) and daphnia (*D. magna*) exposed to pre-illuminated TiO_2 showed that algae were more sensitive with an EC_{50} of 44 mg/l (Hund-Rinke and Simon, 2006). In this experiment enhanced toxicity induced by pre-illuminating TiO_2 particles showed that the photocatalytic activity of nanoparticles can last for a period of time.

Zinc oxide: Phytotoxicity of ENPs has been demonstrated for Zn and ZnO by carrying out inhibition of seed germination and root growth after 2h exposure to ENPs suspensions in deionized water (Lin and Xing, 2007). Five types of nanoparticles (multi-walled carbon nanotubes aluminum, alumina, zinc and zinc oxide) and six plant species (radish, rape, ryegrass, lettuce corn and cucumber) were screened. Fifty percent inhibition of root growth was observed for nano-Zn and nano-ZnO at approx. 50 mg/l for radish and about 20mg/l for rape and ryegrass, whereas other ENPs plant combinations showed weaker inhibition.

Alumina nanoparticles: They are commonly employed in scratch-resistant transparent coating and sunscreen lotions that provide transparent-UV protection. However, a recent study reported that alumina nanoparticles led to phototoxicity by retarding the growth of root in five plant species (corn, cucumber, soybean, cabbage, and carrot) of significant economic value.

e) Ecotoxicity of other nanoparticles:

Quantum dots: Quantum dots are a diverse group of substances with different composition, size and coatings increasingly used in drug targeting and biomedical imaging. Some q-dots have a strong cytotoxic effect (Hardman 2006), but data on ecotoxicity are lacking.

Dendrimers: These are repeatedly branched organic molecules occasionally used to form hollow cage for transport of drug or other therapeutic agents in nanomedicine. There has been some indication that dendrimers display cytotoxicity (Duncan and Izzo, 2005), and that cationic dendrimers were more cytotoxic than hemolytic anionic or PE Gylated dendrimers (Chen *et al.*, 2004). No ecotoxicity studies have so for been performed on dendrimers as ENPs.

Environmental Hazards and Risks

Risk assessment of ENPs has started with identification of hazards and exposure routes for humans in a production setting as this is perhaps the most imminent situation where safety issues have to be resolved to permit other areas of nanotechnology to advance (Kreyling *et al.*, 2006; Lam *et al.*, 2006).

It is premature to attempt the assessment of environmental risks at the present time and it may indeed never be possible to provide any generally valid risk assessment model for ENPs, as they comprise such a broad and continuously developing class of substances (Colvin, 2003). Current approaches for environmental risk assessment of chemicals basically compare environmental concentrations to no adverse effect levels. We may assume that a similar strategy can be applicable to evaluate the environmental impact of released ENPs if speciation and concentration dependent exposure and toxicity are taken into account. Further more ENPs might change the environment, without being toxic in the traditional sense.

A recent workshop co-sponsored by the Nation Science Foundation and the US Environmental Protection Agency has identified a number of critical risk assessment issues regarding ENPs, such as:

i.   Exposure assessment of ENPs

ii. Toxicology of ENPs

iii. Ability to extrapolate ENPs toxicity using existing particles and fiber toxicological databases

iv. Environmental and biological fate, transport, persistence and transformation of ENPs

v.   Recyclability and overall sustainability of ENPs

Environmental Risk Assessment of Nanoparticles

Risk assessment is the task of characterizing a level of risk usually in terms of a relative score or ranking. The goal of performing a risk assessment is to provide information that will help to evaluate alternative and arrive at decisions (Calow 1998).

Traditionally the risk assessment is divided into four steps

i.   Hazard assessment

ii. Dose-response assessment

iii. Environmental exposure assessment

iv. Risk characterization

i) Hazard assessment

In the hazard assessment the potential of the nanoparticle to cause harm is evaluated. Many considerations should be made when characterizing hazards *e.g.* how much is known about the biological uptake and metabolism, what kind of toxicological studies have been performed, which effects have been observed, (mortality, mutations, inflammation irritation, growth reduction, reproduction effects, stress, etc.), are there any other end points of concern, do the data from experiments involve taking into account of realistic exposure mechanisms etc. Hazards, such as toxicity and ecotoxicity, can be measured using different endpoints, including physiological, genetic or functional effects, either acute or chronic etc. But there may also be other potentially negative environmental effects apart from toxicity that need to be considered in a risk assessment paradigm adapted to ENPs.

ii) Dose-response assessment

Dose-response assessment follows the hazard identification in the risk assessment process. The establishment of dose response relationships may involve performing experiments in the laboratory or using mathematical models. However dose-response relationship may not be straight forward for ENPs as dose based on mass concentration may be less relevant than dose based on surface area and also different ENPs preparations may result in differences in surface reactivity and thus will result in different kind of toxicity. As for hazard assessment, dose response assessment need to take into account all relevant end-points (physiological, genetic and functional). Finally, when it comes to predictions of human effects, animal studies or other replacement tests have to be used if results from epidermological studies are not available.

iii) Exposure assessment

The potential for exposure to nanomaterials begins with production of these materials. Thus knowledge on quantitative aspects linked to different production steps, purification, functionalization, conditioning packaging and transport, as well as losses and waste streams associated with each of these are important. Predication of industrial release must be based upon knowledge on the day-to-day operations, including the processes that are likely to be the most important for emission rates, e.g. those involving high temperatures and high pressures, high material flows and all waste streams. When considering environmental exposure it is also important to consider the frequency and magnitude of incidents that may lead to release to air, water and soil (Robichaud *et al.*, 2007).

For environmental exposure it is important to have empirical data or procedures to predict the persistence and mobility in air, soil and water. Examples of parameters that may be needed to make predications on environmental fate are chemical factors like adsorption capacity, degree of aggregation, photolytic degradation, dispersability, interactions with soil particles etc. The tendency of nanoparticles to aggregate is especially important for the environmental persistence and effects. It is still an open question whether we can expect to find individual free of nanoparticles in the environment.

iv) Risk Characterization

Risk characterization is the final step in the risk assessment procedure in which the information from the hazard identification, dose-response and exposure steps are considered together to determine and communicate the actual likelihood of risk to exposed populations. The risk is often characterized by comparing the exposure concentration with an exposure level that is assumed to have no effect.

Important issues in this final step is an evaluation of the overall quality of data, the assumption and uncertainties associated with each step and the level of confidence in the resulting estimates.

Current Status of Environmental Risk Assessment of Enps

According to Robichaud *et al.* (2007) current tools for risk assessment may provide useful guidance for short-term assessment of nanomaterials, as more results from studies on exposure and hazards are waiting. They describe a method for performing a relative risk analysis of nanomaterials using tools from the insurance industry. As an example they mention XL Insurance, a Swiss company that employs a numerical tool to quantify the relative risk of manufacturing processes with respect to financial liability of the insurers. By focusing on environmental pollution and health risks this tool may be useful as a first step for making a risk assessment protocol for ENPs. Five nanoparticle materials (single walled carbon nanotubes, C_{60} fullerenes, one variety of quantum dots, alumoxane nanoparticles and TiO_2) were selected based upon their current or near term potential for large scale production.

The influence of key production parameters such as temperature and pressure in combination with chemical, physical properties of the materials, the relative risks were calculated based upon factors like volatility, carcinogenicity flammability, toxicity and persistence.

Another starting point, of course is to establish a new risk assessment paradigm for ENPs since descriptions of exposure and toxicity can only partly profit from the knowledge based on environmental chemistry, ecotoxicolgy and ecology.

Multi–criteria decision analysis (MCDA) can then be used for a transparent application of WOE based on different disciplines. Critto *et al.,* (2007), Semenzin *et al.,* (2007) and Linkov *et al.,* (2007) reveal that the essential contribution of MCDA is the possibility to link limited information on physical and chemical properties with decision criteria and weightings elicited from scientists and managers, allowing visualization and quantification of the trade off involved in the decision making process.

Below figure summarizes the described path for a risk assessment framework that uses existing risk assessment models and characteristics of ENPs.

ENP-Specific features

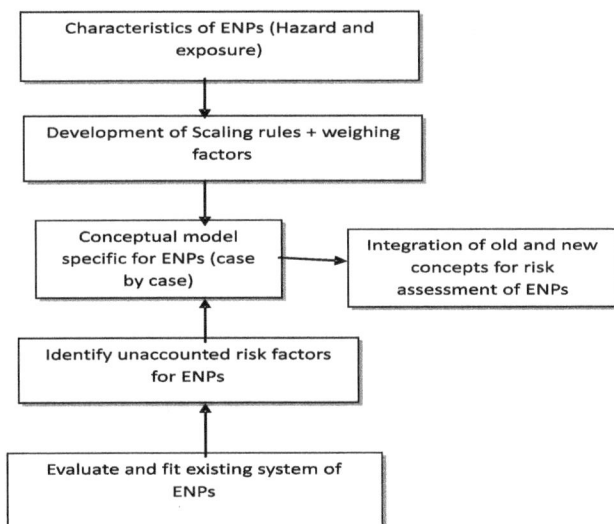

Safety Guidelines for Handling Nanomaterials

Use good general laboratory safety practices as found in your chemical hygiene plan. Wear gloves, lab coats, safety glasses, face shields, closed toed shoes as needed.

i. Be sure to consider the hazards of precursor material in evaluating process hazards.

ii. Avoid skin contact with nanoparticles or with nanoparticles–containing solutions by wearing appropriate personal protective equipment. Do not handle nanoparticls with your bare skin.

iii. If it is necessary to handle nanoparticles powders out side of a HEPA-filtered powered–exhaust laminar flow hood, wear appropriate respiratory protection. The appropriate respirator should be selected based on professional consultation.

iv. Use flame exhaust hoods to expel fumes from tube furnaces or chemical reaction vessels.

v. Dispose of and transport waste nanoparticles according to guidelines of hazardous chemical waste.

vi. Vacuum cleaners used to clean up nanoparticles should be tested for HEPA-filtered units.

vii. Equipment previously used to manufacture or handle nanoparticles should be evaluated for potential contamination prior to disposal or reuse for another purpose.

viii. Lab equipment and exhaust systems should also be evaluated   prior to removal, remodeling or repair.

ix. Given the differing synthetic methods and experimental goals, no blanket recommendation can be made regarding aerosol emission controls. This should be evaluated on a case by case basis.

Conclusion

It must be concluded that current risk assessment methodologies are not suited to the hazards associated with nanoparticles. In particular, existing toxicological and eco-toxicological methods are not up to the mark. Exposure evaluation (dose) needs to be expressed as quantity of nanoparticles and surface area rather than simply mass. Equipment for routine detecting and measuring nanoparticles in air, water or soil is inadequate and very little is known about the physiological responses to nanoparticles.

The heterogeneity of emerging ecotoxicity data for metal oxide nanoparticles shows the need for additional studies involving standardization and modification of the respective test protocols. The toxicity mechanism of metal oxide nanoparticles may be related to either soluble ions or particle properties or both. In all cases, aggregation and chemical speciation play a leading role in their ecotoxicity. Regulatory bodies in the US as well as in European Commission have concluded that nanoparticles form the potential for an entirely new risk and that it is necessary to carry out an extensive analysis of the risk. The out come of these studies can form the basis for government and international regulations.

References

Adams LK, Lyon Dy and Alvarez PJT 2006. Comparative ecotoxicity of nano scale TiO$_2$, SiO$_2$ and ZnO water suspensions. *Water Research* 40, 3527-3532.

Brant J, Lecoanet H, Hotze M and Wiesner M 2005a. Comparison of electrokinetic properties of colloidal fullerenes (n-C-60) formed using two procedures. *Environmental Science and Technology* 39, 6343-6351.

Brant J, Lecoanet H and Wiesner M 2005b. Aggregation and deposition characteristict of fullerence nanoparticles in aqueous system. *Jouranal of nanoparticles Research* 7, 545-553.

Brant J A, Labille J, Bottero J Y and Wiesner M R 2006. Characterizing the impact of preparation method on fullerene cluster structure and chemistry. *Langmuir* 22, 3878-3885.

Braydich-S tolle L, Hussain S, Schlager JJ and Hofmann MC 2005. In vitro cytotoxicity of nanoparicles in mammalian germline stem cells, *Toxicological Sciences* 88, 412-419.

Buffle J 2006. The Key role of environmental colloids/nanoparticles for the sustainability of life. *Environmental Chemistry* 3, 155-158.

Calow p 1998. Hand book of environmental risk assessment and management Blackwell science, Oxford -590 pp.

Chen H T, Neerman M F, Parrish A R and Simanek E E 2004. Cytotoxicity, hemolysis, and acute *in-vivo* toxicity of dendrimers based on melamine, candidate vehicles for drug delivery. *Journal of the American Chemical Society* 126, 10044-10048.

Cheng T, Flahaut E and Cheng S H 2007. Effect of carbon nanotubes on developing Zebrafish embryos. *Environmental Toxicology and Chemistry* 26, 708-716.

cherukuri P, Gannon C J, Leeuw T K, Schmidt H K, Smalley R E, Curley S A and Weisman B 2006. Mammalian pharmacokinetics of carbon nanotubes using intrinsic near-infrared fluorescence. *Proceedings of the National Academy of Science of the United of America* 103, 18882-18886.

Cioffi N, Ditaranto N, Torsi L, Picca R A, De Giglio E, Sabbatini L, Novello L, Tantillo G, Bleve-Zacheo T and Zambonin P G 2005a. Synthesis, analytical characterization and bioactivity of Ag and Cu nanoparticles embedded in poly-vinyl-methyl-ketone films. *Analytical Bioanalytical Chemistry* 382, 1912-1918.

Cioffi N, Torsi L, Ditaranto N, Tantillo G, Ghibelli L, Sabbatini L, Bleve-Zacheo T, D'Alessio M, Zambonin P G and Traversa E 2005b. Copper nanoparticles/polymer composites with antifungal and bacteriostatic properties. *Chemistry of Materials* 17, 5255-5262.

Colvin VL 2003, The Potential environmental impact of engineered nanomaterials. *Nature Biotechnology* 21, 1166-1170.

Critto A, Torresan S, Semenzin E, Giove S, Mesman M, Schouten A J, Rutgers M and Marcomini A 2007. Development of a site-specific ecological risk assessment for contaminated sites: Part I. A multi-criteria based system for the selection of ecotoxicological tests and ecological observations. *Sci. Tot. Environ.*

Deguchi S, Alargova R G and Tsujii K 2001. Stable dispersion of fullerenes, C-60 and C-70, in water preparation and characterization. *Langmuir* 17, 6013-6017.

Duncan R and Ezzo L 2005. Dendrimer biocompatibility and toxicity. *Advanced Drug Delivery Reviews* 57, 2215-2237.

Fortner J D, Kim D I, Boyd A M, Falkner J C, Moran S, Colvin V L, Hughes J B and Kim J H 2007. Reaction of water-stable C60 aggregates with ozone. *Environmental Science and Technology*.

Gimbert L J, Hamon R E, Casey PS and Worsfold P J 2007. Partitioning and stability of engineered ZnO nanoparticles in soil suspensions using flow field-flow fractionation. *Environmental Chemistry* 4, 8-10.

Griffitt R J, Weil R, Hyndman K A, Denslow N D, Powers K, Taylor D and Barber D S 2007. Exposure to copper nanoparticles causes gill injury and acute lethality in zebrafish. *Environmental Science and Technology* 41, 8178-8186.

Guzman K A D, Finnegan M P and Banfield J F 2006. Influence of surface potential on aggregation and transport of titania nanoparticles. *Environmental Science and Technology* 40, 7688-7693.

Handy R D and Shaw B J 2007. Toxic effects of nanoparticles and nanomaterials implications for public health risk assessment and the public perception of nanotechnology. *Health Risk and Society* 9, 125-144.

Hansen S F and Tickner J A 2007. The challenges of adopting voluntary health, safety and environment measures for manufactured nanomaterials: Lessons from the past for more effective adoption in the future. *Nanotechnology Law and Business* 4, 341-359.

Hund-Rinke K and simon M 2006. Eco-toxic effect of photo catalytic active nanoparticles TiO_2 on algae and daphnids. *Environmental Science and pollution Research* 13, 225-232.

Hussain S M, Hess K L, Gearhart J M, Geiss K T and Schlager J J 2005 In-vitro toxicity of nanoparticles in BRL 3A rat liver cells. *Toxicology in-vitro* 19, 975-983.

Jeong S H Hwang Y H and Yi SC 2005. Antibacterial properties of padded PPIPE nonwovens incorporating nano size silver colloidal, *Journal of Materials Science*, 40, 5413-5418.

Kreyling W G, Semmler-Behnke M and Moller W 2006. Health implication of nanoparticles. *Journal of Nanoparticles research* 8, 543-562.

Lam C W, James J T, McCluskey R, Arepalli S and Hunter R L 2006. A review of carbon nanotubes toxicity and assessment of potential occupational and environmental health risk. *Critical Reviews in Toxicology* 36, 189-217.

Lecoanet H F, Bottero J Y and Wiesner M R 2004. Laboratory assessment of the mobility of nanomaterials in porous media. *Environmental Science and Technology* 38, 5164-5169.

Lecoanet H F, and Wiesner M R 2004. Velocity effects on fullerence and oxide nanoparticles deposition in porous media. *Environmental Science and Technology* 38, 4377-4382.

Lee K J, Nallathanby P D, Browning L M, Osgood C J and Xu X-H N 2007. In-vivo imaging of transport and biocompatibility of single silver nanoparticles in early development of zebrafish embryos ACS Nano 1, 133-143.

Lin D and Xing B 2007. Phytotoxicity of nanoparticles: Inhibition of seed germination and root growth. *Environmental Pollution* 150, 243-250.

Linkov I, Satterstrom F K, Steevens J, Ferguson E and Pleus R C 2007. Multi-criteria decision analysis and environmental risk assessment for nanomaterials. *Journal of Nanoparticles Research* 9, 543-554.

Lyon D Y, Thill A, Rose J and Alvarez P J J 2007. Ecotoxicological impacts of nanomaterials. In Environmental Nanotechnology. Application and Impacts of Nanomaterials. Eds. M R Wiesner and J Y Bottero. pp 445-479. McGraw Hill, New York.

Masumoto K, Ohtsuki T, Sueki K, KiKuchi K and Mitsugashira T 1999. Direct synthesis of radioactive carbon labeled fullerenes using nuclear reactions. *Journal of Rioanalytical and Nuclear Chemistry* 239, 201-206.

Mureau N, Mendoza E and Silva S R P 2007. Dielectrophoretic manipulation of fluorescing single-walled carbon nanotubes. *Electrophoresis* 28, 1495.

Murr L E, Soto K F, Esquivel E V, Bang J J, Gurrero P A, Lopez D A and Ramirez D A 2004. Carbon nanotubes and other fullerene-related nanocrystals in the environment: A TEM study. *Journal of Materials;* 56, 28-31.

Nowack B and Bucheli T D 2007. Occurrence, behavior and effects of nanoparticles in the environment. *Environmental Pollution* 150, 5-22.

O'Melia C R 1972. Coagulation and flocculation. In Physicochemical processes for water quality control. Ed. W J Weber jr. pp 61-109. Wiley Interscience, New York.

Oberdoster G, Gelein R M, Ferin J and Weiss B 1995. Association of particulate air-pollution and acute mortality-Involvement of ultrafine particles. *Inhalation Toxicology* 7, 111-124.

Oberdoster G, Oberdoster E and Oberdoster J 2005. Nanotoxicology: An emerging discipline evolving from studies of ultrafine particles. *Environmental Health Perspectives* 113, 823-839.

Oberdoster E 2004. Manufactured nanomaterials (Fullerenes, C60) induce oxidative stress in the brain of juvenile largemouth bass. *Environmental Health Perspectives* 112, 1058-1062.

Oberdoster E Zhu S Q Blikely TM McClellan Green P and HaSchMZ 2006. Ecotoxicolgy of carbon–based ENPs effects of fullerence (C60) on aquatic organism carbon 44, 1112-1120.

Pan Y, Neuss S, Leifert A, Fischler M, Wen F, Simon U, Schmid G, Brandau W and Jahnen-Dechent W 2007. Size-dependent cytotoxicity of gold nanoparticles. *Small* 3, 1941-1949.

Robert A P, Mount A S, Seda B Souther J, QiaoR Lins, Ke P C Rao A M and Klaine S T 2007, *Invivo* biomodification of lipidcoated carbon nanotubes by *Daphnia magna*. *Environmental Science and Technology* 41, 3025-3029.

Robichaud C O, Tanzil D, Weilnmann U and Wiesner M R 2005 Relative risk analysis of several manufactured nanomatrial. An insurance industry context. *Environmental Science and Technology* 39, 8985-8994.

Rose J, Thill A and Brant J 2007 Methods for structural and chemical characterization of nanomaterials. In Environmental Nanotechnology.   Application and Impacts of Nanomatrials. Eds. M R Wiesner and J Y Bottero. pp. 105-154. McGraw Hill, New York.

Semenzin E, Critto A, Rutgers M and Marcomini A 2007. Development of a site-specific ecological risk assessment for contaminated sites: Part II. A multi-criterion based system for the selection of bioavailability assessment tools. *Sci. Tot. Environ.*

Wiesner M R, Lowry G V, Alvarez P, Dionysiou D and Biswas P 2006. Assessing the risks of manufactured nanomaterials. *Environmental Science and Technology* 40, 4336-4345.

Yamago S, Tokuyama H, Nakamura E, Kikuchi K, Kananishi S, Sueki K, Nakahara H, Eonomoto S and Ambe F 1995. In vivo biological behavior of a water-miscible fullerene-C-14 labeling, absorption, distribution, excretion and acute toxicity. *Chemistry and Biology* 2, 385-389.

Zhu X S, Zhu L, Li Y, and Duan Z H, Chen W and Alvarez P J J 2007. Developmental toxicity in zebrafish embryos after exposure to manufactured nanomaterials: Buckminsterfullerene aggregates and fullerol. *Environmental Toxicology and Chemistry* 26, 976-979.

Zhu Y, Zhao Q F, Li Y G, Cai X Q and Li W 2006 The interaction and toxicity of multi-walled carbon nanotubes with *Stylonychia mytilus*. *Journal of Nanoscience and Nanotechnology* 6, 1357-1364.

Chapter 13

Adjudication of Soil Quality by Using Indicators

G. Prakash[1] and B.B. Hosetti[2]

[1,2]*Department of Applied Zoology, Jnanasahyadri, Kuvempu University,
Shimoga – 577 451, Karnataka*

ABSTRACT

Soil contamination occurs by several reasons. The biogeochemical cycles of inorganic biogenic elements and pollutants have been accelerated by human activities. The Industrial Revolution marked a dramatic increase in the release of trace elements to the biosphere. Moreover, locally large concentrations of contaminants may also be related to natural phenomena. The public welfare concern over the hazards of soil pollution has led to legislative action aimed at controlling the major pathways of contamination. Land applications of sewage sludge, combustion of fossil fuel, smelter activities, etc. have been strictly regulated during recent decades.

The main degraded and contaminated soils include highly acid soils, soils deficient in micro-elements, eroded and poorly drained sandy soils, soils suffering from water stress, compacted soils, and rural soils contaminated by trace elements and organic pollutants. Reliable and practical indicators of soil quality are needed to evaluate the condition of degraded and polluted soils. This chapter include indicators are selected for soil quality assessment in sustainable soil management systems, based on the concept of the control chart. The critical level (threshold level, an upper control limit (UCL), and a lower control limit (LCL)) represent the values within which soil quality must be kept for sustainable soil management.

Key words: Soil, Indicators, Soil management, Sustainable management

Introduction

About 45 percent of India's land is degraded, air pollution is increasing in all its cities, it is losing its rare plants and animals more rapidly than before and about one-third of its urban population now lives in slums, says the State of Environment

Report India 2009 brought out by the government (IBN live...). The report, prepared by NGO Development Alternatives under the aegis of the ministry, says 45 percent of India's land area is degraded due to erosion, soil acidity, alkalinity and salinity, water logging and wind erosion. It says the prime causes of land degradation are deforestation, unsustainable farming, and mining and excessive groundwater extraction. On the bright side, the report shows how over two thirds of the degraded 147 million hectares can be regenerated quite easily, and points out that India's forest cover is gradually increasing. Currently, only two percent of India land area is under high density forest cover, while medium density forests cover is about 10 % of the land. Presenting the salient features of the report to the media, Development Alternatives President (Development Enterprises) George C Varughese said one of its most worrisome findings was that the level of respirable suspended particulate matter the small pieces of soot and dust that get inside the lungs--had gone up in all the 50 cities across India. This report was also supported by the All India Institute of Medical Sciences and the Central Pollution Control Board.

"In these 50 cities, with their population of 110 million, the public health damage costs due to this were estimated at Rs.15, 000 crore in 2004," Varughese said. The main causes of urban air pollution were vehicles and factories, he pointed out, appealing for a major boost to public transport.

While India still had some cushion when it came to water use, this scarce resource would have to be managed very carefully, the report says. It identifies lack of proper pricing of water for domestic usage, poor sanitation, and unregulated extraction of groundwater by industry, discharge of toxic and organic wastewater by factories, inefficient irrigation and overuse of chemical fertilisers and pesticides as the main causes of water problems in the country. While India remains one of the world's 35 "Mega-diverse" countries in terms of the number of species it houses, 10 percent of its wild flora and fauna are on the threatened list, Varughese pointed out. The main causes, according to the report, were habitat destruction, poaching, invasive species, overexploitation, pollution and climate change. The report points out that while India contributes only about five percent of the world's greenhouse gas emissions that are leading to climate change, about 700 million Indians directly face the threat of global warming today, as it affects farming, makes droughts, floods and storms more frequent and more severe and is raising the sea level.

Under the section on urbanisation, the report points out that 20 to 40 percent of urban population living in slums. Varghese said there were good projects to upgrade their lives and improve the environment at the same time, but the problem was that most of the money from schemes like the Jawaharlal Nehru National Urban Renewal Mission was taken away by the big cities, "while the major problem is in about 4,000 small and medium towns".

Sustainable Agriculture

The FAO/Netherlands conference on Agriculture and the Environment (FAO 1991) revised the original definition of "Sustainable Agricultural Development" defined by FAO in 1990 and translated it into several basic criteria to measure the sustainability of present agriculture and future trends. These criteria can be listed as follows:

> Meeting the food demands of present and future generations in terms of quantity, quality and the demand for other agricultural products.

> Providing enough jobs, securing income and creating human living and working conditions for all those engaged in agricultural production.

> Maintaining, and enhancing, the productive capacity of the natural resources base as a whole and the regenerative capacity of renewable resources, without impairing the function of basic natural cycles and ecological balance, destroying the socio-cultural identity of rural communities.

> Making the agricultural sector stronger against adverse natural and socio-economic factors and other risks, and strengthening the self-confidence of rural populations.

According to these criteria, the sustainable management of agricultural soils maintain the soil productivity for future generations in an ecologically, economically, and culturally sustainable way.

Multidisciplinary Aspects of Sustainable Soil Management

Sustainable soil management (SSM) must take a multidisciplinary approach. It is not limited only to soil science. Basically, one can consider three aspects of this management system (Steiner, 1996).

> Bio-physical aspects: Sustainable soil management must maintain and improve the physical and biological soil conditions for plant production and biodiversity.

> Socio-cultural aspects: Sustainable soil management must satisfy the needs of human beings in a socially and culturally appropriate manner at a regional or national level.

> Economic aspects: Sustainable soil management must cover all the costs of individual land users and society.

The concept of sustainable land management (SLM) can be applied on different scales to resolve different issues, while still providing guidance on the scientific standards and protocols to be followed in the evaluation for sustainable development in the future (Dumanski, 1997). Based on the above it is clear that sustainable land management, and sustainable land management becomes is the basis of sustainable development (Dumanki , 1997) (Figure. 1).

Land Quality Indicators (LQIs) are being developed as a means of improving coordination when taking action on land-related issues such as land degradation. Indicators are already in regular use to support decision-making at a national level, but few such indicators are available to monitor changes in the quality of land resources. There is a need of more research into LQIs, including:

❖ How to integrate socio-economic (land management) data with biophysical information in the definition and development of LQIs.

❖ How to scale data for application at various hierarchical levels.

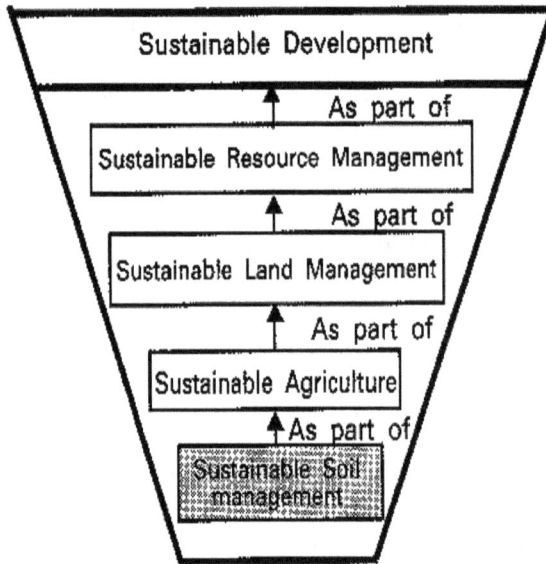

Figure:1 The Relationships among Sustainable Development, Sustainable Land Management, Sustainable Agriculture, and Sustainable Soil Management. (Source: Dumanski, 1997)

Causes of Soil Degradation

The most important challenge in the next century is nutrient depletion, deficiency, and erosion of soils (IBSRAM, 1994). Major soil-related problems for sustainable soil management include:

1. Nutrient depletion and deficiency,
2. Soil erosion and degradation,
3. Socioeconomic prices and marketing,
4. Inefficient water use,
5. Faulty research methods,
6. Unsustainable farming,
7. Soil acidity,
8. Non-adoption by farmers of improved technology,
9. Competing uses for water,
10. Lack of organic matter,

11. Inadequate fertilizer use and management,

12. High compaction,

13. Seasonal drought and,

14. Water stress, water logging and poor drainage.

The rate speed of soil degradation depends on different environmental factors, such as soil type, relief, climate and farming system. The UNEP (United Nations Environment Program) Project and GLASOD (Global Assessment of Soil Degradation) distinguishes four human-induced processes of soil degradation: water and wind erosion, plus chemical and physical degradation (Oldeman *et al.*, 1990). Soil erosion caused by water and wind is the most important form of degradation.

➢ Soil loss due to wind erosion (28%)

➢ Soil loss due to water erosion (56%).

➢ Nutrient depletion due to inadequate fertilizer applications.

➢ Soil acidification.

➢ Salinization due to inadequate irrigation and drainage (12%).

➢ Depletion of organic matter due to fast decomposition and insufficient organic fertilizer and

➢ Compaction, aggravated by the use of heavy machinery (4%).

The most important causes of water erosion are deforestation (43%), overgrazing (29%) and agricultural mismanagement (24%). The most important causes of wind erosion are overgrazing (60%), agricultural mismanagement (16%), over-exploitation of natural vegetation (16%) and deforestation (8%). The most important forms of chemical soil degradation are loss of nutrients and organic matter in South America, and salinization in Asian countries (Oldeman *et al.*, 1990).

Physical Degradation

Compaction, hardpans and crusting are three major causes of physical degradation (Steiner, 1996). Soil compaction is an increase in bulk density caused by external loading, leading to deterioration in root penetration, hydraulic conductivity, and aeration. There are many ways of reducing soil compaction. Hardpans are common in alluvial plains in semi-arid areas with a pronounced rainy season. Crusting is due to the destruction of aggregates in the top soils by rain, and is closely linked to soil erosion. Crusting reduces infiltration and promotes water run-off.

Chemical Degradation

About 36% of tropical soils are low in nutrient reserves. Acidification produces aluminium and ferrous oxides. This intern results in the fixation of phosphorus, which is no longer available for plants. A ferrous oxide/clay ratio of > 0.2 is considered to be the threshold for P fixation, and affects 22% of all tropical soils. This problem also occurs in Andisols, Ultisols, and Oxisols in the humid tropics and tropical highlands.

About 30% of tropical land problems occur in highly acidic soils which contain phyto-toxic aluminum (Al) in the soil solution. This is particularly marked where the Al saturation percentage of total cation exchange capacity (CEC) exceeds 60% in the upper 50 cm of the soil pedon. About 25% of tropical soils are acidic soils with pH values below 5.5 in the upper horizons but without aluminum phyto-toxicity. Salinization can be regarded as a specific form of soil degradation. Salinization is caused by improper irrigation, a high evapo-transportation rate, or changes in hydrological conditions. Maintaining a sufficient level of soil organic matter is very important in tropical countries. The decomposition rate of tropical organic matter is about five times faster in the tropics than in temperate regions.

Biological Degradation

Biological degradation is related to the depletion of vegetation cover and organic matter content in the soils, but also denotes a reduction in beneficial soil organisms and soil fauna. Biological degradation is the direct result of inappropriate soil management. Soil organisms and soil organic matter content can influence and improve the physical structure of the soils, especially with regard to transportation within the soils, mixing mineral and organic materials, and changes in soil micropore volume.

Social-economic Factors

A severe problem in many developing countries is rapid rate of population growth. This increases the demand on natural resources, especially on soil, food and water resources. In many countries, population growth increased the pressure on land.

➢ Degraded and Contaminated Soils in India

➢ Strongly acidic soils (pH<5.6) (30% of total rural soils);

➢ Microelement deficiency of Zn, B, Fe, and Mn.

➢ Salt-affected soils on the western and eastern coast of India;

➢ Soil erosion along the Western ghats and eastern ghats,

➢ Poorly drained soils;

➢ Water stress in deep sandy soils derived from coastal sediments of sandstone and slate;

➢ Compact clay soil in Southern India; and

➢ The soils contaminated by trace elements (Cd, Pb, Cu and Zn).

Most degraded soils have been reclaimed. The reclamation techniques are as follows:

➢ Liming on strongly acid soils;

➢ Application of Zn, B, Fe, and Mn elements for nutrient deficiency soils;

➢ Reclamation of salt-affected soils by natural leaching processes and underground drainage;

> Cover cropping and mulching with grass on slope land soils;

> Regional improvement of drainage canals for poorly drained rice-growing soils;

> Sprinkler irrigation for upland crops and for deep sandy soils; and

> Deep plowing for compact soils.

However, it has not yet been possible to remedy most of the contaminated soils. Only about 10 ha of rural soils contaminated by Cd and Pb have been rehabilitated.

Indicators for Sustainable Soil Management

There are six basic ecological criteria frequently used to evaluate sustainable soil management. These indicators are:

> Soil mass should be conserved for long-term in each small land unit.

> Soil fertility and soil biology should be conserved for long-term, and damage by toxic substances from outside should be minimized.

> Soil use should be stepped up when the marginal return has significantly increased.

> All forms of degradation (erosion, biological, physical, and chemical degradation) should be prevented. In degraded soils, soil formation should be enhanced to improve soil biology and soil fertility.

> Natural biodiversity and the other natural resources of a region should be conserved, to ensure that the extinction of individual species does not endanger the biological community.

> Local land use should not hamper the sustainable development of a zone, especially in social, institutional and economic respects.

Until now, the basic problem of developing and implementing measures for sustainable soil management is that results cannot be transferred and reproduced, because of the multiple factors influencing the process.

Definition of Soil Quality

Soil quality has been defined as "The capacity of a reference soil to function, within natural or managed ecosystem boundaries, to sustain plant and animal productivity, maintain water and air quality, and support human health and habitation." Soil quality can be assessed in terms of the health of the whole soil biological system (Warkentin, 1995). Many penologists feel that any definition of soil quality should consider its function in the ecosystem (Acton and Gregorich, 1995; Kennedy and Papendick 1995; Warkentin, 1995; Doran *et al.*, 1996; Johnson *et al.*, 1997). These definitions are based on monitoring of soil quality (Doran and Parkin, 1994), in terms of the following.

> Productivity: The ability of soil to enhance plant and biological productivity.

> ➤ Environmental quality: The ability of soil to attenuate environmental contaminants, pathogens, and offsite damage.

> ➤ Animal health: The interrelationship between soil quality and plant, animal and human health. Therefore, soil quality can be regarded as soil health (Doran *et al.*, 1996).

Immediately as we can assess human health, we can estimate soil quality and health. Larson and Pierce (1994) proposed that a minimum data set (MDS) of soil parameters should be adopted for assessing the health of world soils, and that standardized methodologies and procedures be established to assess changes in the quality of these factors. These indicators should be useful across a range of ecological and socio-economic situations (Lal 1994, Doran and Parkin 1996).

These Indicators should have the following Advantages

1. Correlate well with natural processes in the ecosystem (this also increases their utility in process-oriented modelling).

2. Integrate soil physical, chemical, and biological properties and processes, and serve as basic inputs needed for estimation of soil properties.

3. Be relatively easy to use under field conditions, so that both specialists and producers can use them to assess soil quality.

4. Be sensitive to variations in management and climate. The indicators should be sensitive enough to reflect the influence of management and climate on long-term changes in soil quality, but not be so sensitive that they are influenced by short-term weather patterns.

5. Be the components of existing soil databases where possible.

Cameron *et al.*, (1998) suggested the use of a simple scoring approach, to help users to decide whether to accept or reject a potential soil quality indicator for degraded soils.

A = sum of (S, U, M, I, R)

> ❖ Where A: Acceptance score for indicator.

> ❖ S: Sensitivity of indicator to degradation or remediation process.

> ❖ U: Ease of understanding of indicator value.

> ❖ M: Cost effectiveness of measurement of soil indicator.

> ❖ I: Predictable influence of properties on soil, plant and animal health, and productivity.

> ❖ R: Relationship to ecosystem processes (especially those reflecting wider aspects of environmental quality and sustainability).

Each parameter in the equation is given a score (1 to 5) based on the user's knowledge and experience of it. The sum of the individual score gives the level of acceptance (A) score which can be ranked in comparison to other potential indicators, aiding the selection of indicators for a site. For example, soil bulk density may receive the following score (S=4, U=4, M=5, I=3, and R=2) giving A values of 18/25 (72%). Particle size, on the other hand, may only get an A value of 10/25 (40%) (S=1, U=3, M=2, I=2, and R=2). In this case, we should select soil bulk density to be one of the indicators for soil quality assessment.

Indicators of Soil Quality

Assessment of soil quality is the basis for assessing sustainable soil management. It is difficult to select factors of soil quality for degraded or polluted soils. Dumanski,(1994) indicated that appropriate sustainable management would require that a technology have five major pillars of sustainability, namely, (1) Be ecological protective, (2) Be socially acceptable, (3) Be economically productive, (4) Be economically viable, and (5) Reduce risk. Appropriate indicators are needed to show whether those requirements are being met. Some possible soil variables which may define resource management domains are soil texture, drainage, slope and land form, effective soil depth, water holding capacity, cation exchange capacity, organic carbon, soil pH, salinity or alkalinity, surface stoniness, fertility parameters, and other limited properties (Eswaran *et al.*, 1998). The utility of each variable is determined by several factors, including whether changes can be measured over time, sensitivity of the data to the changes being monitored, relevance of information to the local situation, and statistical techniques which can be employed for processing information. Doran and Parkin, (1994) have developed a list of basic soil indicators for screening soil quality and health (Table 1).

Table: 1. Proposed minimum data set (MDS) of pssshysical, chemical, and biological indicators for screening the condition, quality, and health of soils

Indicator of soil conditions	Relationships to soil conditions and functions	Unit measured
Physical indicators		
Soil texture	Retention and transport of water and minerals; modelling use and level of soil erosion and estimating variability	Percentage of Sand, Silt and Clay
Depth of soils or toposoil	Estimate of potential productivity and erosion; normalizing landscape and geographic variability	Cm (or Mt)
Infilteration and bulk density	Potential for leaching, productivity and erosion	Minutes/ 2.5 cm water and mg/m^3
Water holding capacity	Related to water retention, transport and erosity	Percentage (cm^3/cm30; cm of available water /30/cm
Chemical indicators		
Soil organic matter (total organic C&N)	Soil fertility, stability and extent of erosion; use in process models and for site normalization	kg C or N/ha?30 cm
pH	Biological and chemical activity thresholds	Compared with upper and lower for plants and microbial activity

Electric conductivity	Threshold of plant and microbial activity	dS/m
Extractable N,P,& K	Available plant nutrients 7 potential for N loss, productivity & environmental quality indicators.	kg/Ha?30 cm
Biological indicators		
Microbial biomass C& N	Microbial catalytic potential and repository for C& N; modelling; effect of organic matter on land management	kg C or N/Ha/30 cm, or CO_2
Potential mineralizable N	Soil productivity & N supplying potential; process modelling	kg N/ ha/ 30 cm
Soil respiration water content, and temperature	Measures microbial activity: process modelling; estimate of biomass activity	kg C/ ha/ day C loss Vs, inputs, Total C pool (kg/ m^2/ 30 cm

- **Physical indicators include** (1) soil texture, (2) depth of soils, topsoil or rooting, (3) infiltration, (4) soil bulk density, and (5) water holding capacity.

- **Chemical indicators include** (1) soil organic matter (OM), or organic carbon and nitrogen, (2) soil pH, (3) electric conductivity (EC), and (4) extractable N, P, and K.

- **Biological indicators include** (1) microbial carbon and nitrogen (2) potential mineralizable nitrogen (anaerobic incubation) and (3) soil respiration, water content, and soil temperature.

Harris and Bezdicek, (1994) stated that soil quality indicators might be divided into two major groups, analytical and descriptive. Experts often prefer analytical indicators, while farmers and the public often use descriptive descriptions. Soil contaminants selected as indicators may be those which have an impact on plant, animal and human health. Soil quality can be viewed from two perspectives: the degree to which soil function is impaired by contaminants, and the ability of the soil to bind, detoxify and degrade contaminants.

Soil Physical Indicators

Doran and Parkin, (1994) have selected some physical indicators for the assessment of soil quality. These indicators include (1) Soil texture, (2) Depth of soils, topsoil, (3) Infiltration, (4) Soil bulk density, and (5) Water holding capacity. Hseu *et al.,* (1999) also selected some indicators for the evaluation of the quality of India's soils. The physical indicators selected included (1) Depth of the a horizon, (2) contents of clay, silt, and sand %, (3) Bulk density, (4) Available water content (%), and (5) Aggregate stability at a depth of 30 cm.

It is easy to understand that measuring the bulk density, soil texture, and penetration of resistance (or infiltration) can provide useful indices of the state of compactness, and the translocation of water, air and root transmission.

Measurements of infiltration rate and hydraulic conductivity are also very useful data, but are often limited because of the wide natural variation that occurs in field soils, and the difficulty of making enough measurements to obtain a reliable average value (Cameron *et al.*, 1998). Measuring the aggregate stability gives valuable data about soil structural degradation, which is affected by pollution (e.g. presence of sodium) and soil degradation (loss of organic matter).

This shows that visual assessment of the soil profile is a very valuable way of assessing the physical condition of the soil, and whether there is a need for soil reclamation or remediation. These physical indicators should include:

1. Soil texture: related to porosity, infiltration, and available water content.

2. Bulk density: related to infiltration rate and hydraulic conductivity.

3. Aggregate stability: related to soil erosion resistance and organic matter content.

Beare *et al.*, (1997) have proposed a quantitative method to show the decline and restoration of soil structure conditions in a typical mixed-cropping rotation system over eight years.

Chemical Indicators

Doran and Parkin, (1994) have also selected chemical indicators for the assessment of soil quality. These indicators include (1) Soil organic matter (OM), or organic carbon and nitrogen, (2) Soil pH, (3) Electric conductivity (EC), and (4) Extractable available N, P, and K. Hseu *et al.*, (1999) selected some chemical indicators for evaluating the quality of India soils. The chemical indicators include (1) soil pH, (2) electric conductivity (EC), (3) organic carbon, (4) extractable available N, P, and K, (5) extractable available trace elements (Cu, Zn, Cd, and Pb).

Standard soil fertility attributes (soil pH, organic carbon, available N, P, and K) are the most important factors in terms of plant growth, crop production and microbial diversity and function. As we know, these parameters are generally sensitive to soil management. For polluted, the soil fertility indicators are regarded as part of a minimum data set of soil chemical indicators.

Biological Indicators

Doran and Parkin, (1994) have selected a number of biological indicators for the assessment of soil quality. These include: (1) Microbial carbon and nitrogen, (2) Potential mineralizable nitrogen (anaerobic incubation) and (3) Soil respiration, water content, and soil temperature. Hseu *et al.*, (1999) also selected some chemical indicators for the evaluation of the quality of India soils. The chemical indicators include (1) Potential mineralization of N, (2) C, N, and P present in the microbial biomass (3) Soil respiration, (4) The number of earthworms, and (5) Crop yield.

Soil biological parameters are potentially early, sensitive indicators of soil degradation and contamination. It follows, the minimum data sets for assessing key soil processes composed of a number of biological (e.g. microbial biomass, fungal hyphae) and biochemical (e.g. carbohydrate) properties (Cameron *et al.*,

1998). Two of the most useful indicators are microbial biomass and microbial activity. Microbial biomass is a sensitive indicator for a long-term decline in total soil organic matter, but does not seem to be a sensitive indicator of the effects of organic pollutants applied to fields.

Assessment of Soil Quality

There are no reliable, practical methods of assessing or evaluating soil quality/ health, although some research reports have established a conceptual framework for assessing this (Karlen *et al.*, 1997). In this chapter, we use the concept of threshold values to evaluate quality of Indian rural soils (Cameron *et al.*, 1998).

Chemical Pollutants Used in Developed Countries

Various criteria for the assessment and abatement of contaminated soils have been developed, especially in industrialized countries, including the United States, Germany, United Kingdom, Australia, Canada, Netherlands, Japan and Taiwan (ICRCL 1987; USEPA 1989; Alloway 1990; Jacobs 1990, Tiller 1992; Ministry of Housing Netherlands 1994; Chen *et al.*, 1996; Adriano *et al.*, 1997; Chen 1998). Many national governments and local authorities who lack their own formal guidelines have used the Dutch standard in assessing contaminated sites, or monitoring sites.

Some have also made modifications to develop their own regulations based on soil qualities they feel are most important. However, the Dutch authorities are continually upgrading their soil quality criteria in the light of new scientific work, especially the ecotoxicology of listed substances and their impact on species. Two values are considered in making decisions on regulating the level of heavy metals in soils, a target value (upper value of natural level) and the intervention value (*i.e.* values which mean that soil needs cleaning up) (Ministry of Housing, Netherlands 1994). The Dutch standards for assessing soil contamination on the basis of the total concentration of heavy metals in the soil. Cameron *et al.*, (1998) suggested that the dynamics of a soil quality value (Q) can be quantified by measuring the changes in soil quality parameters value over time (dQ/dt). This can be done using a quality control chart in which the soil attribute values are plotted as a time series. The control chart may have a critical limit (or threshold level, and a lower control limit (LCL)) which represents the tolerances beyond which soil quality measures of will be severally affected. For example, the UCL and LCL of the total soil copper concentration was proposed as 140 mg/kg as a soil quality guideline and 5 mg/kg for the minimum crops requirement.

Threshold Values of Soil Chemical Pollutants Developed in India

The EPA organized a working group to develop guidelines for assessing sites polluted with heavy metals, and has used these guidelines to monitor these sites since 1990. The guidelines primarily follow the basic soil properties of India, and the effects of heavy metals on:

1. Water quality
2. Activity of soil microorganisims

3. Human health, and

4. Crop productivity and quality;

Final guidelines for soil quality were proposed by this working group over the past few years (Wang *et al.*, 1994, Chen *et al.*, 1996, Chen, 1998). These were primarily based on the effects of heavy metal concentrations on human health, on plant productivity and crop quality, and on guideline values established in other countries. The intervention value for trace elements and threshold phytotoxicity of heavy metals extracted from soil with 0.1 M HC1.

Action Levels

Steiner, (1996) indicated that general conclusions drawn from particular projects can be transferred to other sites only under specific conditions. The solution to the problems of degraded soils must be geared to local needs. Programs should coordinate activity at different levels. Sustainable soil management is part of the effort to achieve sustainable agriculture. Depending on the problems, combined action must be taken at different levels of intervention at the same time (Steiner, 1996).

Agricultural Policy

Maydell, (1994) pointed out that policies to promote sustainable soil management must begin by identifying which aspects should be assisted or can be influenced. Figure.2 depicts the relationships between soil degradation, land use and agricultural policy. The major factors in these relations are cropland area, cropping patterns and cropping techniques.

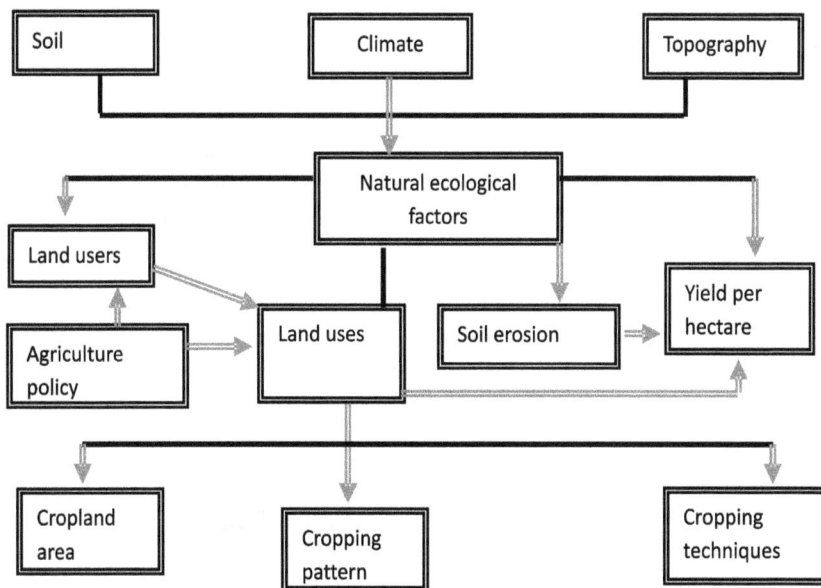

Figure: 2 Relationships between soil erosion, land use and agricultural policy

Source: Maydell, 1994

Cropping Area

Low level of productivity per unit area in less industrialized countries is the main reason why so much new arable land has been cleared in the past, and is still being cleared today. The only land now left is marginal, with low natural soil productivity, unstable soils and a high risk of soil erosion. In the past, prices and subsidies have encouraged people to increase the area under crops. In order to solve these problems in the long term, the most effective way is to raise land productivity by promoting agricultural research, improving agricultural services, and developing high-value special crops for farmers.

Cropping Techniques

Most methods that conserve soil resources involve higher costs. To conserve resources, more research is needed on cropping techniques to protect soil resources.

Approaches and Technical Options

A national strategy for sustainable soil management should be based on the following process.

1. Analyze the background and basic data of degraded and polluted soils.

2. Assemble the components for an effective solution.

3. Produce a set of tools at a national level to meet the needs of farmers and policy makers.

On approach to sustainable soil management at a national level is give in Figure 2 Available data on the background to the problem include the causes of soil degradation, current status of soil quality, numbers and needs of farmers, and agricultural policy. An effective solution for soil problems must include early warning by soil indicators, prevention of soil degradation, rapid assessment of problems, assessment of the economics of production, risk assessment

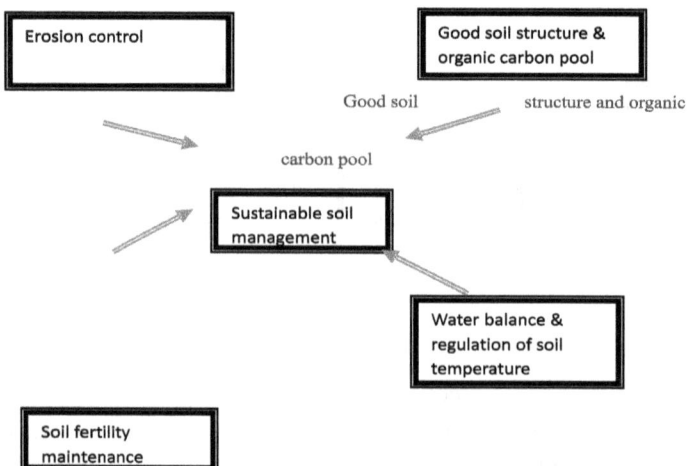

Figure: 3 Components of a sustainable soil management system

Erosion Control

Vigorous ground cover is strongly recommended to avoid soil loss through water run-off. Cropping methods include early sowing, cover crops, mixed cropping, higher seed density, inter-row cropping and planted fallows. Splash erosion can be controlled by mulching. Rill and gully erosion can be controlled by terracing, or by placing other barriers parallel to the slope such as contour strips planted with different species of grass. Contour plowing and minimum tillage are also effective against soil erosion. These methods and technologies are not widespread. They need further development.

Adding crop residues, manure, and compost to the soil is a better way of maintaining soil fertility and soil structure. Another method is mulching, in which people gather organic substances (grass, leaves, litter, branches) from non-agricultural areas and spread it on fields to avoid soil erosion and to increase the fertility of poor soils. The most effective way to maintain soil fertility, soil structure and biological activity is to provide enough soil organic manures, to the soil (Chen and Hseu, 1997). In India, the mean organic carbon content of surface soil is about 2.1 - 2.8%. The mean organic carbon pool in Indian soils is less than 8 kg/m²/m (or less than 5 kg/m²/50 cm). This organic carbon pool is not enough to maintain good soil structure and crop production. An annual application of 20 mt/ha of organic manure or compost is needed to meet the demands of crop production and provide good soil structure and biodiversity in the soil (Gregorish *et al.*, 1995, Studdert *et al.*, 1997, Chen *et al.*, 1998).

References

Acton, D.F., and L.J. Gregorich. 1995. Understanding soil health. In: The Health of our Soils: Toward Sustainable Agriculture in Canada, D.F. Acton and L.J. Gregorich (Eds.). Centre for Land and Biological Resources Research, Research Branch, Agriculture and Agri-Food Canada, Ottawa, Ontario, Canada, pp. 5-10.

Adriano, D.C., J. Albright, F.W. Whicker, I.K. Iskandar, and C. Sherony. 1997. Remediation of soils contaminated with metals and radionuclide-contaminated soils. In: Remediation of Soils Contaminated with Metals, A. Iskandar, and D.C. Adriano (Eds.). *Science Reviews*, Northwood, UK, pp. 27-46.

Alloway, B.J. 1990. Heavy Metals in Soils. Blackie and Son Ltd., London, UK.

Beare, M.H., Tian, M. Vikram, and S.C. Srivastava. 1997. Agricultural intensification, soil biodiversity, and agro-ecosystem function. *Appl. Soil Ecol.* 6: 87-108.

Cameron, K., M.H. Beare, R.P. McLaren, and H. Di. 1998. Selecting physical chemical, and biological indicators of soil quality for degraded or polluted soils. Proceedings of 16th World Congress of Soil Science. Scientific registration No. 2516. Symposium No. 37. Aug. 20-26, 1998. Montpelier, France.

Chen, Z.S. 1998. Management of contaminated soil remediation programmes. *Land Contamination & Reclamation* 6: 41-56.

Chen, Z.S. and Z.Y. Hseu. 1997. Total organic carbon pool in soils of Taiwan. Proc. *National Science Council*, ROC, Part B: *Life Sciences* 21: 120-127.

Chen, Z.S., Z.Y. Hseu, and C.C. Tsai. 1998. Total organic carbon pools in Taiwan rural soils and its application in sustainable soil management system. *Soil and Environment* 1: 295-306. (In Chinese, with English abstract and Tables).

Chen, Z.S., D.Y. Lee, C.F. Lin, S.L. Lo, and Y.P. Wang. 1996. Contamination of rural and urban soils in Taiwan. In: Contaminants and the Soil Environment in the Australasia-Pacific Region, R. Naidu, R.S. Kookuna, D.P. Oliver, S. Rogers, M.J. McLaughlin (Eds.). First Australasia-Pacific Conference on Contaminants and Soil Environment in the Australasia-Pacific Region. Adelaide, Australia, Feb. 18-23, 1996. Kluwer Academic Publishers, Boston, London, pp. 691-709.

Doran, J.W., and T.B. Parkin. 1994. Defining and assessing soil quality. In: Defining Soil Quality for a Sustainable Environment, J.W. Doran, D.C. Coleman, D.F. Bezdicek, and B.A. Stewart (Eds.). *Soil Sci. Soc. Am. Special Publication* No. 35, Madison, Wisconsin, USA, pp. 3-21.

Doran, J.W., and T.B. Parkin. 1996. Quantitative indicators of soil quality: A minimum data set. In: Method for assessing soil quality, J.W. Doran and A.J. Jones (Eds.). *Soil Sci. Soc. Am. Special Publication* No. 49, Madison, Wisconsin, USA, pp. 25-37.

Doran, J.W., M. Sarrantonio, and M.A. Liebig. 1996. Soil health and sustainability. *Advances in Agronomy* 56: 1-54.

Dumanski, J. (Ed.). 1994. Proceedings of the International Workshop on Sustainable Land Management for the 21st Century. Vol. 1: Workshop Summary. Agricultural Institute of Canada, Ottawa.

Dumanski, J. 1997. Criteria and indicators for land quality and sustainable land management. *ITC Journal* 1997-3/4: 216-222.

Eswaran, H., F. Beinroth, and P. Reich. 1998. Biophysical considerations in developing resource management domains. pp. 61-78. In: Proceedings of Conference on Resources Management Domains, J.K. Syers (Ed.). Kuala Lumpur, Malaysia, Published by International Board for Soil Research and Management (IBSRAM). Proceedings No. 16.

FAO (Food and Agriculture). 1991. Issues and Perspectives in Sustainable Agriculture and Rural Development. Main document No. 1 DAO/Netherlands Conference and Agriculture and Environment. S-Hertogenbosch, the Netherlands. April 15-19, 1991. FAO, Rome.

Gregorich, E.G., D.A. Angers, C.A. Campbell, M.R. Carter, C.F. Drury, B.H. Ellert, P.H. Groenevelt, D.A. Holmstrom, C.M. Monreal, H.W. Rels, R.P. Voroney, and T.J. Vyn. 1995. Changes in soil organic matter. In: The Health of Our Soils: Toward Sustainable Agriculture in Canada, D. F. Acton and L.J. Gregorich (Eds.). Centre for Land and Biological Resources Research, Research Branch, Agriculture and Agri-Food Canada, pp. 41-50.

Harris, R.F., and Bezdicek, D.F. 1994. Descriptive aspects of soil quality/health. In: Defining Soil Quality for a Sustainable Environment. J.W. Doran, D.C. Coleman, D.F. Bexdicek, and B. A. Stewart. (Eds.). *Soil Sci. Soc. Am. Special Publication* No. 35. Madison, Wisconsin, USA, pp. 23-35.

Hseu, Z.Y., Z.S. Chen, and C.C. Tsai. 1999. Selected indicators and conceptual framework for assessment methods of soil quality in arable soils of Taiwan. Soil and Environment. 2: 77-88. (In Chinese, with English abstract and tables).

IBN live, A network 18 venture. Published on Tuesday, August, 2009, Indian section.

IBSRAM (International Board for Soil Research and Management). 1994. Soil, Water, and Nutrient Management Research: A new agenda. IBSRAM position paper, Bangkok, Thailand.

ICRCL 1987 Guidance on the assessment and redevelopment of contaminated land. ICRCL paper 59/83. Department of the Environment, London, UK.

Jacobs, L.W. 1990. Potential Hazards When Using Organic Materials as Fertilizers for Crop Production. FFTC Extension Bulletin No. 313, Food and Fertilizer Technology Center of Asia and Pacific Regions (FFTC/ASPAC). 20 pp.

Johnson, D.L., S.H. Ambrosce, T.J. Bassett, M.L. Bowen, D.E. Crummey, J.S. Isaaxson, D.N. Johnson, P. Lamb, M. Saul, and A.E. Winter-Nelson. 1997. Meanings of environmental terms. *J. Environ. Quality* 26: 581-589.

Karlen, D.L., M.J. Mausbach, J.W. Doran, R.G. Cline, R.F. Harres, and G.E. Schuman. 1997. Soil quality: A concept, definition, and framework for evaluation. *Soil Sci. Soc. Am.* J. 61:4-10.

Kennedy, A.C., and R.I. Papendick. 1995. Microbial characteristics of soil quality. *J. of Soil and Water Conservation* 50: 243-248.

Lal, R. 1994. Data analysis and interpretation. In: Methods and Guidelines for Assessing Sustainable Use of Soil and Water Resources in the Tropics, R. Lal (Ed.). Soil Management Support Services Technical. Monograph. No. 21. SMSS/SCS/USDA, Washington D.C, pp. 59-64.

Larson, W.E., and F.J. Pierce. 1994. The dynamics of soil quality as a measure of sustainable management. In: Defining Soil Quality for a Sustainable Environment. J.W. Doran, D.C. Coleman, D.F. Bezdicek, and B.A. Stewart (Eds.). Soil Sci. Soc. Am. Special Publication No. 35. Madison, Wisconsin, USA, pp. 37-51.

Maydell, O.V. 1994. Agrarpolitische Ansatze zur Erhaltung von Bodenressourcen in Entwicklungslandern. Landwirtschaft und Umwelt. Schriften zur Umwdltokonomik, Band 9. Wissenschaftscerlag Vauk. Kiel. Germany.

Ministry of Housing - Netherlands. 1994. Dutch intervention values of heavy metals and organic pollutants in soils, sediments, and ground water. *Physical Planning and Environmental Conservation Report* HSE 94.021.

Oldeman, L.R., V.W.P. Van Engelen, and J.H.M. Pulles. 1990. The extent of human induced soil degradation. Annex 5 of L.R. Oldeman, R.T.A. Hakkeling, and W.G. Sombrock. World Map of the Status of Human-Induced Soil Degradation: An Exploratory Note. 2nd Rev. ISRIC (International Soil Reference and Information Centre) (Ed.). Wageningen, Netherlands.

Steiner, K.G. 1996. Causes of Soil Degradation and Development Approaches to Sustainable Soil Management. (English version by Richard Williams). CTZ, Margraf Verlag, Weilersheim, Germany.

Studdert, G.A., H.E. Echeverria, and E.M. Casanovas. 1997. Crop-pasture rotation for sustaining the quality and productivity of a Typic Argiudoll. *Soil Sci. Soc. Am. J.* 61: 1466-1472.

Tiller, K.G. 1992. Urban soil contamination in Australia. *Australian J. Soil Research* 30: 937-957.

USEPA (United States Environmental Protection Agency). 1989. Standards for the disposal of sewage sludge: Proposed rules. *Federal Register* 54: 5778-5902. USA.

Wang, Y.P., Z.S. Chen, W.C. Liu, T.H. Wu, C.C. Chaou, G.C. Li, and T.T. Wang. 1994. Criteria of soil quality: Establishment of heavy metal contents in different categories (Final reports of four years projects). Project reports of EPA/ROC (Grant No. EPA-83-E3HI-09-02). 54p. (In Chinese, with English abstract and tables).

Warkentin, B.P. 1995. The changing concepts of soil quality. *J. Soil and Water Conservation.* 50: 226-228.

Chapter 14

Impact of Arbuscular Mycorhiza Fungi on Plants Grown Under Salt Stressed Conditions

H.C. Lakshman, K. Sandeepkumar, A. Channabasava and M.A. Khuli

*DOS in Botany, Karnatak University, Pavate Nagar,
Dharwad – 580 003, Karnataka*

ABSTRACT

Salinity has become a serious agricultural problem in arid and semi arid zones where rainfall is not sufficient to leach down salts from the plant root zone to downward depth. In India, about 7.2 million hectares of the land has been affected by accumulation of salts. This area cannot be left unused in view of the new obligations of enhanced production and sustainability. The reclamation and utilization of these soils can lead to additional production of 35 to 50 million tones every year. A glimpse of studies on plants showing tolerance to high salinity level carried out by various workers has been highlighted in this chapter.

Key words: *AM fungi, salinity, stress, tolerance*

Introduction

Salt affected soils consist of saline, saline-sodic and sodic soils. Millions of hectares (about 952 Millions hectares) of land through out the world is affected by salinity and sodicity and this area is increasing year after year because of salt accumulation. The problem of salinity has become a serious agricultural problem in arid and semi arid zones where rainfall is not sufficient to leach down salts from the plant root zone to downward depth. In India, about 7.2 million hectares of the land has been affected by the accumulation of salts. This area can not be left unused in view of the new obligations of enhanced production and sustainability.

By the end of 21ˢᵗ century, the population of country is expected to be more than 150 millions. To support this population, the food grains requirement will be about 250 million tones. The requirement for vegetable oils, cotton and milk would be about 8.5 to 10.5 million tones, 10.7 to 17.8 million bales and 49.4 to 64.5 million cubic meters, respectively. Therefore, the formers and scientists will have to find a solution for the restoration of productivity of the salt affected soils.

The agricultural production can be increased either by multicroping on the existing land or by bringing the additional area under cultivation. So naturally ones attention is drawn towards the reclamation of salt affected soils. The reclamation and utilization of 7.2 million hectares of salt affected soils can lead to additional production of 35 to 50 million tones every year. As there is no good land available which can be used for cultivation, the only alternative is to increase the cultivated area by using waste and barren lands. Even in these waste and barren lands also some parts can not be reclaimed either due to technical or economical reasons. Such lands can be used for fuel, timber and fodder etc.

Soil salinity may affect the plant growth in two distinct ways: (1) decreased osmotic potential of the soil solution which causes the physiological dryness and (2) accumulation of one or more elements, likely to influence the uptake of one or more of the important plant nutrients. Among various edaphic factors, the salt stress, sodicity and boron are important factors which limit the growth and development of plants and even cause premature termination of life cycle by altering their morphological, physiological and biochemical attributes. There are certain plants which grow vigorously and produce more dry matter under such conditions than the other plants. Plants differ in their capacity to tolerate salinity and alkalinity. Even the different varieties of particular species may exhibit differential tolerance behavior in this regard (Joshi, 1976). Apparently, salinity research has acquired many dimensions.

Different biotic and abiotic stress such as drought, salinity, low temperature, high irradiance and nutritional factors have a great impact on various physiological and biochemical processes of plants and they show a rapid molecular response to changing environmental conditions (Lidreman, 1988; Lichenthaler, 1996). These environmental conditions are major factors limiting the plant productivity and plant distribution. As plants have only limited mechanisms of stress avoidance, they require flexible means of adaptations to changing environmental conditions. (Polle and Rennenberg, 1993). Out of various abiotic stresses, salinity (both of soil and water) is a serious problem in the arid and semiarid regions of the world where the precipitation is not enough to leach the excess soluble salts from root zone. It is estimated that out of about 0.34×10^9 million hectare (23% of cultivated land) world over (Szabolcs, 1989) and approximately 7 million hectare of land in India (Sehgal and Abrol, 1994) is affected by salts. This area is gradually increasing year after year because of salt accumulation. The reclamation of 7 million hectare of salt affected soils can lead to additional production of 35-50 million tones every year.

One of the best known responses of plants to drought and other stresses is accumulation of soluble, low molecular- mass solutes such as proline (i.e. proline concentration is more in stressed plants) (Paleg *et al.*, 1984). Photosynthesis is also greatly affected by salinity because of stomatal closure (Taleisnik, 1987). Dawson and Gibbs (1987) reported salinity induced decrease in the photosynthetic rates of wheat and barley, while the water use efficiency (WUE) was marginally affected. The transpiration per unit leaf area in several crops has been shown to decrease in response to increasing salinity (Hoffman and Phene, 1971). Osmotic adjustment, as well as increased photosynthesis, may enable salinized plants to maintain better turgor and growth during of stress periods.

The introduction of arbuscular-mycorrhiza (AM) fungi to the sites with saline soils may improve early plant tolerance and growth (Jain *et al.*,1989). Although improved salt tolerance of mycorrhizal plants can be related to enhanced mineral nutrition, particularly phosphorus (P) (Graham, 1986), the effect of these fungi on salt tolerance may not only be limited to this mechanism. Hence, in a study of the interaction between AM fungi and salinity, it is essential to determine if the effect fungus on "P" uptake is the primary mechanism by which it is able to increase salt tolerance or whether this effect is due to additional or alternative mechanisms. Arbuscular mycorrhizal fungi may influence plant hormones (Danneberg *et al.*, 1992) or improve water uptake (Ruiz-Lozano an Azcon, 1995). Other mechanisms include osmotic adjustment, which assist in the maintenance of leaf turgor and effects on physiological processes such as photosynthesis, transpiration, conductance and water use efficiency.

In the present chapter, the effect of different levels of soil salinity with inoculation of AM fungi on seedling vigor of plants and their tolerance to soil salinity have been discussed. In saline soils improved growth of onion and pepper inoculated with AM has been observed (Hirrel and Gerdemann, 1980). Also changes in biochemistry of the host plant with respect to phenols in roots (Dehne and Schonbeck, 1979) have been correlated with resistance to pathogens (Krishna and Bagyaraj, 1984; Lakshman, 2007, 2008).

Mycorrhizal Fungi to Salt Stress

The salt tolerance of selected plant genera have been evaluated in controlled experiments (Dixon *et al.*, 1993). However, in many of these studies, plants were supplied with adequate mineral nutrients and grown hydroponically or in artificial soil media. The effects of salinity on some symbiotic relationships (e.g. Actinorrhizal or legume-*Rhizobium* symbiosis) have been assessed in a preliminary manner. The role of AM in tree host water relations, gas exchange, and photosynthesis may be significant (Dosskey, *et al.*, 1990).

Plants relying on symbiotic relationships for adequate mineral nutrition and water uptake may differ in salt tolerance, depending on host and symbiont tolerance to soil salinity (Osonuki, *et al.*, 1991). The relative sensitivity of symbiotic fungi and host feeder roots to sodium salts, individually or in their composite mycorrhizal association, has not been evaluated. In addition to negative impacts on the host plant, extreme soil salinity could adversely affect mycorrhizal propagules in the rhizosphere, fungal colonization of feeder roots, and/or mycorrhizal structure and function.

Preliminary investigations suggest that rhizosphere organisms may be influenced by salt accumulation in soils. For example, the Basidiomycotina, intolerant to salt stress in *vitro*. Salt tolerance may be important to the long-term survival, reproduction, and spread of mycorrhizal fungi in saline soils.

High salt concentration inhibits germination of AM fungal spores as well as growth of hyphae resulting in decreased growth and development of AM fungal density in soil (Juniper and Abbott 1993; Al-Karaki and Clark 1998; Al-Karaki *et al.*, 2001). McMillen *et al.*, (1998) found that increasing concentration of NaCI inhibits either the hyphal growth or the infectivity of hyphae and AM colonization of plant roots. This may be due to adverse effect of NaCI on the hyphal growth as well as altered supply of carbohydrates from the plant to the fungus.

The extent of AM response on growth as well as root colonization varied with the species and salinity levels. Mycorrhiza-inoculated plants accumulated greater amount of P, and K while N uptake lowered as salinity increased. The greater nutrient acquisition, change in root morphology and electrical conductivity of soil in response to AM colonization was observed during course of the study and this may be the possible mechanism to protect the plant from salt stress. Inoculation of *Sesbania grandiflora* and *Sesbania aegyptiaca* with AM fungus *Glomus macrocarpum* had significant increase in growth and biomass production (Giri and Mukerji, 2004).

Under saline condition, *Sesbania* spp. had higher amount of Mg and reduced Na content in shoot tissues; the increased Mg uptake and reduced sodium uptake helped in chlorophyll synthesis. AM fungus also increased establishment and survival of tree plants. Both the tree species were highly dependent on *G. macrocarpum* (Giri and Mukerji, 1999; Giri, *et al.*, 1999).

The response of mycorrhizal and non-mycorrhizal *Olea europaea* under saline conditions with or without supplemental Ca resulted in less depolarization at cellular level (cell transmembrane electropotential) in roots of mycorrhizal than non-mycorrhizal plants. Supplement of Ca in the saline treatments had protective effect on membrane integrity canceling or reducing the differences in depolarization between mycorrhizal and non-mycorrhizal plants. Mycorrhizal roots accumulated greater quantities of Na, K, and Ca and exhibited a lower K: Na ratio, but in leaves, mycorrhizal plants had greater Potassium and Sodium ratio than non-mycorrhizal plants (Rinaldelli and Mancuso, 1996).

Mycorrhizal colonization brought about a noticeable improvement in salt-tolerance in Olive plants, which was clearly demonstrated by trends in impedance parameters (Mancuso and Rinaldelli, 1996). In saline conditions, the electrical impedance parameters in shoots and leaves of Olive plants were studied to understand variations in extracellular resistance, intracellular resistance and the state of membrane in mycorrhizal and non-mycorrhizal plants.

There was a reduction in extra- and intra cellular resistance for non-mycorrhizal plants with increased NaCl concentration (Mancuso and Rinaldelli, 1996). Ezz and Nawar (1994) found that sour orange seedlings irrigated with saline water of 450 ppm salt showed reduced total leaf chlorophyll, peroxidase activity, starch and total carbohydrate concentration in leaves and roots. Inoculation with *Glomus intraradices* increased total chlorophyll, polyphenol activity, leaf and root sugars, and carbohydrate concentrations but peraxidaes activity was not altered.

Complexity of Salt Tolerance

Most crop plants do not fully express their original genetic potential for growth, development, and yield under salt stress, and their economic value declines as salinity levels increase (Lauchli, and Epstein, 1990; Maas 1990). Although numerous attempts have been made to improve the salt tolerance of crops by traditional breeding programmer, commercial success has been very limited due to the complexity of the trait: salt resistance is genetically and physiologically complex (Flowers, 2004). At present, major efforts are being directed towards the genetic transformation of plants in order to raise their tolerance (Borsani *et al.*, 2003).

Improving salt resistance of crop plants is of major concern in agricultural research. A potent genetic source for the improvement of salt resistance in crop plants resides among wild populations of halophytes (Glenn, *et al.*, 1999; Serrano *et al.*, 1999). These can be either domesticated into new, salt-resistant crops, or used as a source of genes to be introduced into crop species by classical breeding or molecular methods.

Mechanisms of Salt Resistance with AM Fungi

Salt tolerance involves physiological and biochemical adaptations for maintaining protoplasmic viability while cells compartmentalize electrolytes. Salt avoidance involves structural and physiological adaptations to minimize salt concentrations of the cells or physiological exclusion by root membranes. In principle, salt tolerance can be achieved by salt exclusion or salt inclusion with AM fungal inoculation as mentioned (Fig. 1).

At the whole plant level, plant resistance may be the process of salt regulation, but at the cellular level it may be the salt tolerance of the cytoplasm (Breckle, 2002). Physiological and biochemical research has shown that salt resistance in halophytes depends on a range of adaptations embracing many aspects of plants physiology, including regulation of H_2O/CO_2 gas exchange in the leaves, osmotic adaptation, selective transport and uptake of ions (salt balance), exclusion of NaCl from the symplast (compartmentalization) to maintain homeostasis, and enhanced synthesis of organic solutes (osmolyte production) (Marschner, 1995; Debez *et al.*,2006; Megdiche *et al.*, 2007; Lakshman 2008*).

Metabolite control of gene expression

Genes involved in salt stress perception and response are essential for the salt tolerance phenotype of plants. The degree of salt tolerance rather has to be attributed to control of gene expression as well as regulation of their products (Wang *et al.*, 2003).

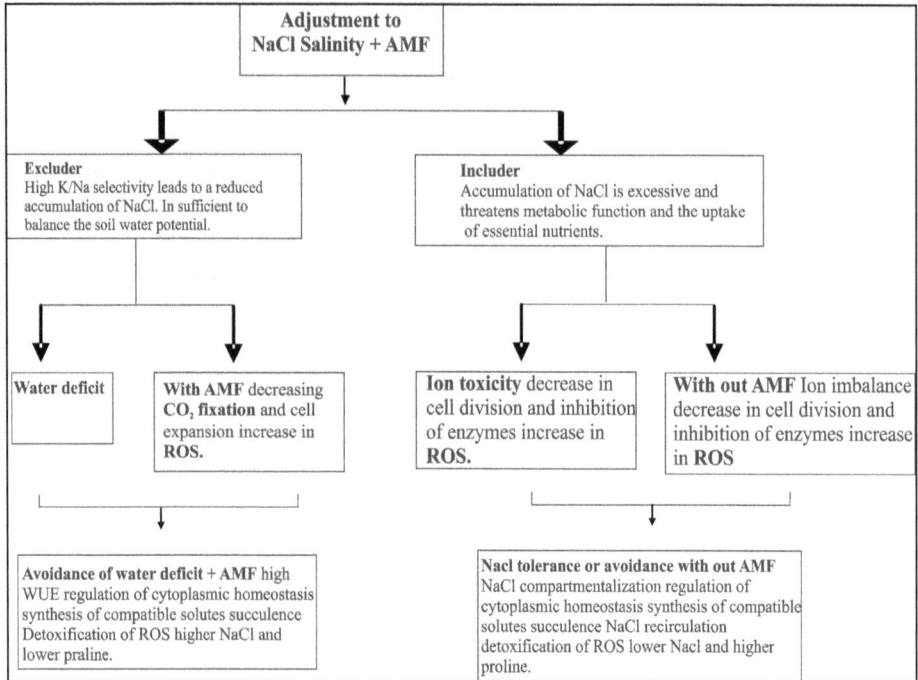

Figure: 1. Flow chart showing the possible mechanism of some plants with AMF to adjust to high external NaCl salinity.

Another aspect, crucial for any experimental approach aiming at elucidation of salt tolerance mechanisms, is based on interrelations between metabolic and physiological mechanisms allowing plants to adapt to ecological parameters of the individual area they are found (Lawlor, 2002; Flowers, 2004). Plants' salt tolerance is an additional parameter within this adaptation spectrum. An individual adaptation mechanism leading to salt tolerance of one plant species at one site does not need to be advantageous for this plant at another site. Moreover, it is not easy to predict the effect of over expression of genes involved in a specific tolerance mechanism in a different plant species. A very obvious prerequisite is, for instance, that metabolism of the transgenic plant has to be capable of feeding in ample amounts educts to be metabolized by an over expressed enzyme (Nuccio *et al.*, 1998). Nevertheless, as outlined below, in case of salt tolerance brought about by single genes, examples for improvement of salt tolerance of glycophytes has been published in the literature.

Proline and Glycinebetaine

Proline accumulation has been correlated with improved tolerance to salt and drought stresses. The accumulation of proline under dehydrated conditions is caused both by activation of the proline biosynthesis and inactivation of the proline degradation. As it has already been suggested by many studies, rapid catabolism of proline upon relief from stress might provide reducing equivalents that support mitochondrial oxidative phosphorylation and the generation of ATP for recovery from the stress and repair of stress induced damage (Hare and Cress, 1997; Hare *et al.*, 1998). Transgenic tobacco cells (Bright Yellow 2), which were silenced at their tobacco Prodehydrogenase (NtProDH) gene, accumulated more Prodehydrogenase than wild type cells and showed enhanced osmotolerance (Tateishi *et al.*, 2005). Proline has been shown to protect membranes and proteins against damaging effects of high concentrations of inorganic ions and temperature extremes by many researchers (Pollard and Wyn Jones, 1979; Nash *et al.*, 1982; Santoro *et al.*, 1992; Rajendrakumar *et al.*, 1994; Jain *et al.*, 2001). Prodehydrogenase also functions as a protein-compatible hydrotrope and radical scavenger (Smimoff and Cumbes, 1989; Chen and Dickman, 2005).

The exact mechanism by which system tolerance to unbalanced ionic conditions in salty soils is not known. The non-micorhizal plants under high salinity level have greater concentrations of potassium, sodium and phosphorus (Ojala *et al.*, 1983) where as micorhizal plants maintained a steady Na/K balance as micorhizal plants. It is reported in *Parthenium aggentatum* inoculated with *Glomus intradices* that, mycorrhizal advantage occurs through the increased concentration of phosphate and decreased concentration of sodium in shoot tissue compared to non- inoculated plants (Pfeiffer and Bloss, 1988).

Various workers have reported tolerance of plants to high salinity levels (Table 1).

Table 1: List of plants showing tolerance to high salinity level with inoculation of AM fungi

Plants	AM Fungi	pH	References
Bell pepper	*Glomus fasciculatum*	1-12	Hirrel and Gerdemann, 1980.
Bell pepper	*Gigaspora margarita*	1-12	Hirrel and Gerdemann, 1980.
Onion	*G. fasciculatum*	1-12	Hirrel and Gerdemann, 1980.
Onion	*Gigaspora margarita*	1-12	Hirrel and Gerdemann, 1980.
Indian ricegrass	*Entrophospora infrequences*	1.6-2.0	Stahl and Williams, 1986.
Indian ricegrass	*G.fasciculatum*	1.6-2.0	Stahl and Williams, 1986.
Indian ricegrass	*Glomus microcarpum*	1.6-2.0	Stahl and Williams, 1986.
Indian ricegrass	*Glomus mosseae*	1.6-2.0	Stahl and Williams, 1986.
Yellow sweetclover	*E. Infrequences*	1.6-2.0	Stahl and Williams, 1986.
Yellow sweetclover	*G. fasciculatum*	1.6-2.0	Stahl and Williams, 1986.
Yellow sweetclover	*G. microcarpum*	1.6-2.0	Stahl and Williams, 1986.
Yellow sweetclover	*G. mosseae*	1.6-2.0	Stahl and Williams, 1986.
Big sagebrush	*G. fasciculatum*	0.6	Stahl *et al.*, 1988.
Big sagebrush	*G. microcarpum*	0.6	Stahl *et al.*, 1988.
Big sagebrush	*E. Infrequences*	2.6-3.8	Stahl *et al.*, 1988.
Big sagebrush	*G. fasciculatum*	2.6-3.8	Stahl *et al.*, 1988.
Big sagebrush	*G. macrocarpum*	2.6-3.8	Stahl *et al.*, 1988.

Table 1: contd.

Big sagebrush	G. microcarpum	2.6-3.8	Stahl et al., 1988.
Big sagebrush	G. mosseae	2.6-3.8	Stahl et al., 1988.
Acacia auriculiformis	G. fasciculatum	1-10	Giri et al., 2003b.
Acacia auriculiformis	G. macrocarpum	1-10	Giri et al., 2003b.
Sesbania aegyptiaca	G. macrocarpum	1-5	Giri and Mukerji, 2004.
Sesbania randiflora	G. macrocarpum	1-5	Giri and Mukerji, 2004.
Erythrina indica Lam.	Scutellospora erythropa	3.1 -3.3	Lakshman, 1999.
Jatropa curcus L.	Scutellospora erythropa	3.1-3.3	Lakshman, 2005.
Solanum nigrum L.	G. Citrocola	2.9-3.1	Lakshman and Hiremath, 2009.
Capsicum annum L.	G. fasciculatum	3.0-3.2	Lakshman and Hiremath, 2009.

Conclusion

Plant exposure to salinity is rapidly followed by a decrease in growth rate. Most of the physiologists interpreted this phenomenon as a direct consequence of metabolic perturbations induced by NaCl. The concept of NaCl toxicity is generally derived from studies on plant response to lethal NaCl concentrations. It is evident that these conditions provoke some irreversible perturbations in plant metabolism. However, it is not clear whether a linear gradient in physiological disturbance exits between exposure to sublethal and lethal NaCl concentrations.

Recent studies revealed the role of the root as a sensing organ of soil environment (reviewed by Davies and Zhang, 1991; Poljakoff-Mayber and Lerner, 1993). The hormonal message sent by the root is able to reduce shoot growth before appearance of the first damages caused by the stress. Therefore, the adaptation of the whole plant with mycorrhizal inoculation provides a potential for selection of genotypes with different agricultural trait, within a common pool of plants with acquired potentially stable salinity tolerance.

Many microorganisms especially nitrogen fixers have shown to change the properties of salt-affected soils, which cause a bioremediation of the salinity (Kaushik and Ummat, 1992). The major group of salt tolerant microorganisms reported from many salt affected soils include free living Bacteria, e.g., *Klebsiella, Azotobacter, Xanthobacter, Alcaligens, Leptocola* and *Azospirillum*; cynobactria, *Anabaena, Gloeocapsa, Oscillotoria, Nodularia, Nostoc*, etc., symbiotic arbuscular mycorrhizal fungi and *Rhizobia* (Lakshman, 2010). However, more intensive studies are required to understand the molecular mechanisms and common signaling systems in plants to the stresses. Further, in many cases, especially during short term exposure to the stress specific plant responses which are even species, cultivar and organ specific have been reported. The regulatory mechanisms for these stress specific responses also need proper attention.

References

Al-Karaki, G. N., Hammad, R and Rusan, M (2001). Response of two tomato cultivars differing in salt tolerance to inoculation with mycorrhizal fungi under salt stress. *Mycorrhiza*. 11: 43-47.

Apse, M.P, Aharon, G.S, Smedden W.A. and Blumwald, E. (1999). Salt tolerance conferred by over expression of vacuolar Na+ / H+ antiport in Arabidopsis. Science 285 (5431); 1256-1258.

Borsani. O., Valpuesta, V. and Botella, M. A. (2003). Developing salt tolerant plants in a new century: a molecular biology approach. *Plant cell tissue organ cult.* 73:101-115.

Breckle, S.W. (2002). Salinity, halophytes and salt affected natural ecosystems. In Lauchli A, Luttge U, editors. Salinity: Environment, plants, molecules. Kluwer academic publishers, pp. 53-77.

Chen, H., An R., Tang, J.H., Cui, X.H., Hao, F.S., Chen, J., Wang, X.C. (2007). Over-expression of a vacuolar Na+ / H+ antiporter gene improves salt tolerance in upland rice. *Mol. Breed.* 19: 215-225.

Danneberg, G., Latus, C., Zimmer, W., Hundeshagen, B., Schneider-Poetsch, H. J and Bothe, H (1992). Influence of vesicular-arbuscular mycorrhiza on phytohormone balances in maize (*Zea mays* L.). *J. Plant Physiol.* 141:33-39.

Davies, W.J and Zhang, J. (1991). Root signals and the regulation of growth and development of plants in drying soil. *Ann Rev Plant Physiol*, 42:55-73.

Dawson, J. O and Gibson, A. M (1987). Sensitivity of selected isolates from *Casuarina, Auocasuarina* and North American host plants to sodium chloride. - *Physiol. Plant.* 70:111-211.

Debez, A., Saadaoui. D., Ramani, B., Ouerghi, Z., Koyro, H.W Huchzermeyer, B. and Abdelly C. (2006). Leaf H+ - ATPase activity and photosynthetic capacity of Cakile maritime under increasing salinity. *Environ. Exp. Bot.* 57: 285-295.

Ezz, T., Nawar, A. (1994). Salinity and mycorrhizal association in relation to carbohydrate status, leaf chlorophyll and activity of perooxidase and polyphenol oxidase enzyme in sour orange seedlings. *Alexandria J Agri Res* 39:263-280.

Flowers, T.J. (2004). Improving salt tolerance. *J. Exp. Bot.* 55: 307-319.

Giri, B. and Mukerji, K.G (2004). Mycorrhizal inoculant alleviates salinity stress in *Sesbania aegyptiaca* Pers and *Sesbania grandiflora* Pers. Under field conditions: Evidence for improved magnesium decreased sodium uptake. *Mycorrhiza* 14:307-312.

Giri, B. and Mukerji, K.G. (1999). Improved growth and productivity of *Sesbania grandiflora* Pers. Under salinity stress through mycorrhizal technology. *J Phtol. Res.* 12: 35-38.

Giri, B., Kaur, M. and Mukerji, K.G. (1999). Growth response and dependency of *Sesbania aegyptiaca* on vasicular arbuscular mycorrhiza in salt stress soil. *Ann. Agric. Res.* 20: 109-112.

Glenn, E., Brown, J.J. and Blumwald, E. (1999). Salt tolerance and crop potential of halophytes. *Crit. Rev. Plant Sci.* 18: 255-277.

Graham, J. H. (1986). Citrus mycorrhizae: Potential benefits and interactions with pathogens. –*Hort. Science.* 21: 1302-1306.

Gusbert, C., Rus, A.M., Bolarin, M.C., Lopez-Coronado, J.M., Arrilaga, I., Motesinos, C., Caro, M., Serrano, R. and Moreno, V. 2000. The yeast HAL1 gene improves salt tolerance of transgenic tomato. *Plant Physiol.* 123: 393-402.

Hare, P.D and Cress, W.A. (1997). Metabolic implications of stress-induced praline accumulation in plants. *Plant growth. Reg* 21: 79-102

Hayasi. M., Takahasi, H., Tamura, K., Huang, J., Yu, L.H., Kawai-Yamada, M., Tezuka, T. and Uchimiya, H. (2005). Enhanced dihydroflovonol-4-reductase activity and NAD homeostasis leading to cell death tolerance in transgenic rice. *Proceed. Natl. Acad. Sci. USA.* 102: 7020-7025.

Hoffman, G. J and Phene, C. J (1971). Effect of constant salinity levels on water-use efficiency of bean and cotton. - *Trans. Am. Soc. Agric, Eng.* 117: 1103-1106.

Jain, M., Mathur, G., Koul, S. and Sarin, N.B. (2001). Ameliorative effects of proline on salt stress induced lipid peroxidation in cell lines of ground nut (*Arachis hypogaea* L.). *Plant cell Rep.* 20: 463- 468.

Jain, R. K., Paliwal, K., Dixon. R. K and Gjerstad, D. H. (1989). Improving productivity of multipurpose trees on substandard soils in India. - *J. For.* 87: 38-42.

Juniper, S. and Abbott, L. (1993). Vesicular arbuscular mycorrhizas and soil salinity. *Mycorrhiza* 4:45-57.

Kasuga, M., Liu, Q., Miura S, Yamaguchi-Shinozaki K. (1999). Improving plant drought, salt and freezing tolerance by gene transfer of a single stress-inducible transcription factor. *Nature Biotechnol.* 17: 287-291.

Kaushik, B.D and Ummat, J. (1992). Reclamation of salt affected soils with blue green algae (Cyanobacteria). A technology development. *In Biofertilizer technology Transfer* (Ed).

Lakshman, H.C. and Hiremath, S.G. (2007). Survey of AM fungi in plants growing costal beaches and their effect on *Solanum nigrum* L. *Nature Env..Polt. tech.* 6(1);81-84.

Lakshman, H.C. and Ratageri. (2010). Combined inoculation of AM fungi and Azotobacter beneficial to *Triticum astivum* L. grown in alkaline soil. *Int.J. Plant .Prop.* 2 (2): 267-277. s

Lakshman, H.C. and Ratna, A. (2008). Development of AM fungi during growth of *Cyanomopsis tetragonolaba* (L.) Tanb. (clusterbean) grown on polluted soil. *Bioinflet.* 5 (4): 375-377.

Lauchli, A. and Epstein, E. (1990). Plant responses to saline sodic conditions. In: Tanji KK editor. Agricultural salinity assessment and management. NY, ASCE, manual No. 71, pp. 113-137.

Lawlor, D.W. (2002). Limitation to Photosynthesis in water-stressed leaves: stomata vs. metabolism and the role of ATP. *Ann. Bot.* 89: 871-885.

Lichenthaler, H. K (1996). Vegetation stress: an introduction to stress concepts in plants. *J.plant physiol.* 148: 4-14.

Maas, E.V. (1990). Crop salt tolerance. In Tanji KK, editor. Agriculture salinity assessment and management. NY, ASCE manual No. 71, pp. 262-304.

Marschner, H. (1995). Mineral nutrition of higher plants. London, Academic press, p. 889.

Mc Millien, B. G., Juniper, S. and Abbott, L. (1998). Inhibition of hypal growth of vesicular arbuscular mycorrhizal infection from fungus spores in soil containing sodium chloride. *Soil Biol Biochem* 30:1639-1646.

Megdiche W, Ben Amor, N., Debez A, Hessini, K., Ksouri, R., Zuily-Fodil, Y. and Abdelly, C. (2007). Salt tolerance of the annual halophyte *Cakile maritima* as affected by the provenance and the developmental stage. *Acta physiol. Plant.* 29: 375-384.

Nash, D., Paleg, L.G. and Wiskich, J.T. (1982). Effect of proline, betain and some other solutes on the heat stability of mitochondrial enzymes. *Aust. J. Plant physiol.* 9: 47-57.

Nuccio MLm Russel, B.L., Nolte, K.D., Rathinasabapati, B., Gage, D.A., Hanson, A.D. (1998). The endogenous choline supply limits glycine betaine synthesis in transgenic tobacco expressing choline mono-oxygenase. *Plant J.* 16: 487-498.

Ojala, J. C., Jarrell, W. M., Menge, J. A and Johnson, E. L. V (1983) Influence of mycorrhizal fungi on the mineral nutrition and yield of onion in saline soil. *Agronomy Journal.* 75: 255-259.

Paleg, L.G., Stewart, G.R. and Bradbeer, J.W. (1984). Proline and giycine betaine influence protein salutation. - *Plant Physiol.* 75: 974-978.

Pfeiffer, C. M. and Bloss, H. E. (1988). Growth and nutrition of guayule (*Parthenium argentatum*) in a saline soil as influenced by vesicular-arbuscular mycorrhiza and phosphorus fertilization. *New Phytol.* 108: 315-321

Pojakoff-Mayber, A. and Lerner, H.R. (1993). Plants in saline environments. In hand book of plant and crop stress (Ed. Pessakali, M) Marcel Dekker Inc, New york, pp. 65-96.

Pollard, A. and Wyn Jones, R.G. (1979). Enzyme activities in concentrated solutions of glycinebetain and other solutes. *Planta.* 144: 291-298.

Polle, A. and Rennenberg, H. (1993). Significant of antioxidants in plant adaptations to environmental stress. In: Fowden L., Mansfield, T. and Stoddart, T. (eds). *Plant adaptation to environmental stress.* Champman and Hall, pp. 263-273.

Rajendrakumar, C.S.V., Rddy, B.V.D., Reddy, A.R. (1994). Proline-protein interactions: protection of structural and functional integrity of M_4 lactate hydrogenase. *Biochem. Biophys. Res. Commun.* 201: 957-963.

Rinaldelli, E. And Mancuson, S. (1996). Response of young mycorrhizal and non-mycorrhizal plants olive tree (*Olea europaea* L.) to saline conditions. I. short term electro physiological and long term vegetative salt effects. *Adv. Hort. Sci* 10:126-134.

Ruiz-Lozano, J. M and Azcon, R. (1995). Hyphal contribution to water uptake in myconfiizal plants as affected by the fungal species and water status. - *Physiol, Plant.* 95: 472-478,

Santoro, M.M., Liu, Y., Khan, S.M.A., Hou L-X., Bolen, D.W. (1992). Increased thermal stability of proteins in the presence of naturally occurring osmolytes. *Biochemistry.* 31: 5278-5283.

Sehgal, J. and Abrol, I. P. (1994). Soil degradation in India: status and impact. Oxford and *IBH Publication*. Co. Pvt Ltd. New Delhi, India. pp. 80.

Serrano, R,, Rodriguez-Navarro, A. (2001). Ion homeostasis during salt stress in plants. *Current opinion in cell Biol*. 13: 399-404.

Serrano, R., Mulet, J.M., Rios, G., Marquez, J.A., de Larrinoa, I.F., Leube, M.P., Mendizabal, I., Pascual-Ahuir, A., Proft, M., Ros, R. and Montesinos, C. (1999). A glimpse of the mechanism of ion homeostasis during salt stress. *J. Exp. Bot.* 50: 1023-1036.

Tarczynski, M.C., Jensen, R.G., Bohnert, H.J. (1993). Stress protection of transgenic tobacco by production of the osmolyte manitol. *Science*. 259: 508-510.

Tateishi, Y., Nakagawa, T. and Esaka, M. (2005). Osmotolerance and growth stimulation of transgenic tobacco cells accumulating free praline by silencing praline dehydrogenase expression with double-stranded DNA interference technique. *Physiol. Plant*. 125: 224-234.

Wang, X.W., Vinocur, B. and Altman, A. (2003). Plant responses to drought, salinity and extreme temperature: towards genetic engineering for stress tolerance. *Planta*. 218 (1): 1-14.

Xu, D., Duan, X., Wang, B., Hong, B., Ho, T.H.D and Wu, R. (1996). Expression of a late embryogenesis abundant protein gene, HVA1, from barley tolerance to water deficit and salt stress in transgenic rice. *Plant Physiol*. 110: 249-257.

Chapter 15

Trends in Management of Environmental Issues

T.C. Taranath[1], H.C. Lakshman[2], S.S. Kamble[3], A. Channabasava[4] and K.P. Kolkar[5]

[1,2,3]*DOS in Botany, Karnatak University, Pavate Nagar,*
Dharwad – 580 003, Karnataka
[4]*Department of Botany, Shivaji University, Vidyanagar,*
Kolhapur-416 004, Maharashtra
[5]*Department of Botany, Karnatak Science College,*
Dharwad – 580 007, Karnataka

ABSTRACT

India's rapid growth is causing its rapid environmental destruction. Due to the increasing population pressure, India is pushing itself ahead way too hard with aggressive industrial development, producing enormous amounts of untreated toxic waste, which often end up in our rivers, lakes, forest and landfills.

Unplanned and unmanageable growth of urban areas is causing tragic environmental degradation. Tropical deforestation is responsible in part for the increase in carbon dioxide in the atmosphere. Of all the human induced land degradation problems, the permanent loss of soil productivity due to erosion is worst on the global scale. Accelerated erosion occurs where agriculture is practiced and it is irreversible in nature. Loss of nutrient rich fine soil not only reduces productivity, but also results on silting of water bodies and streams, and induces release of soil carbon from particulate organic material, which contributes global warming. Over fishing is a problem with many marine stocks, physical alterations to coastal habitats increased the susceptibility of coastal populations to flooding and erosion. Sea level will rise as climate change pushes planetary temperatures high. Environmental changes, due to increase of green house gas (GHG) concentration has impacts on agro-forestry ecosystems, biodiversity, regeneration, biomass growth rate and geographic distribution of plant species. Thus this review article of the most highly cited papers in this journal shows significant contributions across five broad themes: the drivers and impacts of systemic and cumulative change, cross cutting concepts such as vulnerability and resilience, approaches to management, control and policy, and different perspectives of climate change.

Key words: *Pollution, Management, Restoration, Sustainability, Act*

Introduction

Environment consists of both biotic and abiotic substances and provides favorable conditions for the existence and development of living organisms. In fact, environment is the sum of all social, economical, biological, physical or chemical factors, which constitute the surroundings of man who both creates and moulds his environment. Our immediate concern is the quality of surrounding and space we live in, the purity of air we breathe in, the water we drink, the food we eat and the resources we draw from our environment to support our economy. However, environment is viewed from different angles by different environmentalists. The environment of any living organism has never been constant or static. It has always been changing, sometimes slowly and sometimes rapidly or drastically. Thus, like other organisms drastically organizing, man is also affected by his environment and these changes in environment may benefit the man or other organism in it. Many species on earth could not cope with change in environment, as a result of which they have since vanished (*e.g.* Dinosaurs) and many others are on the brink of extinction.

Every living species such as plants or animals is influenced by its environment and in turn gets influenced by it. The magnitude of such influence is not usually high in these species because of the fact that due to natural checks their population cannot raise beyond certain limits and they also cannot modify their own way of life. However, man is an exception; with increasing medical and scientific knowledge, man is able to modify the environment to suit his inundated needs much more than any other organism. This enables man to improve the quality of his life. After the scientific and industrial revolution in the recent past there has been immense impact of man on his environment.

The natural environmental system operates through self-regulating mechanism called homeostatic environment mechanism i.e., any change in natural ecosystem brought about by natural process is counter-balanced by change in other components of the environment. Thus, there exists a reciprocal relationship among the various components of the environment like air, water, soil, radiation, land, forest, wild life, flora and fauna, etc. The other forms of pollutions are induced by human activities (anthropogenic) and the effect may range from slight to heavy. Actually man is the most powerful environmental agent spearheaded by modern technology capable of modifying the environment according to his needs to a great extent. Man made environment includes technologies, industrial revolution, agricultural implements, transportation, housing, dam-building, channelization of energy sources like hydro, thermal and atomic energy, etc.

Understanding the Causes and Change in the Environment

Present India is facing many important environmental challenges which currently threaten both the development of India and the outlook for its future (Divan and Rosencranz, 2001). The state of India's environment is upset at the hands of uncontrolled human activities, and these ecological ailments are affecting social growth potential. Decrease of land quantity, increasing air pollution, increasing population growth rate, depletion of water resources, loss of indigenous species of flora and fauna and the background of overwhelming poverty are depicted in the report to detract from the positive growth of Indian people and the country as a whole (Geetanjoy, 2008). Thus, India's rapid growth is driving equally rapid environmental destruction and therefore, this paper brings out some of the important causes that lead to change in the environments, which are as follows:

i. Increasing Human Population and Its Impact on Environment

The earth's expanding human population and industrial growth have been known to cause serious environmental disasters. At the end of 2011, India's population reached 1.21 billion and its economy is growing at 8.5%, the fastest after China (Govindasamy, 2012). Due to the population pressure, India pushes ahead with aggressive industrial development. Consequently, thousands of industrial clusters nationwide produce enormous amounts of untreated toxic waste that often end up in rivers, lakes, forests, and landfills (Govindasamy, 2012). Even though India has sufficient environmental laws, weak enforcement and the lack of funds and manpower are most often the stumbling blocks for the pollution control boards (Govindasamy, 2012). The issues of environment are the effects of human activities that have no civic conscience and only think of the profit, without any concern about the impact towards the environment and their future of life. The long term effect from the environmental pollution can be seen when the ecosystem is not able to endure the pollution. According to the relevant literature, the major cause of this ecological crisis is regarding the value and belief in shaping human relations with the surrounding and the lifestyle itself (Bajaj, 1996).

Population growth is outpacing agricultural production and food-output per person has fallen, despite increase in food production. Annually, about 18 million people, mostly children, die of hunger or starvation, malnutrition and dietary deficiency (Hinrichsen, 1997). As per Technical Atlas of the FAO, undernourished is defined as lacking access to enough food to meet dietary energy supply requirement (2200 calories per day for adults). On this basis, presently about 20-30% of the Indian population is undernourished. In some countries, intensive cultivation has been undertaken even in areas with dry, sandy or thin and rocky soils, but demanding too much from the soil simply causes it to lose nutrients and ground water with consequent erosion and thus, the food producing capacity is decreased (FAO Report, 1996). While adopting better land management and conservation measures to protect the environment, the countries are to develop and introduce new agricultural technologies (Green, 1992). This has been mainly due to adoption of crop rotation, production and use of fertilizers and pesticides, expanded irrigation and genetically superior, high-yielding and disease resistant cultivated crop varieties etc.

Already the urban population is increasing so much that by 2015 AD about 60% of the world's population would be town-dwellers (Pudasaini 1997). UNDP calculates that presently Tokyo has the largest number (27.2 million) of urban residents while that of the second largest one (Mexico City) has about 16.9 million of urban dwellers. Country wise, India has the fourth largest urban population after

USA, USSR and China (about 350 million). Having less access to the civic amenities and housing etc., many of them are forced to reside in what is called slum or ghetto. Rising urban population densities are sure to promote the environmental damage by way of air and water pollution and the associated socio-economic conflicts of various natures. (Brown and Jacobson, 1987; Shaw, 1989). While slowing the urban population growth alone cannot solve such problems, it may help relieve the underlying pressure and may provide some time to provide more services. It may be mentioned that one of the characteristic features of human society is vulnerability to conflict over the limited and available natural resources and privileges. Presently, over 800 million people are unemployed or underemployed in the high fertility developing countries which need to provide about 40 million new jobs per year – a phenomenon that dampens economic growth. But when population growth slows; the per capita income increases (Brander and Dowrick, 1993). Higher earnings in town generate more savings and investment. Besides, countries with rapid population growth can afford to spend less per child on education, resulting in declining educational level. But with declined fertility a higher percentage of young people may be educated. A steady work force with higher educational attainment enhances the labor productivity and income also rise (Mason, 1997).

The notion of population pressure is central to the environmental security perspective (Homer-Dixon, 1999; Homer-Dixon and Blitt, 1998). Rapid population growth may lead to a reduced per-capita access to subsistence resources as resource reproduction is unable to keep up with the growing demand. This is typically taking place in rural contexts where the dependence on renewable resources is great. Overpopulation may also lead to a decline in the overall supply of certain resources, for example due to pollution, deforestation, overgrazing, unrestrained fishing, and clearing of land for housing. Rwanda, a densely populated and predominantly rural country, is a classic case in point. Resource scarcity may also spark or escalate inter-group competition. Under unfavorable economic and political conditions, such competition may take the form of violent conflict. Poor countries are argued to be particularly susceptible to violent resource conflicts as they have limited capacities to adapt to changing environments and often lack institutional arrangements for peaceful conflict resolution (Barnett and Adger, 2007) even if a conditional relationship is yet to be robustly verified (Benjaminsen *et al.*, 2012; Devitt and Tol, 2012; Gizelis and Wooden, 2010; Hendrix and Salehyan, 2012; Raleigh, 2010; Raleigh and Kniveton, 2012; Theisen, 2012; Theisen *et al.*, 2011–2012).

While, some individuals and communities manage to adapt to forms of resource scarcity, substitute resources or use them more efficiently, others will exit and settle in more promising environments, including in urban areas. In this way, rural–urban migration could act as a safety valve, relieving the countryside of the impending population pressure.

ii. Growing Urbanization and Its Impact on Environment

The developing countries like India, since the beginning of industrial revolution, towns and cities have grown rapidly both in size and power. The transition from an agriculture based society to an industrial one has been accompanied by urbanization. It is predicted that by the end of 21st century, 80-90% population will live in the cities. The major problem of urbanization is providing resources to a large and fast increasing population and thus, the unplanned and unmanageable growth of these cities is causing tragic and severe urban environmental problems.

- Degradation of water resources
- Sanitation , Air pollution and Water pollution
- Traffic and congestion due to increased vehicles/automobile industries.
- Loss of agricultural lands, Loss of landscape and wild life.
- Depletion of energy resources
- Floating population, increasing huts/slums lead to housing problems.

Edward .R. Carr examines the role of environmental change as a driver of migration, drawing on political ecology and a Foucauldian conceptualization of power. The concept demonstrated that environment is rarely a sufficient basis for the decision to migrate, but cannot be excluded from migration decision making because it is a key element of local power relations and local knowledge. A similar approach is adopted by Black *et al.*, (2011). The area of explanation of migration focused on more social 'causes' of migration, and particularly the role of social networks (Boyd, 1989). It has been made clear from the empirical evidence that many movements over shorter distances are associated with social factors, for example migration at the point of marriage. There is a large body of evidence that demonstrates the importance of social networks in perpetuating migration flows.

Government policy can play a key role in stimulating economic development that leads to migration. The creation of Special Economic Zones in China from 1978 onwards led to rapid urbanization in areas such as the Pearl and Yangtze River deltas, including the cities of Guangzhou, Dongguan, Foshan, Shenzhen, Shanghai, Changzhou, Hangzhou and Suzhou. In these areas, urban population is predicted to increase from fewer than 10 million people in1990 to a projected total of more than 65 million by 2025. Similar rural urban shifts are evident in other Asian mega-deltas such as the Red River and Mekong deltas in Vietnam and the Chao Praya delta in Thailand. Whilst migration to these urban areas can lead to increased income and to improved living standards, there can also be a greater vulnerability to the effects of environmental hazards, including floods, hurricanes and coastal erosion (Seto, 2011).

Income and wage differentials alone, however, cannot explain the specifics of migration. More broadly, migration is not a general process of people moving from poorer to richer places. It is a highly specific process as people move from one relatively poor area to another specific relatively rich area. The scale and direction of movement is linked to the personal circumstances of migrants, such as class, ethnicity, religion, language, education levels and connections with people in planned destinations, mitigated by the intervening effects of migration policies. Those who are most exposed to conflict may actually lack the resources to move and may remain exposed to high levels of danger in their home towns and villages.

Conflict can also interact with other drivers to create conditions where political tensions, poverty, environmental hazard and a relatively young population all contribute to migration and displacement, as has recently been the case in Pakistan (Raleigh, 2011). Political uncertainty, even in the absence of actual conflict, may also be push factors for migration. More positively, perceived political stability may be a pull factor that attracts immigrants, or at least encourages people not to leave. Government policies to relocate people can also be seen as a political driver of migration, this time primarily focusing on mobility. For example, policies for the creation of growth hubs or new urban development can act as a pull factor for migration. Similarly, policies for the management of rural land can act as a push, as can specific types of development projects (such as the construction of dams and reservoirs).

The specificity of migration is grounded in the connections that develop between places as a result of histories and cultures of migration. Social drivers of migration include family and cultural expectations, cultural practices regarding inheritance, the need to acquire funds for dowries or bride payments, and the search for educational opportunities. Migration networks can be formal through the operation of agencies, or more informal through kith and kin networks. Past migration and its direction can therefore be a good predictor of future migration (Massey, 1990). The creation of links between sending places and destinations can open 'transnational spaces' within which remittance flows can potentially contribute to economic development (Vertovec, 2008). New social media and communications technologies have the potential to reduce the social and psychological costs of migration and provide images and representations of destination countries. They also allow connections to be maintained between migrants and their families. Migration can thus provide resources to sustain livelihoods, but social drivers help us to understand how and why opportunities to migrate are not evenly distributed.

iii. Carbon Emissions and Its Impact on Environment

Tropical deforestation is responsible in parts, for the increasing concentration of CO_2 in the atmosphere. Estimates of net release of carbon (C) at the global level are highly uncertain, ranging from 0.4 – 1.6 Gt C/year to 1.1-3.6 Gt C/year, whereas net annual C emission estimates for India is 0.67×10^4 Gt C/year. Carbon emissions can be from deforested areas as well as from degradation. FSI (1988) has estimated that annually, 27×10^6 of firewood is removed from the forests through clear cutting. The carbon emission estimates in India are based on the standing biomass determined using the crown cover estimates and a few published studies on standing biomass. Similarly, the C-uptake is estimated to be in the range of 1.25 to 2.85% of the standing biomass. Therefore, any changes in the estimates of crown cover, standing biomass and annual productivity will also change the net emission rates.

iv. Impact of Carbon sequestration on environment

Tree legumes are often important components of arid and semi-arid ecosystem. Many of these trees and shrubs have the ability to develop a deep root system and symbiotically fix atmospheric nitrogen. Woody legume trees, which have potential to fix significant quantities of nitrogen, can have a possible influence on yield in the nitrogen deficient soils. Some nitrogen fixing trees have proved to be very efficient in phosphorus uptake and can effectively serve as nutrient pumps for this. Other mineral nutrients, predicted the possibility of nitrogen fixation in woody legumes to the order of 100 kg N ha⁻¹in areas receiving 500 mm annual rainfall. Nitrogen fixing trees can achieve rates of fixation comparable with those of leguminous crops. The major recognized avenue for addition of organic matter to the soil from the tree standing on it is through the fall of litter, twigs, branches and fruits. In some cases levels of these nutrients approached or exceeded those found in the nearby rain forest.

v. Soil Erosion and Its Impact on Environment

Of all the human induced land degradation problems, the permanent loss of soil productivity due to erosion is the worst on the global scale. Accelerated erosion occurs in nearly every place where agriculture is practiced and is irreversible in nature. Loss of nutrient rich fine soil not only reduces productivity, but also results on silting of water bodies and streams, and induces release of soil carbon from particulate organic material, which contributes to global warming. Asia has suffered the most from human induced soil erosion than any other continent; of which agricultural activities account for 25 per cent (Wani *et al.* 2001).

On average, nationwide soil loss rate is 16 tones /ha/year with 29 percent being permanently lost to sea and another 9 percent deposited into major reservoirs reducing their capacity by 1-2 percent annually. This adds up to 5334 million tones of soil eroded every year (Dhruvanarayana and Ram Babu, 1983). This is much higher than the acceptable rate of erosion which is at 11 tonnes/ha/ year (Al-Swaify *et al.*, 1985). The annual water erosion rate values range from less than 5 tones /ha/yr for dense forest, snow clad cold desert, and the arid region of western Rajasthan to more than 80 tones /ha/yr in the Shiwalik hills. Ravines along the banks of the Yamuna, Chambal, Mahi, Tapti and Krishna rivers and in the shifting cultivation regions of Orissa and the North Eastern states also revealed soil losses exceeding 40 tones /ha/yr. The annual erosion rates in Western Ghats coastal regions vary from 20-30 tones /ha (Singh *et al.*, 1992). Erosion rates in the black soil region (vertisols) of the country were reported at 20 tones /ha /yr while red soil of Chhotnagpur plateau recorded a soil loss of 10-15 tones /ha/ yr. The hilly regions have rates up to 20 tones /ha/yr. Soil erosion rates on alluvial Indo-Gangetic Plains of Punjab, Harayana, Uttar Pradesh, Bihar and West Bengal are moderate (5-10 tones/ha/yr).

vi. Soil related Externalities Arising from Indian Agriculture:

The physical processes that constitute land degradation are mainly water and wind erosion, compaction, crusting and water logging. The chemical processes include salinization, alkalization, acidification, pollution and nutrient depletion. The biological processes on the other hand are related to the reduction of organic matter content in the soil, denudation of vegetation and impairment of activities of microorganisms and fauna.

Salinity is a severe problem affecting mainly irrigated areas. Lands with nutrient loss are small, in spite of the fact that large areas of rain fed soils are naturally deficient in nutrients and also receive only a small application of fertilizer and organic matter (Kerr, 1996). In economic terms the country loses a phenomenal amount of Rs 285.5 billion annually (at current prices) at the current level of total land degradation. At constant (1979-82) prices the annual economic loss is Rs. 89.38 billion. In per hectare terms, the economic loss due to land degradation is Rs. 1521 at current prices and Rs. 476 at constant prices annually. State-wise the magnitude of loss is high in AP, Gujarat, Karnataka, Maharashtra, MP, Rajasthan, TN, and WB. These states, together, account for nearly 73 per cent of the total loss in the country (Singh *et al.*, 2003).

vii. Oceanography and Its Impact on Environment

Oceanic and costal water circulation actively transport pollutants across geopolitical biometrics with source and impact separated on various spaces and time scales (Ratageri *et al.*, 2006). However, coastal zone ecosystem serves important socio-economic activities, such as offshore drilling for all, fisheries, aquaculture, marine transport, defense activities and tourism. Over fishing is a problem with many marine stocks. Physical alterations to costal habitats have increased the susceptibility of costal populations to flooding and erosion. In addition, the oil spill and oil leakage has increased considerably over the last three decades. Therefore, there is a need of field research that has to supplement the results of laboratory experiments. Future research needs to be done mainly on marine organisms resulting from exposure to oil effect on population and the interaction of oil with other contaminants.

The level of the ocean has fluctuated by more than 100 m over the past 100,000 years as ice stored on land has changed in volume. During the last ice age, sea level was 120 m below compared to where it is today. During the transition out of this period of glaciations, sea level changed at an average rate of 10 mm/yr (some periods, rates as high as 40 mm/yr). During the inter-glacial periods, rates of sea level rise have been much slower (0.1 – 0.2 mm/yr over the past 3000 years; Church *et al.* 2001). Changes in sea level have had major impacts on the abundance and particularly the distribution of both marine and terrestrial diversity.

Sea level will rise as climate change pushes planetary temperatures higher. This occurs due to the thermal expansion of ocean water (responsible for about 70 percent of the increase), the melting of glaciers, and changes in the distribution of Greenland and Antarctic ice sheets (responsible for the remainder). The expected increase in sea level is approximately 9-29 cm over the next 40 years, or 28- 98 cm by 2090 (Church *et al.* 2001; IPCC 2001). These changes may sound trivial but are expected to have major impacts on coastal regions and human infrastructure. Vast areas of the world's coastal regions are expected to be inundated. A 25- cm rise, for example, would displace a large number of people from the delta regions of major rivers such as the Nile, Ganges and Yangtze as well drowning Pacific and Indian Ocean nations such as the Maldives, Kiribati and Tuvalu (Church *et al.* 2001).

In concert with the direct effects of coastal inundation are the impacts of storm surge, which could result in a fivefold increase in displaced people by 2080 (Nichols *et al.* 1999). Sea temperature, in turn, has an influence on a broad range of other critical features of the marine environment. Due to its direct effects on the density of seawater, changes in global temperatures can play directly upon the rates and directions of ocean water movement. Most global circulation models indicate that the thermohaline circulation of the planet, for example, is likely to weaken as greenhouse warming continues. Dickson and colleagues (2002) have produced convincing data that indicate a rapid and sustained freshening (decreased salinity) of the deep Atlantic Ocean. Though these changes may appear small (0.03 ppm salinity change over the past 40 years), they could indicate that major changes may be in store for the heat budget and functioning of the earth's oceans.

viii. Biological Responses to Climate Change in the Ocean

Not surprisingly, organisms in the ocean respond to changes in the physical and chemical makeup of their environment. These responses may be mild, as organisms adjusting to their physiological processes to the new conditions sicken or experience higher mortality rates as their thresholds for particular types of change are exceeded. The latter may result in a shift in the genetic structure (adaptation) and / or geographic range of a population. There is strong evidence that both types of changes have already occurred among marine organisms in response to climate change.

Coral Reefs

Tropical intertidal and subtidal regions are dominated by ecosystems that are characterized by a framework of scleractinian corals. Coral reef ecosystems deserve separate mention in this chapter for two reasons. First, they stand out as a coastal ecosystem, that has an order of magnitude greater biodiversity than all other coastal ecosystems (Bryant *et al.*1998). Second, they have undergone major changes over the past 20 years, much of which has been associated with climate change (Hoegh- Guldberg 1999) and other stresses (Bryant *et al.* 1998).

Changes in reef-building coral communities are likely to have huge impacts on marine biodiversity. Corals form the essential framework within which multitudes of other species make their home. How these communities and interactions will change as coral health and abundance decline is still largely unassisted. Fish that depend on corals for food, shelter, or settlement cues may experience dramatic changes in abundance or go extinct. Thousands of other organisms are also vulnerable. For example, over 55 species of decapods crustacean are associated with living colonies of a single coral species.

Fish Population

Coastal fisheries are critical recourses for hundreds of millions of people. Many scientists now point to the dramatic over exploitation of fisheries and the subsequent decline in fish stocks as the major factor in ecosystem change over the past two centuries (Jackson *et al.* 2001). Overfishing is one driving pressure that has had devastating impacts on coral reefs. A perfect example would be that of the Great Barrier Reef. The biodiversity of reefs supports the aquarium and aquaculture industries, biomedical industry and other commercial industries.

The management of coral reef fishers falls across several groups, including National Oceanic and Atmospheric Administration (NOAA) through the regional fishery management councils, and state, territory commonwealth and local agencies. Because over-fishing has become a huge problem in the Great Barrier Reefs, the Australian government has had to make plans in order to help future projections. Overfishing of important herbivores has only been increasing over the past few decades. Direct overexploitation of different fishes and invertebrates by recreational, subsistence, and commercial fisheries has resulted in the rapid decline in populations. The NOAA has proof that overfishing affects fish size, abundance, species composition and genotypic diversity. Also, overexploitation of marine organisms contributes to the degradation of coral reef ecosystems as a whole.

ix. Environmental change and its impact on Agro-Forestry

Environmental changes, due to increase of green house gas (GHG) concentration has impacts on forests / agro forestry ecosystems, biodiversity, regeneration, biomass growth rate and geographic distribution of plant species. Sustainable restoration of resources and conservation compulsions has compelled to discover the role of alternative land use systems in recent years. Growing of trees and agricultural crops as well as rearing of animals on the same piece of land simultaneously or sequentially is now being viewed as optimum strategies. As against the recommendations of National Forest Policy (1988) to keep 33% geographical area under forest, India has only 19% forest. Mounting demographic pressure on natural resources is further deteriorating the situation due to increasing pressure on forests for firewood, fodder, timber, etc.,. Although area-wise India stands as the 7th largest country of the globe (2.3% land area), but accommodates 1027 and 4708 million of human and livestock population, which ranks 2nd (16.6% of Globe population) and first in the world. The projection for food grains, fodder and firewood requirement is quite huge of the tune of 295 million tons (by 2020 AD), 676 million tons and 46 MM^3 (by 2025) respectively. Per capita land has decreased and rain fed farming areas cover 70% of cropped area but contributes hardly to 40% of national food basket. Shrinking crop area, fast degenerating natural resources, depleting soil fertility, declining use efficiencies of agricultural inputs, dwindling output input ratio have rendered crop production less and less remunerative. Nearly 60% population of the country is dependent on agriculture in which rain fed agriculture supports 40% of our population and 75% of poor in the country. Therefore, maintenance of health of production stands as prime goal to achieve for sustainable production. Hence, tree culture is necessitated highly for amelioration of climate, production of firewood, fodder, timber, fibre, fruit, fertilizers, gum etc., and for this agro-forestry is the viable option particularly in Indian tropics and subtropics. Sustainable land use is the key for successful management of resources to satisfy the changing human needs, while maintaining or enhancing the quality of environment and conserving natural resources.

X. Enhanced Green House Effect and Its Impacts on Environment

After all, the greenhouse effect may not lead to a very unpleasant situation for all of us. People battling harsh winters every year, may look forward to a warmer climate. In the past, temperature increase due to greenhouse effect had helped mankind in more than a single way. In the eleventh and twelfth centuries, an increase of about only 0.5°C beyond today's temperature was sufficient to allow farming on the coast of Greenland, for vineyards to flourish extensively in England and for the Vikings to travel the North Atlantic and settle in new found land.

The term green house effect simply means that the average air temperatures will increase as a result of the buildup of different greenhouse gases in the atmosphere termed as global warming and is responsible for the temperature increase of about 0.6^0 C that has occurred since 1800.

Clouds absorb about one-fifth of solar radiation striking them but unless they are very thin, they are almost opaque to IR radiation. The presence of cirrus clouds after a period of clear sky at night is enough to increase the surface air temperature rapidly by several degrees because of long wave radiation emitted by the clouds. Green house effect is most marked at night and usually keeps the diurnal temperature range below 20°F. This natural phenomenon is called greenhouse effect and is responsible for the average temperature at the earth's surface being +15°C. The earth's surface is much more warmed by this phenomenon than it is by solar energy it receives directly.

According to projections made in 1992 by the Inter-Governmental Panel for Climate Change (IPCC), a group sponsored by the United Nations Environmental plan, if no step is taken to reduce emission of CO_2 and the other green house gases, then by about the year 2040 the average global atmospheric temperature will be 1^0 C higher than at present. By 2100 it will increase still by 1.5^0 C. The temperature increase will probably be greater towards the Polar Regions.

Increase in temperature may also result in a higher amount of evapo-transpiration which may lead to increased frequency of droughts. There could be significant changes in the rainfall pattern which will aggravate the situation and will make the predictions more difficult to make. For example, in India in 1987, lower than average rainfall reduced the food grains production from 152mt to 134mt, lowering the buffer stocks from 23 to 9 mt.

In arctic region, algae, bacteria and other microscopic organisms grow on the underside of the sea ice as longer spring days bring increasing amounts of sunlight. As arctic days get warmer with the change of seasons, the ice breaks up and the organisms are released into the water, where they support an escalating food chain that ends with seals, whales and polar bears. If global warming melts the polar ice, the number of microorganisms grown under the ice would be reduced and the marine animals they support will also suffer.

Globally, agriculture may remain stable at levels which are sufficient for meeting human food requirements but the cost of the measures to maintain these levels are unclear. Serious negative impacts may occur in some regions, which are currently climate dependent. These may include the semiarid and subtropical areas of Africa (a 30-70% reduction in crop yield may be expected in some regions with a warming up to 3.5^0 C) and South America, as well as the tropical and equatorial parts of South east Asia and Central America (crop reduction up to 7%

and 5-25% respectively). Crop yields in some countries may increase (Philippines, Japan, etc.). In the South and South East of European USSR, where the annual total precipitation may drop by up to 20% by the end of 20th century, mainly during winter, the increased frequency of droughts may cause 10-20% reduction in yield of cereal crops.

The earth's expanding human population and industrial growth have been known to cause serious environmental disasters. At the end of 2011, India's population reached 1.21 billion and its economy is growing at 8.5%, the fastest after China (Govindasamy, 2012). Due to the population pressure, India pushes ahead with aggressive industrial development. Consequently, thousands of industrial clusters nationwide produce enormous amounts of untreated toxic waste that often end up in rivers, lakes, forests, and landfills (Govindasamy, 2012). Even though India has sufficient environmental laws, weak enforcement and the lack of funds and manpower are most often the stumbling blocks for the pollution control boards (Govindasamy, 2012). The issues of environment are the effect from the human's activities that have no civic conscious and only think the profit without concern about the impact towards the environment and their future of life. The long term effect from the environmental pollution can be seen when the ecosystem is not able to endure the pollution. According to the relevant literature, the major cause of this ecological crisis is regarding the value and belief in shaping human's relation with the surrounding and the lifestyle itself (Bajaj, 1996). Present India is facing many important environmental challenges which currently threaten both the development of India and the outlook for its future (Divan and Rosencranz, 2001). The state of India's environment is in upset at the hands of uncontrolled human activities, and these ecological ailments are affecting social growth potential. Decrease of land quantity, increasing air pollution, depletion of water resources, loss of indigenous species of flora and fauna and the background of overwhelming poverty are depicted in the report to detract from the positive growth of Indian people and the country as a whole (Geetanjoy, 2008). Thus, India's rapid growth is driving equally rapid environmental destruction.

xi. The Loss of Genetic Resources Due to Change In the Environment

Genetic resources are the building blocks with which new varieties are made. Hence, their availability to plant breeders is absolutely necessary for the further development of crop to meet the ever diversifying needs of man (Singh, 1989). Extensive forest clearance and the large scale adaptation of uniform modern varieties are eroding the range of genetic resources. An all-out effort must be made to conserve and utilize genetic resources in a rational manner to avoid a conflict between development and conservation.

For development to be sustainable there is a need to conserve and manage the biological resource base, much of which has unknown potential. Future needs for breeding and crop diversification cannot be predicted in the face of changing climates, new agricultural production systems and unknown human requirements. Not only has man already lost part of the genetic resource base, he is also subjecting the production system to high risk by electing to use only a narrow genetic base for many of his most important crops. Consequently, the genetic base of crop varieties is becoming increasingly narrower before drastic degradation of the resource occurs and productivity falls even below its present level.

Possible Control Measures for Environmental Management

1. Restoration Ecology and Management of Natural Resources

Degradation of the natural resources has been a global problem. Conversion of forest lands into arable land or for other development activities such as urbanization and industrialization, intensive agriculture, over-exploitation, overgrazing, pollution of various kinds, mining and other anthropogenic activities have resulted in degradation of both the land and water resources. About 173.64 million ha of land are degraded in India with one or other problems and about 260 million ha are drought prone (India Agriculture in Brief, 2000).

The suitable improvement of farming system is possible by halting further degradation of natural resource base and maintaining soil health through alternative agricultural practices such as adoption of efficient soil and water conservation measures, crop rotation, diversification of crops, reduction of tillage, integrated nutrient management and efficient crop/livestock systems. There should be fertility management strategies for specific soil problems viz. acid soils, saline and alkali soils, waterlogged soils, and hilly, arid and coastal soils.

- There should be better development of sound data base on water resource availability and utilization.
- Faulty and uncontrolled methods of irrigation have to be removed.
- There has to be an improvement in the hydro-biological productivity needs adequate attention.
- Encourage the involvement of women in policy making and participatory issues in watershed development programmes.

Adaptations of Technology by Farmers to Restore Environmental Changes

The following guidelines are considered important in the adaptation of technology by farmers:

- Technology should be tested under farm conditions and recommendations must be made only after consulting with the farmers concerned. They must take fully into account the farmer's circumstances, focusing on the local farming system and the related processing and marketing components.
- There is an urgent need to develop and promote technology that increases or sustains productivity at lower cost and for a lower labor input and which does not harm the environment (Lakshman, 1999 and 2000).
- Soil and water conservation and other measures to achieve sustainable development should, where possible be designed to benefit the farmer in the same year they are carried out; otherwise they are unlikely to be widely adapted by other farmers. Simple water harvesting methods can meet this requirement as can certain forms of minimum tillage.

- Emphasis should be placed on low external input farming systems, such as the integration of crop and livestock production systems to provide manure and draught animal power, to reduce residue problems from fertilizers and pesticides and lower the cost of external production inputs (Lakshman and Channabasava, 2013).

2. Use of Wastewater in Irrigation system for conservation of water resource

Based on information from the countries providing data on irrigated areas, it is estimated that more than 4-6 million hectares (ha) are irrigated with wastewater or polluted water. A separate estimate indicates 20 million ha globally an area that is nearly equivalent to 7 per cent of the total irrigated land in the world[10].

The resulting agricultural activities are indeed most common in and around cities[6]. But, can also be seen in rural communities located downstream of where cities discharge unless treatment or self purification processes take place. Much of this use is not international and is the consequence of water sources being polluted due to poor sanitation and waste disposal practices in cities. It was suggested from a survey across the developing world that wastewater without any significant treatment is used for irrigation purposes in four out of five cities.

Few studies have quantified the aggregate contribution of wastewater to food supply. In Pakistan about 26 per cent of national vegetable production is irrigated with wastewater, while in Hanoi, Vietnam which is much better than Pakistan about 80 percent of vegetable production is from urban and peri-urban areas irrigated with diluted wastewater[3]. In major cities in West Africa between 50% and 90% of vegetables consumed by urban dwellers are produced within or close to the city. Where much of the water used for irrigation is polluted. Faisalabad is the third largest city of Pakistan, a detailed description of the city and its wastewater use practices is given by Ensink.

Sewage wastewater has source of different solid wastes. Hence ionic concentrations and hardness in wastewater is higher than in ground water. Sewage wastewater proved superior to ground water for the growth parameters, giving higher shoot length, shoot fresh weight and dry weight, leaf number and leaf area. The role of certain nutrients in the wastewater is well known. For example, nitrogen is essential for cell division and expansion. Sulphur is useful in providing certain amino acids and vitamins and the role of Phosphorus is known for energy transfer compounds.

The physiological parameters i.e. total chlorophyll content and photosynthetic rate were significantly enhanced by the application of wastewater as compared to ground water. It may also be due to the presence of many important nutrients in the wastewater in surplus amount. For e.g., Mg. which is an important constituent of chlorophyll and K has a role in photosynthesis as a co-factor for many photosynthetic enzymes[16,17]. Earlier reports have also indicated the beneficial effect of waste water on chlorophyll contents[17]

3. Water Harvesting in Small Catchments

In most of the semi-arid regions of India, where dry farming is practiced, the annual rainfall is 600-1000 mm or more. But the rainfall is erratic and confined to a short duration. Consequently, despite water availability, there is deficiency of water for crops. In recent years efforts have been made to revive the old practice of conserving water in tanks (Sinha *et al.*, 1985). The collected water can be used for drinking, for "life saving" irrigation of crops, or as a pre sowing irrigation for a Rabi crop. The provision of this irrigation calls for a carefully planned management of rain water. Interestingly, the International Crops research Institute for the Semi-arid Tropics (ICRISAT)., Hyderabad, has rediscovered the utility of farm tanks as an innovation for south India, and has now advocated water harvesting and tank construction as a major recommendation for the semi-arid regions. The following problems need attention in this respect.

- In view of soil topography and the small holdings of farmers, large reservoirs of water may be impracticable.

- The ownership of the reservoir, whether it should belong to an individual or the community, has to be decided.

- Suitable methods should be devised for the control of seepage and evaporation.

- Application of water should be with minimal investment in energy.

- Relative cost of development of water resources should be worked out in comparison with the medium and the major irrigation projects.

- Consideration should be given to pollution problems.

- Resources should be mobilized for the project.

It is clear that it would not be possible to have a uniform pattern of water-harvesting techniques suited to different parts of the country. Depending upon the soil, rainfall and socio- economic conditions, independent or complementary approaches would have to be developed. A successful project could serve as catalyst in this respect. One such project is an operation in Sukhomajri (Tehsil Kalla) in Haryana. Here protective irrigation for 16.14 hectares has been developed at an approximate cost of Rs 5000 per hectare. The cost of irrigation development through large projects was about Rs. 15000 at that time.

4. Environmental Education is the Need of the Hour

India is the country of villages. 70% of India is rural. In villages, conditions of life are very difficult and villagers are unaware of the importance of environmental education. Hence, if this education is not rooted in village itself, all external help is unproductive. Environmental education helps individuals and groups to understand the concept of a sustainable environment. The ultimate goal of such education is to help young people in developing, caring and committed attitudes and the desire to act responsibly for the environment and towards one another. Therefore, environment education is concerned not only with teaching conceptual knowledge and skills for monitoring and measuring environmental quality, but also with the development of the values, attitudes and skills which will motivate and empower young people to work, both individually and with others to help in promoting the sustainability of natural and social environment (Macleod, 1992).

Environmental education constitutes a comprehensive life-long education, one responsive to changes in a rapidly changing world. It prepares the individual and communities for life through an understanding of the major problems of the contemporary complex world, the problems resulting from the interaction of the biological, physical, social, economic and cultural aspects of the individual and the communities. The National Curriculum framework says that for the primary grades, the natural and social environment will be explained as integral parts of languages and mathematics. Children should be engaged in activities to understand the environment through illustrations from the physical, biological, social and cultural spheres. The ECEEN (European Community Environmental Education Primary Network) aims at an improvement of the quality of environmental education in the schools involved, by means of mutual cooperation and learning from each other's experiences; and gathering and dissemination of teaching materials on environmental education.

5. There is A Need of Change Towards Sustainability

In evolving more sustainable production systems, agriculture and rural development effort should be directed towards three essential goals.

- Food security
- Employment and income generation in rural areas, in order to eradicate poverty.
- Natural resource conservation and environmental protection.

It has been recognized that, the root causes of environmental degradation are social and institutional in nature. Measures to address the problem will require integrated strategies which involve an adjustment of policies, values and institutional structures. The major thrust of the strategy for sustainability for the poor must aim at eradicating poverty (FAO, 1991). Measures to promote sustainability should include the following:

- There must be active participation by rural people in the development of integrated farming systems, by means of organizations such as agricultural cooperatives. Such groups will help prevent an increase in influence of the middleman.
- There must be decentralization and recognition of the role of farmers, their families and local authorities in decision making, including incentives for initiatives by local communities.
- Clear and fair legal rights and obligations must be allocated with regard to the use of land and other natural resources, including land reform where necessary. Such allocations should pay particular attention to the important role of women as decision maker, food producer and food provider.
- Pressure on natural resources should be relieved by investment into improving, rehabilitating and conserving them so, that they can be used safely and productively.

- Agricultural policies should be adjusted to promote production systems that can help attain the objective of sustainability. This includes promoting the demand for crops and livestock which can be produced sustainability.

- More attention should be paid to safeguard human health and environmental quality in relation to the use of dangerous pesticides and other chemicals.

- More sources of off-farm income such as food processing and handicrafts are needed in rural areas to prevent the migration of farmers to urban centers.

6. Environmental Legal Issues Framed by Government of India

The Indian government has framed many acts and regulations to protect environment and natural resources. Some of the important legal issues are mentioned as follows:

- **The River Boards Act, (1956)** enables the states to enroll the central government in setting up an Advisory River Board to resolve issues in inter-state cooperation.

- **The Merchant Shipping Act, (1970)** aims to deal with waste arising from ships along the coastal areas within a specified radius.

- **The Water (Prevention and Control of Pollution) Act, (1974)** establishes an institutional structure for preventing and abating water pollution. It establishes standards for water quality and effluent. Polluting industries must seek permission to discharge waste into effluent bodies. The CPCB (Central Pollution Control Board) was constituted under this act.

- **The Water (Prevention and Control of Pollution) Cess Act, (1977)** provides for the levy and collection of fees on water consuming industries and local authorities.

- **The Water (Prevention and Control of Pollution) Cess Rules, (1978)** contains the standard definitions and indicate the kind of and location of meters that every consumer of water is required to affix.

- **The Air (Prevention and Control of Pollution) Act, (1981)** provides for the control and abatement of air pollution. It entrusts the power of enforcing this act to the CPCB.

- **The Forest (Conservation) Act and Rules, (1981),** provides for the protection of and the conservation of the forests.

- **The Air (Prevention and Control of Pollution) Rules, 1982** defines the procedures of the meetings of the Boards and the powers entrusted to them.

- **The Environment (Protection) Act, (1986)** obligates the central government to protect and improve environmental quality, control and reduce pollution from various sources, and prohibit or restrict the setting and /or operation of any industrial facility on environmental grounds.

- **The Environment (Protection) Rules, (1986)** lay down procedures for setting standards of emission or discharge of environmental pollutants.

- **Factories Act, 1948 and its Amendment in 1987:** The Act contains a comprehensive list of 29 categories of industries involving hazardous processes, which are defined as a process or activity where unless special care is taken, raw materials used therein or the intermediate or the finished products, by-products, wastes or effluents would: i.) Cause material impairment to health of the persons engaged. ii.) Result in the pollution of the general environment.

- **The Factories Act, 1948 and Amendment in 1987** was the first to express concern for the working environment of the workers. The amendment of 1987 has sharpened its environmental focus and expanded its application to hazardous processes.

- **The Air (Prevention and Control of Pollution) Amendment Act, (1987)** empowers the central and state pollution control boards to meet with grave emergencies of air pollution.

- **The objective of Hazardous Waste (Management and Handling) Rules, (1989)** is to control the generation, collection, import, storage, handling and treatment of hazardous waste.

- **The Manufacture, Storage, and Import of Hazardous Rules, (1989)** define the terms used in this regard, and sets up an authority to inspect yearly, the industrial activity connected with hazardous chemicals and its storage facilities.

- **Public Liability Insurance Act (PLIA), 1991:** The PLIA was amended in 1992, and the central government was authorized to establish the environmental relief fund, for making relief payments.

- **The Coastal Regulation Zone Notification, (1991)** puts regulations on various activities, including construction,. It gives some protection to the backwaters and estuaries.

- **The National Environmental Tribunal Act, (1995)** was created to award compensation for damages to persons, property, and the environment arising from any activity involving hazardous substances.

- **The National Environment Appellate Authority Act, (1997)** was established to hear appeals with respect to restrictions of areas in which classes of industries etc. are carried out or prescribed, subject to certain safeguards under the EPA.

- **The Environment (Sitting for Industrial Projects) Rules, (1999)** lay down detailed provisions relating to areas to be avoided for sitting of industries, precautionary measures to be taken for site selecting as also the aspects of environmental protection which should have been incorporated during the implementation of the industrial development projects.

- **The Municipal Solid Wastes (Management and Handling) Rules, (2000)** apply to every municipal authority responsible for the collection, segregation, storage, transportation, processing, and disposal of municipal solid wastes.

- **The Ozone Depleting Substances (Regulation and Control) Rules, (2000)** have been laid down for the regulation of production and consumption of ozone depleting substances.

- **The Biological Diversity Act, (2002)** is an act to provide for the conservation of biological diversity, sustainable use of its components, and fair and equitable sharing of the benefits arising out of the use of biological resources and knowledge associated with it.

Conclusion

Since 1990, global population has grown from roughly 5.3 to 6.8 billion and sustained global economic growth, accompanied by total and per capita increase in consumption in many parts of the world has also increased in many countries including Brazil, Russia, India and China. However, our world remains driven by differences in access to resources and per capita consumption both between and within countries. A review of the most highly cited papers in this journal shows significant contributions across 5 broad themes: the drivers and impacts of systemic and cumulative change, cross-cutting concepts such as vulnerability and resilience, approaches to management, control and policy, and different perspectives on climate change. The scientific community has clearly documented and quantified global environmental change with increasing precision and improved models to understand the future consequences of our actions, although large uncertainties still remain. The community has also developed tools to quantify our footprints and the effects of our lifestyles beyond our immediate surroundings (Rees, 1992; Hoekstra and Hung, 2005) and we have far greater potential to understand our interconnectedness across scales, in both biophysical and socio-economic terms, which as Rifkin (2009) suggests may cultivate increased empathy. But it is perhaps at the interface between individual and collective perceptions and action that research has progressed the least but where there is the greatest potential to address the challenges.

References

AATSE-Australian Academy of Technological Sciences and Engineering, 2004. Water Recycling in Australia, AATSE, Victoria, Australia.

Al-Swaify ,S .A., P. Pathak, T. J. Rego, and S. Singh, 1985. Soil Management for Optimized Productivity under Rain fed Conditions in the semi- arid Tropics', *Journal Article* No 383, ICRISAT.

Attrill, M. andPower, M. 2002. Climatic influence on a marine fish assemblage. *Nature* 417:275-278.

Babcock Hollowed, A., Hare, SR., Wooster, W.S 2001. Pacific Basin Climate variability and paterns of North Easte Pacific marine fish production *Progress in Oceanography* 49: 257-282.

Bajaj R. 1996. CITES and the wildlife trade in India, New Delhi: Centre for Environmental Law, WWF – India, 182.

Barnett, J., Adger, W.N., 2007. Climate change, human security and violent conflict. *Political Geography* 26, 639–655.

Benjaminsen, T.A., Alinon, K., Buhaug, H., Buseth, J.T., 2012. Does climate change drive land-use conflicts in the Sahel? *Journal of Peace Research* 49, 97–111.

Black, R., Adger, W.N., Arnell, N.W., Dercon, S., Geddes, A and Thomas, D. S.G. 2011. The effect of environmental change on human migration. *Global Environmental Change*. 21S: S3-S11.

Black, R., Kniveton, D., *et al.*, 2011. Migration and climate change: towards an integrated assessment of sensitivity. *Environment and Planning*A43, 431–450.

Boyd, M., 1989. Family and personal networks in international migration: recent developments and new agendas. *International Migration Review* 23 (3), 638–670.

Brandier, J.A. and Downick, S. 1993. The role of fertility and population in economic growth : Empirical results from aggregate cross-national data. National Bureau of Economic Research, Cambridge, Massachussett. Working paper No.4270, pp.40.

Brown, L.R. and Jacobson, J.L. 1987. The future of urbanization: Lacing the ecological and economic constraints, Worldwatch Institute, Washington D.C., Paper No.77, pp.58.

Bryan t, D., Burke, L., McManus, J., Spalding, M. 1998 . Reefs at Risk: A Map-Based Indicator of Threats to the World's Coral Reefs. Washington, D. C: World Resources Institute.

Buhaug, H and Urdal, H. 2013. An urbanization bomb?, population growth and social disorder in cities. *Global Environmental Changes*. 23: 1-10.

C. Baird, Environmental Chemistry, 1995. pp: 149-155. (W.H. Freeman and Company, New York).

Church, J. A., Gregory, J. M., Huyberchts, P., Kuhn, M., Lambeck, K., Nhuan, M. T., Qin, D., and Woodworth, P.L. 2001. Changes in sea level. In Climate Change (2001). The scientific Basis. Contribution of working Group1 to the Third Assessment Report of intergovernmental Panel on climate change, Houghton, J. T., Ding, Y.,Griggs , D. J., Noguer, M., Van der Linden, P. , Dai, X., Maskell, K., and Johnson, C. I., eds., Cambridge: Cambridge University Press, 639-694.

Devitt, C., Tol, R.S.J., 2012. Civil war, climate change, and development: a scenario study for Sub-Saharan Africa. *Journal of Peace Research* 49, 51–64.

Dhruva Narayana, V.V. and Ram babu, 1983 'Estimation of soil erosion in India', Journal of Irrigation and Drainage Engineering, *American Society of Civil Engineering*, Vol. 109.

Dickson, R., Yashayaev, I., Meincke, J., Turrell, W., Dye, S., Holfort, J .2002. Rapid freshening of the deep North Atlantic Ocean over the past four decades. *Nature* 416:832-837.

Divan S. and Rosencranz A. 2001. Environmental law and policy in India, cases, materials and status, 2nd edition, New York, Oxford University Press.

FAO. 1991. Social and institutional aspects of sustainable agriculture and rural development. Background document No. 5, FAO/NETHERLANDS Conference on agriculture and the environment. S-Hertogembosch, The Netherlands, 15-19 April.

Geetanjoy S. 2008. Implications of Indian Supreme Court's Innovations for Environmental Jurisprudence, Law. *Environment and Development Journal*, 4(1):1–19.

Gizelis, T.-I., Wooden, A.E., 2010. Water resources, institutions, and intrastate conflict. *Political Geography* 8, 444–453.

Govindasamy, A. 2012. India's Pollution Nightmare: Can It Be Tackled?, Environ. Sci. Technol., 46, 1305–1306.

Hendrix, C., Salehyan, I., 2012. Climate change, rainfall, and social conflict in Africa. Journal of Peace Research 49, 35–50.

Hinrichsen, D. 1997. Winning the food race. Population Reports – Series M, No.13 (Population Information Programme, Baltimore). pp. 24.

Hoegh- Guldberg, O.1999. Coral bleaching, climate change and the future of the World's coral reefs. Marine and Freshwater Research Mar. Freshwater Res. 50: 839-866.

Hoekstra, A.Y., Hung, P.Q., 2005. Globalisation of water resources: international virtual water flows in relation to crop trade. *Global Environmental Change* 15,45–56.

Homer-Dixon, T., 1999. Environment, Scarcity, and Violence. Princeton University Press, Princeton, NJ.

Homer-Dixon, T., Blitt, J. (Eds.), 1998. Ecoviolence: Links among Environment, Population and Security. Rowman and Little field, Lanham, MD.

IAB,2000, *Indian Agriculture in brief*, Agri. Statistics Division, Ministry of Agric., Govt. of India, New Delhi.

IPCC 2001. Intergovernmental panel on climate change. Summary for policy makers. A report of working group I of the intergovernmental panel on climate change, pp. 1-20. R.T.Watson, ed. Cambridge: Cambridge University press.

J.Jager and H.L. Ferguson (Eds). Climate change: Science, Impacts and policy. 1991. Pp: 83-109. (Cambridge University Press, UK).

Jackson, J.B.C., Kriby, M.X., Berger, W. H., Bjorndal, K . A., Botsford, L.W., Bourque, B. J., Bardbury, R.H., Cooke R., Erlandson, J., Estes, J. A., Hughes, T.P. , Kidwell, S., Lange, C. B., Lenihan, ?H.S., Pandolfi, J.M., Peterson, C.H., Steneck, R. S., Tegner, 2001. Historical overfishing and the recent collapse of coastal ecosystem. *Science* 293: 629- 638.

Keraita, B., B. Jimenez and P. Dreschsel, 2008. Extent and implications of agricultural reuse of untreated partly treated and diluted wastewater in developing countries, Agriculture, Veterinary Science, Nutrition and National Resources, 3, 58, 15.

Kerr, John M. 1996. 'Sustainable Development of Rainfed Agriculture in India', EPTD Discussion Paper No.20, International Food Policy Research Institute, Washington D. C.,

Klyashtorin , L. 1998. Long-term climate change and main commercial fish production in the Atlantic and Pacific. *Fisheries Research* 37: 115- 125.

Lakshman,H.C. and Channabasava A. 2013. Impact of climate change on Agriculture in India. *Inter. J. of Science in Society.* (In press).

Lakshman, H.C. 1999. Aggregation of sand dune soil by arbuscular mycorrhizal fungi and its use in revegetation practices. *Indian J. of Environ. Ecoplanning.* 2(3): 247-252.

Lakshman, H.C. 2000. Occurance and tolerance of VAM plants growing on polluted soils with sewage and industrial effluents. *J of Nat. Con.* 12(1): 9-18.

Raschid-Sally L., and P. Jayakodi, 2008. "Drivers and characteristics of wastewater agriculture in developing countries: Result from a global assessment, Colombo, "Sri Lanka' IWMI Research report 127, International Water Management Institute, Colombo.

Ratageri, R.H., Taranath, T.C. and Lakshman, H.C. 2006. Toxicity of dimethoate on primary productivity of lentic aquatic ecosystem: A microcosm approach. *Bulletin of Environmental Contamination and Toxicology.* 76(3): 373-380.

Macleod, H. 1992. Teaching for ecologically sustainable development. Queensland Dept. of Education, Brisbane.

Mason, A. 1997. Will population change sustain the "Asian economic miracle", Analysis from the east-West Centre, 33: pp. 1-8.

Massey, D., 1990. Social structure, household strategies and the cumulative causation of migration. Population Index 56 (1), 3–26.

McGraw Hills Encyclopedia of Science and Technology, 1992, pp: 225-226. (McGraw Hills, Inc., US)

Nichols, R., Hoozemans, F. M.J., Marchand, M .1999. Increase food risk and wetland losses due to global sea-level rise: Regional and global analysis. *Global Environmental Change* 9: S69-S87.

Pudasaini, S.P. 1997. Population, Environment and Sustainable development in a global perspective. In, Demographic Transition: The Third World scenario. Eds. Ahmad, Noin and Sharma. Rawat Publications, Jaipur, India, pp. 307-321.

Raleigh, C., 2010. Political marginalization, climate change, and conflict in African Sahel states. *International Studies Review* 12, 69–86.

Raleigh, C., Kniveton, D., 2012. Come rain or shine: an analysis of conflict and climate variability in East Africa. *Journal of Peace Research* 49, 51–64.

Raleigh,C., 2011.The search for safety :the effects of conflict, poverty and ecological influences on migration in the developing world. *Global Environmental Change* 21, S82–S93.

Rees, W.E., 1992. Ecological footprints and appropriated carrying capacity: what urban economics leaves out. *Environment and Urbanisation* 4, 121–130.

Rifkin, J., 2009. The Empathic Civilization: The Race to Global Consciousness in a World in Crisis. Tarcher, New York.

Seto,K.,2011.Non-environmentaldriversofmigrationtocities in Asian and African mega-deltas. *Global Environmental Change* 21, S94–S107.

Singh, G., Ram Babu, P.Narayanan and I. P. Abrol.1992. Soil Erosion Rates in India', *Journal of Soil and Water Conservation*, Vol. 47, No. 1.

Singh, R. P., Vasisht, A. K. and Mathur, V. C. 2003. Quantitative Assessment of Economic Losses of Degraded Land in India. Agricultural Economics Research Association, New Delhi, India.

Singh, R.B., 1989. Conservation of fast eroding plant genetic resources. Environment and Agriculture. FAO Publication, Bangkok Thailnad.

Sinha, R.P. 1983. Farm ponds. Project bulletin 6, All –India Co-ordinated Research Project for Dryland Agriculture, Hyderabad.

Theisen, O.M., 2012. Climate clashes? Weather variability, land pressure, and organized violence in Kenya, 1989–2004. *Journal of Peace Research* 49, 81–96.

Theisen, O.M., Holtermann, H., Buhaug, H., 2011–2012. Climate wars? Assessing the claim that drought breeds conflict. *International Security* 36, 79–106.

Vertovec, S., 2008. Transnationalism. Routledge, London.

Wani,S.P., P. Pathak, L. S. Jangawad, H. Eswaran, and P. Singh. 2001. Improved management of vertisols for increased productivity with increased carbon sequestration in the semiarid tropics WWW.capri.cgiar.org /projects/ projects_043.asp.

WHO 2006. Guidelines for the safe use of wastewater, excreta, and Grey water, volume 2: Waste Water use in agriculture, World Health Organization, Geneva.

Index

www.ingramcontent.com/pod-product-compliance
Lightning Source LLC
Chambersburg PA
CBHW050524190326
41458CB00005B/1656